*Wheat Genetic Resources: Meeting Diverse Needs*

# *Wheat Genetic Resources: Meeting Diverse Needs*

Edited by *J. P. Srivastava* and *A. B. Damania*

A Wiley-Chayce Publication

**JOHN WILEY & SONS**
Chichester - New York - Brisbane - Toronto - Singapore

Published by John Wiley & Sons Ltd.
Baffins Lane, Chichester
West Sussex PO19 1UD, England



*Other Wiley Editorial Offices*

John Wiley & Sons, Inc., 605 Third Avenue, New York, NY 10158-0012, USA

Jacaranda Wiley Ltd, G.P.O. Box 859, Brisbane, Queensland 4001, Australia

John Wiley & Sons (Canada) Ltd, 22 Worcester Road, Rexdale, Ontario M9W 1L1, Canada

John Wiley & Sons (SEA) Pte Ltd, 37 Jalan Pemimpin 05-04, Block B, Union Industrial Building, Singapore 2057

**Library of Congress Cataloging-in-Publication Data**

Wheat genetic resources : meeting diverse needs / editors, J.P. Srivastava and A.B. Damania.
p. cm.
'[Papers presented at the] International Symposium on the Evaluation and Utilization of Genetic Resources in Wheat Improvement, Aleppo, Syria, 18–22 May 1989'—P.
Includes bibliographical references and index.
ISBN 0 471 92880 1
1. Wheat—Germplasm resources—Congresses. I. Srivastava, J. P., 1940– . II. Damania, A. B. III. International Symposium on the Evaluation and Utilization of Genetic Resources in Wheat Improvement (1989 : Aleppo, Syria)
SB191.W5W514 1990
633.1′123—dc20 90-45855
CIP

**British Library Cataloguing in Publication Data**

International Symposium on the Evaluation and Utilization of Genetic Resources in Wheat Improvement (1989: Aleppo, Syria).
Wheat genetic resources.
1. Wheat. Breeding
I. Title II. Srivastava, J. P. (1940) III. Damania, A. B.
633.113

ISBN 0 471 92880 1

Printed in Great Britain by Biddles Ltd, Guildford.

# *Contents*

# *List of Contributors*

**M. Abdennedhar**, National Institute of Agronomy of Tunisia, Department of Agronomy and Plant Breeding, 43 Ave. Charles Nicolle, 1002 Tunis-Belvedere, Tunisia

**Y. J. Adham**, International Board for Plant Genetic Resources (IBPGR), Via delle Sette Chiese 142, 00145 Rome, Italy

**K. Ammar**, National Institute of Agronomy of Tunisia, Department of Agronomy and Plant Breeding, 43 Ave. Charles Nicolle, 1002 Tunis-Belvedere, Tunisia

**K. Attary**, Seed and Plant Improvement Institute (SPII), Mard-Abad Avenue, Karadj, Iran

**D. Barnhart**, Department of Botany and Plant Sciences, University of California, Riverside, California 92521, USA

**S. Benedettelli**, Agroselviculture Research Institute, Centro Nazionale di Ricerche, Villa Paolina, 05010 Porano (TR), Italy

**A. Biesantz**, Institute for Research on Useful Plants, Technical University of Berlin, Albrecht-Thaer-Weg 5, D-1000 Berlin 33, West Germany

**G. Boggini**, Experimental Research Institute for Cereal Crops (Catania Section), Via Varese 43, Catania 95123, Italy

**F. Calcagno**, Experimental Research Station for Grain Crops, Via Rossini 1, Caltagirone, Catania 95123, Italy

**C. Ceoloni**, University of Tuscia, Via S.C. de Lellis, 01100 Viterbo, Italy

**T. T. Chang**, International Rice Germplasm Center, International Rice Research Institute, P.O. Box 933, Manila, Philippines

**M. Ciaffi**, University of Tuscia, Via S.C. de Lellis, 01100 Viterbo, Italy

**J. M. Clarke**, Research Station, Agriculture Canada, Swift Current, Saskatchewan, S9M 3X2, Canada

**L. Cseuz**, Cereals Research Institute, P.O.B. 391, H-6701 Szeged, Hungary

**A. Daaloul**, National Institute of Agronomy of Tunisia, Department of Agronomy and Plant Breeding, 43 Ave. Charles Nicolle, 1002 Tunis-Belvedere, Tunisia

**A. B. Damania**, International Center for Agricultural Research in the Dry Areas, P.O. Box 5466, Aleppo, Syria

**Tong Daxiang**, Plant Introduction and Exchange Laboratory, Institute of Crop Germplasm Resources, CAAS, 30 Bai Shi Qiao Road, Beijing, China

**V. Delre**, University of Tuscia, Via S.C. de Lellis, 01100 Viterbo, Italy

**F. J. El-Haramein**, International Center for Agricultural Research in the Dry Areas (ICARDA), P.O. Box 5466, Aleppo, Syria

**J. E. Foster**, Department of Agronomy, Purdue University, West Lafayette, Indiana 47907, USA

**R. J. Giles**, Agriculture and Food Research Council, Institute of Plant Science Research, Maris Lane, Trumpington, Cambridge CB2 3JB, UK

**J. Gorham**, Center for Arid Zone Studies, University College of North Wales, Bangor, Gwynedd LL57 2UW, UK

**G. Haile Mariam**, Plant Genetic Resources Centre/Ethiopia, Institute of Agricultural Research, P.O. Box 30726, Addis Ababa, Ethiopia

**M. Harrabi**, National Institute of Agronomy of Tunisia, Department of Agronomy and Plant Breeding, 43 Ave. Charles Nicolle, 1002 Tunis-Belvedere, Tunisia

**B. O. Humeid**, International Center for Agricultural Research in the Dry Areas, P.O. Box 5466, Aleppo, Syria

**S. Jana**, Department of Crop Science and Plant Ecology, University of Saskatchewan, Saskatoon S7N 0W0, Canada

**A. A. Jaradat**, Jordan University of Science and Technology, P.O. Box 3030, Irbid, Jordan

**D. Knott**, College of Agriculture, University of Saskatchewan, Saskatoon S7N 0W0, Canada

**M. N. Koppar**, National Bureau of Plant Genetic Resources, Pusa Campus, New Delhi
110 012, India

**N. Kyzeridis**, Cereal Institute, P.O. Box 10514, 54110 Thessaloniki, Greece

**D. Lafiandra**, University of Tuscia, Via S.C. de Lellis, 01100 Viterbo, Italy

**P. Limberg**, Institute for Research on Useful Plants, Technical University of Berlin, Albrecht-Thaer-Weg 5, D-1000 Berlin 33, West Germany

**M. C. Mackay**, Australian Winter Cereals Collection, P.M.B. 944, Tamworth 2340, Australia

**B. Malinovsky**, N.I. Vavilov All-Union Scientific Research Institute of Plant Industry, 42-44 Herzen Street, 190000 Leningrad, USSR

**B. Margiotta**, Germplasm Institute, Via G. Amendola 165/A, Bari 70126, Italy

**J. Matuz**, Cereals Research Institute, P.O.B. 391, H-6701 Szeged, Hungary

**J. G. McLeod**, Agriculture Canada Research Station, Swift Current, Saskatchewan, S9M 3X2, Canada

**H. Mekbib**, Plant Genetic Resources Centre/Ethiopia, Institute of Agricultural Research, P.O. Box 30726, Addis Ababa, Ethiopia

**S. K. Mithal**, National Bureau of Plant Genetic Resources, Pusa Campus, New Delhi 110 012, India

**M. Obanni**, Department of Agronomy and Plant Breeding, IAV Hasan II, B.P. 6202, Rabat, Morocco

**K. Obari**, Genetic Resources Unit, Agricultural Research Center, P.O. Box 113, Douma, Syria

**H. W. Ohm**, Department of Agronomy, Purdue University, West Lafayette, Indiana 47907, USA

**R. D'Ovidio**, University of Tuscia, Via S.C. de Lellis, 01100 Viterbo, Italy

**C. De Pace**, University of Tuscia, Via S.C. de Lellis, 01100 Viterbo, Italy

**R. M. De Pauw**, Research Station, Agriculture Canada, Swift Current, Saskatchewan, S9M 3X2, Canada

**M. Palumbo**, Experimental Research Institute for Cereal Crops, Catania Section, Via Varese 43, Catania 95123, Italy

**R. Paolini**, University of Tuscia, Via S.C. de Lellis, 01100 Viterbo, Italy

**H. Pashayani**, International Center for Agricultural Research in the Dry Areas, P.O. Box 5466, Aleppo, Syria

**F. L. Patterson**, Department of Entomology, Purdue University, West Lafayette, Indiana 47907, USA

**L. Pecetti**, International Center for Agricultural Research in the Dry Areas, P.O. Box 5466, Aleppo, Syria

**P. Perrino**, Germplasm Institute, Via G. Amendola 165/A, Bari 70126, Italy

**E. Porceddu**, University of Tuscia, Via S.C. de Lellis, 01100 Viterbo, Italy

**C. O. Qualset**, Genetic Resources Conservation Program, University of California, Davis, California 95616, USA

**S. Rajaram**, Centro Internacional de Mejoramiento de Maíz y Trigo, Apdo. Postal 6-641, 06600 Mexico, D.F.

**M. N. Sankary**, Range and Arid Zone Ecology Research Unit, University of Aleppo, P.O. Box 6656, Aleppo, Syria

**A. Sayegh**, International Center for Agricultural Research in the Dry Areas, P.O. Box 5466, Aleppo, Syria

**G. T. Scarascia Mugnozza**, University of Tuscia, Via S.C. de Lellis, 01100 Viterbo, Italy

**D.V. Seshu**, International Rice Germplasm Center, International Rice Research Institute, P.O. Box 933, Manila, Philippines

**J. A. Shamkhe**, Cereals Research Institute, P.O.B. 391, H-6701 Szeged, Hungary

**V. Shevelukha**, Vashknil, 21 Bolshoy Kharitonyevsky Pereulok, B-78, 107814 Moscow, USSR

**B. Skovmand**, Centro Internacional de Mejoramiento de Maíz y Trigo, Apdo. Postal 6-641, 06600 Mexico, D.F.

**M. van Slageren**, International Center for Agricultural Research in the Dry Areas, P.O. Box 5466, Aleppo, Syria

**D. H. van Sloten**, International Board for Plant Genetic Resources, Via delle Sette Chiese 142, 00145 Rome, Italy

**P. L. Spagnoletti-Zeuli**, Institute of Agricultural Biology, University of Basilicata, 85100 Potenza, Italy

**J. P. Srivastava**, International Center for Agricultural Research in the Dry Areas, P.O. Box 5466, Aleppo, Syria (*as from 1 June 1990* Agricultural and Rural Development Department, World Bank, 1818 H Street N.W., Washington DC 20433, USA)

**M. Tahir**, International Center for Agricultural Research in the Dry Areas, P.O. Box 5466, Aleppo, Syria

**O. A. Tanzarella**, University of Tuscia, Via S.C. de Lellis, 01100 Viterbo, Italy

**C. Tomassini**, University of Tuscia, Via S.C. de Lellis, 01100 Viterbo, Italy

**J. G. Waines**, Department of Botany and Plant Sciences, University of California, Riverside, California 92521, USA

**P. Williams**, Canadian Grain Commission, Grain Research Laboratory, Winnipeg, Manitoba R3C 3G8, Canada

**R. C. Yang**, Department of Crop Science and Plant Ecology, University of Saskatchewan, Saskatoon, S7N 0W0, Canada

**D. Zamzam**, Beniflor, Casablanca, Morocco

**Hongtao Zhang**, College of Agriculture, University of Saskatchewan, Saskatoon S7N 0W0, Canada

# *Preface*

This book is based on the presentations given by a multidisciplinary group of scientists at an international symposium on the Evaluation and Utilization of Genetic Resources in Wheat Improvement, held at the International Center for Agricultural Research in the Dry Areas (ICARDA), Aleppo, Syria. The symposium attracted 58 participants from five continents and was sponsored jointly by ICARDA and the University of Tuscia in Viterbo, Italy.

Plant breeders now have a greater range of genetic diversity available for utilization than in the past. However, the usefulness of collections depends on the extent to which they are geographically and ecologically representative and on the presence of genes which are not commonly found in conventional breeding germplasm. Various reasons have been given for the prevailing reluctance of breeders to make more use of this germplasm. There is a tendency for breeders to use materials with which they are familiar and which are already adapted to their particular environment, as opposed to using alien germplasm, which requires carrying out a laborious programme of pre-adaptation trials and the evaluation of the germplasm for desirable traits. To enable researchers to identify potentially useful accessions, information on collections needs to be more readily available.

The presentations discussed factors which are likely to limit or facilitate germplasm evaluation and provided information on some of the current uses being made of wheat genetic resources. They also give an overview of genetic resource conservation activities in some of the national programmes.

The text is divided into five parts. Part 1, 'Evaluation of Biodiversity', deals with various aspects of germplasm description and documentation. The first three papers, written by recognized authorities on the subject, put forward the general principles and considerations of germplasm conservation. The following six papers give the results of evaluation studies for traits of specific interest to users.

In Part 2, 'Constraints to Evaluation and Germplasm Networks', the first three papers look at some of the problems faced in carrying out germplasm evaluation; the fourth paper makes a case for the establishment of a wheat genetic resources network. Part 3, 'Research at National Genebanks', consists of country reports on genetic resource conservation and evaluation activities. Part 4, 'Utilization for Diverse Needs', is the largest section in the book. In the final part of the book, 'Summary and Recommendations', we have attempted to bring together most of the ideas expressed in the papers and the major points raised by participants during the symposium. These conclusions were read out during the closing session and were endorsed by the participants.

During the symposium, most presentations were followed by a short discussion. A verbatim report of each discussion is included at the end of the relevant papers. In some cases, however, there was no discussion; this was generally because of time constraints. It is also worth pointing out that while every effort has been made to include in this book all the papers presented at the symposium, some authors found that their professional commitments did not allow them enough time to prepare their papers for submission to the editors.

We are grateful to the Government of Italy for providing financial support for the symposium and for the production of this book. The success of the symposium was due largely to the interest and encouragement of Dr Enrico Porceddu, Chairman of the ICARDA Board of Trustees and Professor of Agricultural Genetics at the University of Tuscia, and Dr Nasrat Fadda, Director General of ICARDA.

We thank Kay Sayce and Simon Chater of Chayce Publication Services in the United Kingdom for editing the papers and preparing the book for production. We are also immensely indebted to many members of staff at ICARDA whose enthusiasm and hard work helped ensure a well-organized and successful symposium.

J. P. SRIVASTAVA and A. B. DAMANIA

# Foreword

The usefulness of germplasm collections to plant breeders, crop evolutionists, plant pathologists, taxonomists and other experimental biologists depends upon the amount and quality of information available. In recent years, collections have been enriched by several exploration missions, many of which were organized by the international agricultural research centers supported by the Consultative Group on International Agricultural Research (CGIAR).

Collections of wheat, the most important crop in many parts of the world, are substantial, but a more thorough evaluation of these collections and a wider distribution of information on their characteristics needs to be undertaken to promote their effective utilization in breeding programmes.

The International Center for Agricultural Research in the Dry Areas (ICARDA) has a joint mandate with the Centro Internacional de Mejoramiento de Maíz y Trigo (CIMMYT) for wheat improvement in the West Asian and North African (WANA) region. The utilization of genetic resources such as landraces, wild progenitors and primitive forms of wheat to improve and stabilize crop production in the face of biotic and abiotic stresses is a key component in the strategy to develop appropriate germplasm for the dry areas. Environmental factors which reduce productivity, such as drought, temperature extremes, salinity, infertile soils, diseases and pests are common in the WANA region. Plant breeders are searching for desirable genes in the exotic germplasm assembled by genebanks, including the Genetic Resources Unit at ICARDA.

This book explores factors that are likely to present constraints to the evaluation and utilization of wheat genetic resources. It is intended to stimulate discussion on the handling of germplasm in order to optimize its utilization in crop improvement programmes.

The papers in this book are the work of germplasm collectors, evaluators, plant breeders and scientists from other related disciplines, presented at a symposium on the Evaluation and Utilization of Genetic Resources in Wheat Improvement convened by the University of Tuscia, Viterbo, Italy and ICARDA, at Aleppo, Syria, from 18 to 22 May 1989.

I am confident that the knowledge, interest and ideas generated by this symposium will go a long way towards developing strategies for more effective evaluation and utilization of germplasm, as an important component of the effort to improve the welfare of the farmers in the WANA region and elsewhere.

Director General, ICARDA

# *Introduction*

PROFESSOR G. T. SCARASCIA MUGNOZZA

*University of Tuscia, Viterbo, Italy*

It was just over 10 years ago that the decision was taken to accept the offer from the Government of Syria to establish the main experimental station of the International Center for Agricultural Research in the Dry Areas (ICARDA) at Tel Hadya, near Aleppo. Having represented Italy on the committee set up in 1975 to select the location and crop commodity programmes for this new centre in the Consultative Group on International Agricultural Research (CGIAR) system, I served on the ICARDA Board of Trustees from the centre's inception until 1982. Many changes have taken place since those difficult early days in the history of ICARDA. Among the most significant changes was the establishment of the plant genetic resources unit, which has contributed to the leading role ICARDA is now playing in the field of genetic resources.

The papers contained in this book, written by leading international experts in genetic resources, were presented at a symposium sponsored by ICARDA and the University of Tuscia in Viterbo, Italy. This symposium was the fifth in a series of conferences organized by the two institutions. The first was held in Viterbo in 1982 and focused on breeding problems in durum wheat and Triticale. The second meeting, held on the island of Capri, dealt with drought tolerance in winter cereals. At the third meeting, held in Viterbo, the discussion revolved around the strategic plans for ICARDA. The fourth meeting, also in Viterbo, was devoted to the evaluation of durum wheat germplasm.

The principal objective of the 1989 symposium, held at ICARDA to coincide with the opening of the centre's new Genetic Resources Unit building, was to emphasize the importance of plant genetic resources research, not only for the collection and conservation of genetic traits but also for the evaluation of these traits and their utilization in breeding. Although the presentations concentrated mainly on wheat and its wild relatives, many of the findings could also be applied to other crops. The joint ICARDA/University of Tuscia project on the Evaluation and Documentation of

Durum Wheat Germplasm is one of the largest exercises of its kind in the world in terms of the number of traits examined and the number of locations involved. The results of this project could act as a model for the transfer of technology from advanced institutions to national programmes.

Over the past 20 years, the collection and conservation of the most important food crops in the Mediterranean and South-West Asian regions have been largely accomplished. Much of the credit for this, particularly in relation to cereals and food legumes, must go to the International Board for Plant Genetic Resources (IBPGR), which has carried out the work as part of its Mediterranean Program; for several years, this Program was based at the Germplasm Institute in Bari, Italy. The IBPGR has also promoted the establishment of cold storage units for germplasm conservation and the computerized documentation of collections in the region. With the opening of ICARDA's new Genetic Resources Unit for the Mediterranean and South-West Asia, these activities will be strengthened and we look forward to further progress in the conservation and utilization of genetic resources.

# PART 1

# Evaluation of Biodiversity

Wheat Genetic Resources: Meeting Diverse Needs
Edited by J. P. Srivastava and A. B. Damania

Published by John Wiley & Sons

# 1

# *Evaluation, Documentation and Utilization of Durum Wheat Germplasm at ICARDA and the University of Tuscia, Italy*

E. PORCEDDU and J. P. SRIVASTAVA

Of the estimated 1 710 million ha of land in the West Asia and North Africa (WANA) region, about 75 per cent is desert and only 13 per cent is arable. Much of this arable land (about 73 per cent) is rainfed, and most of the rain falls in winter. The rainfall is highly erratic, causing substantial fluctuations in crop productivity. Temperatures in the region rise sharply in spring, and this is followed by a hot dry summer. Thus, winter cereals in the region are highly vulnerable to the cold in winter and to drought and heat later in the crop-growing season. The crop-growing season is also highly variable, ranging from 70 to 200 days depending on soil moisture and depth, rainfall, temperature and altitude.

Worldwide, durum wheat covers only about 19.2 million ha, or about 8 per cent of the land under wheat production. However, it is one of the most important food crops in the WANA region. Although this region accounts for nearly 45 per cent of durum wheat cultivation in the world and about 80 per cent of all durum wheat grown in developing countries, durum wheat is also the region's major food import item.

Faced with the difficult task of developing durum wheat germplasm for this vast area, with its diverse and variable environments, crop scientists at the University of Tuscia in Viterbo, Italy and at ICARDA recognized that the most appropriate long-term strategy would be to develop a genetically broad-based germplasm pool by assembling elite materials from the global genepool of tetraploid wheats.

In the WANA region, durum wheat is grown primarily under rainfed conditions, mainly in areas where the annual precipitation is 250-450 mm and abiotic stresses, such as cold, drought and high temperatures, prevail during the crop-growing period.

This is also where primitive cultivars and landraces are still in cultivation, many of which are well adapted to harsh environments. Considerable genetic diversity is known to be present in these populations, and this diversity can be used in durum wheat improvement. Hence it is important not only to collect but also to evaluate and document these genetic resources for their immediate and long-term use in the improvement of durum wheat, and possibly of bread wheat.

Recognizing the role of ICARDA as a world centre for the preservation of genetic diversity in durum wheat, a collaborative project between ICARDA and the University of Tuscia for the Evaluation and Documentation of Durum Wheat Germplasm was initiated in 1985. This project is supported by the Government of Italy, and involves the Department of Agrobiology and Agrochemicals of the University of Tuscia and the Genetic Resources Unit and Cereal Improvement Program of ICARDA. The major objectives of the project are to:

- evaluate all the durum wheat accessions available at ICARDA;
- document and disseminate the information accumulated;
- test elite germplasm for characters of practical importance in durum wheat breeding;
- duplicate durum wheat accessions of the Germplasm Institute, National Research Council, Bari, Italy;
- provide training for national programme staff in the evaluation, documentation, conservation and utilization of crop genetic resources.

In this paper we outline some of the important features of the project, its achievements and constraints, and future directions for research to make the best possible use of the results obtained to date.

## IMPORTANT FEATURES OF THE PROJECT

About 15 000 accessions of durum wheat from 60 countries have been assembled for evaluation. The major sources of these accessions are: the United States Department of Agriculture (USDA) Small Grains Collection, Beltsville, Maryland, USA; the Germplasm Institute in Bari, Italy; and various national programmes in the WANA region. Because of the large number of accessions, the evaluation work has been spread over three seasons — 1985-86, 1986-87 and 1987-88. Various aspects of this work are reported in Jana et al. (1990) and Pecetti et al. (in prep.)

This project is one of the most ambitious and comprehensive programmes on the evaluation and documentation of crop germplasm undertaken since conscious efforts to collect and preserve the genetic resources of economic species began. Important features of the project are:

- evaluation of about 34 characters, most of which are relevant to the breeding and improvement of durum wheat in the WANA region; these include seed quality characters and agronomic characters such as plant height and grain-filling period, which are important for grain production under drought conditions;

- multilocation evaluation, whereby each location was chosen for the evaluation of a set of characters that were considered appropriate for it; for example, ICARDA's main research station at Tel Hadya, near Aleppo in Syria, was chosen for general evaluation, while Breda, 20 km south-east of Tel Hadya, was selected for drought-related traits, Viterbo for seed-storage proteins and Tunis for resistance to Septoria leaf blotch;
- use of an experimental design appropriate for large-scale field evaluation, namely the modified augmented design;
- emphasis on the characterization of germplasm for its tolerance to drought, the major physical stress that threatens current and future crop productivity;
- emphasis on important biotic stresses in the WANA countries, such as yellow rust, common bunt and Septoria leaf blotch.

## ACHIEVEMENTS, CONSTRAINTS AND FUTURE DIRECTIONS

The major achievements of the project to date are:

- identification of promising accessions for nine characters considered important in durum wheat breeding in the WANA region;
- identification of promising accessions with desirable multi-character associations (for example, earliness, long grain-filling period and tallness);
- identification of the geographical distribution of desirable traits;
- identification of the geographical distribution of several desirable multi-character associations;
- assessments of diversity for a wide range of characters;
- population studies on seed-storage proteins.

Inevitably, in an evaluation project of this scale, there have been a number of difficulties. These include:

- inadequate passport information;
- misclassification of accessions;
- evaluation and characterization of genetically heterogeneous accessions;
- choosing optimal evaluation sites;
- choosing appropriate characters and character states;
- choosing appropriate experimental designs;
- inaccuracy of using discrete ordinal scores for continuously varying traits;
- effects of seasonal fluctuations in growing conditions on evaluation data;
- inadequacies of statistical measures of diversity;
- extracting meaningful information from the wealth of data.

The major future directions for research in this project will be:

- assessment of genetic diversity for characters associated with drought tolerance, disease resistance and grain quality;
- replicated studies repeated in different years and locations to confirm desirable attributes of promising accessions;

- studies on the repeatability of performance and general combining ability of elite germplasm;
- correlation of environmental factors with the performance of promising accessions;
- assessment of the usefulness and efficiency of large-scale germplasm evaluation;
- comparison of short- and long-term adaptation of plant populations to arid conditions;
- assessment of the influence of environment on the preservation and erosion of genetic diversity within and between accessions;
- development of mixed populations with optimal frequencies of desirable accessions — a modern equivalent of landrace populations.

## CONCLUSION

Within a relatively short period — only 4 years — considerable progress has been made in screening and evaluating durum wheat accessions for characters associated with resistance or tolerance to the biotic and abiotic stresses that prevail in the WANA region. The evaluation data have been documented so as to facilitate the use of promising accessions in breeding, as well as in basic or strategic research.

With its emphasis on tolerance to drought, the project is attempting to demonstrate that a genetic solution to the devastating consequences of the so-called 'greenhouse effect' is possible through the carefully planned evaluation and exploitation of genetic diversity. It is our opinion that the project should continue in a modified form, for several reasons.

First, the data accumulated needs to be analyzed and synthesized to extract the maximum amount of useful information. Second, national agricultural research programmes are increasingly interested in using the promising accessions already identified. Evaluation should continue so as to maximize the returns on our investment in time and resources. Third, additional problem-specific evaluations are necessary. We should establish an international network of scientists for this purpose, and this effort should be strengthened and expanded with the active cooperation of the International Board for Plant Genetic Resources (IBPGR) and other international agencies. Finally, continued research and development are essential for developing an effective system of information transfer and increased use of wheat genetic resources. We foresee an important leadership role for IBPGR in these two areas.

## *Discussion*

A. B. DAMANIA: Do you think that variability, within populations, is important in landraces and wild species?

E. PORCEDDU: I would like to mention some points here on the evaluation and characterization of heterogeneous accessions. In a crop such as wheat, which is self-pollinating, there are highly variable genotypes within the same accession, and sometimes even different subspecies of wheat constitute a single accession. Regarding further research, I assessed the influence of environment on the preservation and erosion of genetic diversity between and within accessions. Breeders have been selecting lines with favourable traits, but there is still much scope for the study of distribution of variability, between populations from different environments and within the same populations. These kinds of studies are grouped under ecological genetics and have been conducted for several species from the Triticeae, especially *Avena* species. ICARDA is in a good position to study this variability in relation to wheat and barley. It would not only be a great contribution to scientific knowledge but would also throw light on the adoption of *in situ* conservation methods of both wild and cultivated species.

A. BIESANTZ: You made no mention of *in situ* conservation. All material is now stored in genebanks but its evolution is literally 'frozen'. It has no chance of evolving under natural conditions. Could not some farmers be asked to grow these primitive forms and landraces, so as to maintain genetic variability under natural conditions?

J. P. SRIVASTAVA: We have been giving this a lot of thought at ICARDA. Is the storage of germplasm in a cold room the best solution? It is probably desirable to identify one or two locations where material could be allowed to grow and multiply. In a genebank, as you correctly pointed out, the material stagnates. Storage of material in genebanks, therefore, is very important, but it has its limitations. Perhaps we should discuss this further and if there is general support for this type of conservation it should be put forward as a recommendation emanating from this symposium and we, at ICARDA, will pursue the issue further.

A. A. JARADAT: Between 1984 and 1988 we collected wild emmer with temporal and spatial variability and it is interesting to note how certain characters differ in their diversity indices over locations and in the same location over the years. Therefore there is a case for the *in situ* conservation of emmer wheat in Jordan which could be supported by international agencies.

M. N. SANKARY: In my work at the University of Aleppo and at the Arab Centre for the Studies of Arid Zones and Drylands (ACSAD) we have planted species of wild relatives with mixtures of landraces to create of hybrid swarms for colonization of new habitats. Dr J. Harlan reported such hybrid swarms in Turkey. In Syria, our heterogeneous landraces have disappeared or have become pure varieties, so we need to create variability in this manner.

H. MEKBIB: At the Plant Genetic Resources Centre in Ethiopia, there are plans to conserve areas where many landraces and primitive species are found in their diverse forms. The Government of Canada has provided funds for this *in situ* conservation project; the project began earlier this year in the Ethiopian provinces of Wollo, Shoa and Tigray.

E. PORCEDDU: I would like to add to what Dr Srivastava has said. We are talking about two different situations here: *in situ* conservation of landraces, and that of wild relatives. We should distinguish between the two. For wild species, there are fairly easy solutions. For example, there are some Mediterranean islands which are totally protected by the navy.

People are not allowed on these islands, and thus wild species could be preserved without disturbance, although some evolution could take place. There is no import or export of genes or material to and from these islands, and a degree of equilibrium could be established. When we look at the cultivated material, the situation is quite different. Should we use traditional agricultural practices, or should we preserve the cultivated forms using modern agricultural practices? It is an issue that merits some consideration.

Wheat Genetic Resources: Meeting Diverse Needs
Edited by J. P. Srivastava and A. B. Damania

Published by John Wiley & Sons

# 2

# *Evaluation of Exotic Germplasm for Genetic Enhancement and Plant Breeding, with Special Reference to Rice*

T. T. CHANG and D.V. SESHU

Progress in plant breeding requires a broad genetic base. A rich and diverse germplasm collection is the backbone of every successful crop improvement programme. Recent advances in biotechnology have created new opportunities for using exotic germplasm beyond the traditional limits within a biological species or genus.

In this paper the term 'exotic germplasm' is used to refer specifically to wild relatives, landraces and cultivars adapted to special ecological niches or possessing unusual characteristics. When plant breeders use this term they usually mean germplasm that is already used elsewhere in large-scale commercial production but is not adapted to their target environment. However, Hallauer and Miranda (1981) define exotic germplasm as 'all germplasm that does not have immediate usefulness without selection for adaptation to a given area.'

Chang et al. (1977), Huaman (1984) and Goodman (1985) have described the major weaknesses of exotic rice, potato and maize germplasm. However, not all landraces and wild species carry undesirable gene complexes. Many landraces have contributed to yield increases (Frankel and Soulé, 1981; Frey et al., 1984), pest resistance (Chang et al., 1975a, 1982a; Harlan and Starks, 1980) and tolerance to adverse soil conditions (Ikehashi and Ponnamperuma, 1978). Wild species have provided genes for pest resistance, yield increases, physiological efficiency, tolerance to adverse environments and improved grain quality (Hawkes, 1977; Stalker, 1980; Harlan, 1984; Chang, 1985c, 1989).

This paper discusses the basic principles that should be applied when evaluating exotic germplasm, and some recent advances in the methods used. We draw on our own experience with rice to provide examples. The International Rice Germplasm Center (IRGC), located at the International Rice Research Institute (IRRI), is responsible for the collection, multiplication/regeneration, characterization, preservation and dissemination of genetic resources belonging to the genus *Oryza*. The International Rice Testing Program (IRTP) coordinated by IRRI serves as a mechanism for the exchange and evaluation of rice germplasm in rice-growing countries (Chang et al., 1988).

## PRINCIPLES OF EFFECTIVE AND EFFICIENT EVALUATION

### Increase of seed or plant parts

Genebanks should maintain enough materials, whether in the form of seed or plant parts, to provide for the needs of evaluation experiments. Even so, seed increase by the user is sometimes also needed.

Because of the heterogeneous and heterozygous nature of exotic germplasm, a sample size larger than that normally used for pure-line cultivars is needed in critical tests. Wild species generally produce little seed and are difficult to handle. For users of wild species of the genus *Oryza*, the IRGC provides guidance on how to break strong seed dormancy, raise slow-growing seedlings, provide appropriate radiation levels, follow a taxonomic key for re-identification, and collect the few but precious seeds (Chang, 1976, 1988). The IRTP maintains ample seed stocks of parent donors for assessing varietal performance under biotic, climatic and ecoedaphic stresses.

### Maintenance of the population structure

It is imperative to preserve the population structure of an exotic accession during cycles of seed increase. Provisions such as maintaining a large sample size for the seed stock, controlled pollination or cross-pollination in sib-mating, avoidance of subjective selection or rigorous rogueing, use of a controlled photoperiod, and seed increase under an environment similar to the home habitat, are essential for maintaining the composition of the original population to the maximum extent possible. Workers should recognize the existence of ecostrains, morphological variants and spontaneous mutants under a single variety name. On the other hand, obvious duplicates should be excluded.

Seed samples collected from farmers' fields are commonly heterogeneous (Chang, 1976, 1985a, 1985b). Distinct genotypes with respect to disease and insect resistance have been found repeatedly in single samples of rice (Chang et al., 1975a; Chang, 1985b) and in other cereals (Harlan, 1984). Once a resistant plant is identified, it should be separated from the original population and maintained as a sub-accession (Chang, 1976).

## Germplasm exchange and dissemination

Evaluators, researchers and breeders depend on genebanks to supply seed or plant parts, or both. The most frequently encountered constraints to germplasm exchange and dissemination are:

- insufficient stock of viable seeds for distribution;
- cost ceilings on seed multiplication, preparation, packaging and mailing;
- government restrictions on seed exchanges;
- plant quarantine restrictions;
- obstacles posed by legislation on plant breeding rights.

In addition, the knowledge and experience of genebank managers also affect the quality of the service they offer (Chang et al., 1989; Palmer, 1989; Marshall, 1989).

It has often been pointed out that the use of exotic germplasm has been impeded by the lack of information exchanged on conserved materials and evaluation data (Chang, 1983, 1985b; IBPGR, 1983; Peeters and Williams, 1984; Marshall, 1989; Palmer 1989).

Users themselves contribute to ineffective or delayed dissemination when their requests to genebank managers are vague or over-ambitious. In our own experience, many seed requests involve two or more exchanges of letters following the initial request before the user's needs and capacity for handling large seed shipments are ascertained. Moreover, when crops are to be grown at a marginal or new site, the specific needs often cannot be matched with the information contained in the computerized files of genebanks.

## Multidisciplinary approach to evaluation

Most evaluation operations now require expertise beyond that of one or two closely related disciplines. A multidisciplinary approach is essential in the planning and implementation of an evaluation exercise and the assessment of its results. The Genetic Evaluation and Utilization (GEU) programme at IRRI provides a successful example of such an approach (Brady, 1975; Chang, 1984b, 1985b). Effective evaluation experiments lead to more refined and rewarding subsequent research (Chang, 1985b). The IRTP is the international extension of the GEU and involves about 800 rice scientists in 70 countries.

Genetic resources specialists have a crucial role to play in evaluation. Not only do they hold access to the seed/plant materials, but they are also highly knowledgeable about the origins, breeding history, ecogeographic background and particular characteristics of accessions. They should be involved at the initial stages of planning, kept informed of progress in evaluation, and included in the assessment and re-planning phases. Unfortunately, many genetic resources specialists tend to play a passive role in such collaborative activities (Williams, 1989); this attitude leads to diminishing recognition of their potential contribution. Often they are handicapped by meagre support and inadequate facilities. Weaknesses in such areas need to be remedied before effective evaluation can be realized. In general, evaluation systems are more effective when the genetic resources specialists, evaluators and breeders all belong to the same institution (Chang et al., 1975b; Frankel, 1989).

## Planning and implementing evaluation tests

The usual sequence in evaluation is from mass screening to confirmatory tests, leading to critical experiments or research (Chang, 1985b). The categories of traits to be included in evaluation programmes — observable versus variable or complex — dictate not only the scientific approach but also the division of labour and joint planning among the various disciplines concerned (Chang 1985b; Frankel, 1989).

Every evaluation test should be treated as a scientific investigation. It should embody well-defined objectives; the experimental design should be efficient, with a judicious choice of treatments and control varieties; the environmental factors should be controlled, if necessary; and a complete statistical analysis should be carried out. Data should be entered on computerized data files for easy retrieval and monitoring. Results from repeated tests and those from multiple sites should be compared with previous data before updating the information on the files (Chang, 1985b).

The IRTP nurseries were planned and developed by rice workers in national research programmes who share a similar research interest, be it in grain yield or in other economic traits. Other international agricultural research centres, including the International Institute of Tropical Agriculture (IITA), the Centro Internacional de Agricultura Tropical (CIAT) and the West African Rice Development Association (WARDA) also participate in planning and implementing evaluation exercises.

## Choice of a representative environment or site

It is imperative to carry out evaluation experiments in environments that represent the major production systems targeted for improvement. For many adverse environments, off-station test sites in stressed environments are necessary. To experience a heavy epidemic, disease and insect nurseries need to be planted at 'hot spots' where the pest is endemic. Controlling temperature, relative humidity and light intensity in greenhouses or growth chambers is necessary, although this may pose cost problems to the researcher. Cost-saving designs are needed to maximize use of available resources. For instance, at IRRI we screen thousands of rice accessions for drought reaction in the dry season and assess their agronomic promise in the wet season. We select hybrid progenies by using correlated responses to water stress (Chang and Loresto, 1986).

The sites for IRTP screening nurseries were chosen on the basis of the endemic production constraints and the availability of local scientists to implement the experiments. Many IRTP nurseries are located in hot spots.

## Preventing exotic plants from becoming serious pests

The high seed-shattering and strong rhizomatous characteristics of wild or primitive rice germplasm pose a threat if plants or their seeds escape into water canals or farmers' fields. Measures should be taken to prevent such accidents. At IRRI we grow the wild rices in large pots to control their spread, preserve their identity, and obtain ratoon crops. Plant remains and dropped seeds are collected and destroyed after harvest.

## Verification of evaluation data

Susceptibility to diseases and insects observed in initial tests may be recorded as such without retesting, but resistant or tolerant populations need to be retested under more controlled conditions in order to establish the validity of the preliminary evaluation. Appropriate control varieties or treatments should be included in each experiment.

Retesting should also be applied to the evaluation of other traits, especially those being subjected to the effect of variable environments or planting density in field trials. Testing for environmental variance and genotype x environment (G x E) interactions add considerable scope to evaluation experiments.

## Communication among conservationists, problem-area scientists and plant breeders

On-going communication among conservationists, problem-area scientists such as entomologists and physiologists, and plant breeders is essential to sustain evaluation efforts (Chang, 1984a). At IRRI the three groups of scientists and the statisticians of the GEU programme meet frequently to exchange information. In addition, there is an annual internal programme review and a 5-year external review. The crop advisory committees of the IBPGR and the national institutes also meet periodically to sustain cooperation (Chang, 1985b). For the IRTP participants, findings are presented and discussed at the annual International Rice Research Conference, during monitoring tours and at the IRTP Advisory Committee meetings (IRRI, 1980).

## Documentation

Efficient documentation is needed during all phases of the conservation and evaluation effort. The availability of evaluation data is crucial to arousing the interest of users in evaluated germplasm. Free exchange of evaluation results leads to enhanced exchange of germplasm (Chang, 1985b).

Evaluation results from the IRTP nurseries are compiled, analyzed and summarized each year, and the findings disseminated to all collaborators and their institutions.

## ADVANCES IN EVALUATION METHODS

A number of examples, taken primarily from rice, will illustrate some of the recent advances in evaluation methods and in our understanding of the basic mechanisms, both genetic and physiological, that control resistance or tolerance.

### Physiological mechanisms underlying drought resistance and other complex traits in rice

By approaching the complex subject of drought resistance on a component basis, we have a fuller understanding of the physiological mechanisms responsible: escape, via early maturity; avoidance, via a deep, thick root system and water conservation by leaf reactions; tolerance, via the ability of tissues to withstand desiccation; and recovery, via vegetative growth vigour (Chang et al., 1982b; O'Toole, 1982; Chang and Loresto, 1986). Plant reactions also differ between vegetative and reproductive growth phases and between varieties of different height groups (Loresto and Chang, 1981).

Other successful applications of component analysis to elucidate complex traits in rice include investigations of varietal differences in lodging resistance (Chang, 1967) and of the different physiological phases contributing to growth duration (Vergara and Chang, 1985). More recently, a reliable biochemical procedure was developed to assess cold tolerance (Mazumder et al., in press).

### Criteria for assessing varietal reactions to leaf-sucking insects

A plant's resistance to an insect involves preference, antibiosis and tolerance. IRRI's entomologists have used a variety of tests to distinguish between true resistance, pseudo-resistance and the physiological mechanisms (Pathak, 1975; Saxena and Pathak, 1977). Use of different insect biotypes, and bulk testing versus field testing were compared. Studies were made on: feeding sites and frequency of sucking on individual tester plants; egg-laying volume; survival of nymph populations; relative growth rate; and excretion of amino-acid-containing honeydew by leafhoppers and planthoppers (Paguia et al., 1980; Khan and Saxena, 1984; Heinrichs et al., 1985). A no-choice method of testing seedlings has led to a more reliable differentiation of varietal reactions by eliminating the preference factor (Saxena and Khan, 1984).

### Insect vector and plant pathogen interactions in the virus diseases of rice

IRRI's virologists and entomologists were able to separate varietal resistance to a virus from resistance to the related insect vector. It is the interaction between the two that determines the final varietal response (Ling, 1972). Host reaction to the tungro virus is determined largely by reaction to the insect vector (Hibino, in press). The tungro virus has been recognized as a composite of the bacilliform (B) form and the spherical (S) form; B is the more virulent form, while S serves as the 'helper' or carrier (Hibino, 1987).

## Linking rice seedling reactions to leaf-sucking insects with adult plant reactions

Inoculating young rice seedlings in seed boxes with a heavy inoculum of leafhoppers and planthoppers generally led to the categorization of either resistant or susceptible varieties. Comparatively few intermediate types were found. By using older (32- to 48-day-old) plants and a lighter dose of insects, resistant (R), moderately resistant (MR) and susceptible (S) categories could be more clearly differentiated and the findings could be compared with field tests in terms of plant damage and yield reduction. The tolerance of Utri Rajapan, an Indonesian variety, to heavy infestation was revealed in such a comparison (Panda and Heinrichs, 1983). Field resistance of 35- to 40-day-old plants was also differentiated when an infestation level of 3 to 5 second-instar nymphs/ seedling was used instead of very heavy insect pressure (Velusamy et al., 1986).

## Using environmental data at the site of collection to accelerate evaluation

Information on environmental conditions which is gathered at the site of collection can be profitably used to direct the evaluation test towards specific factors and thus accelerate the evaluation process. For instance, wild tomato, *Lycopersicon cheesmanii*, was collected from Isabela Island in an area exposed to salt spray and with a high salt content in the soil. Later tests in hydroponic culture showed that the accessions could be grown at a soil salt level of one-third (Epstein, 1977; Rick, 1979). Similarly, an Indonesian rice variety, Silewah, collected by a Central Research Institute for Food Crops (CRIFC)-IRRI team at 1300 m above sea level on Sumatra Island, has shown higher tolerance to coolness at the reproductive stage than have Japanese varieties on Hokkaido Island (Satake and Toriyama, 1979). At IRRI we make special seed increases of accessions which have been collected from distinctive environments and channel the seeds for specific evaluation tests ahead of the routine schedule (Chang, 1980, 1985b). In this way, collectors' observations and records can be put to rapid and profitable use.

Yield data obtained from IRTP nurseries indicate varietal differences in performance and stability. Varieties with wide adaptiveness or site-specificity were identified from different hydroedaphic regimes. The relationships between yield on the one hand and solar radiation and temperature on the other were elucidated (IRRI, 1980; Seshu and Cady, 1984; Seshu, 1985). Collaborative testing also showed the geographical differentiation of biotypes among the brown planthopper, gall midge and bacterial blight pathotypes, as well as revealing promising sources of resistance (Seshu and Kauffman, 1980; Seshu, 1985).

## CONSTRAINTS TO USING EXOTIC GERMPLASM

Lack of adaptation to commercial production and poor agronomic characters are the common complaints aired by breeders concerning exotic germplasm. Low crossability and sterility are additional drawbacks. Confusion between taxonomical ambiguity and fertility relationships adds to the reluctance to use exotic germplasm. However, these problems are not as serious when exotic materials are used solely as donor parents of specific traits. Many techniques, including genetic engineering, are available to overcome barriers in distant crosses (Harlan, 1984; Chang, 1985c). Backcrosses, multiple-parent crosses and bridging parents can help to produce usable progenies. Successful examples of these approaches have been summarized by Riley and Lewis (1966), Stalker (1980), Hermsen (1984), Peloquin (1984), Rick (1984) and Chang (1985c).

New biotechnology innovations have overcome some of the sexual barriers to wide hybridization. Desirable genes in unrelated genera may now be manipulated to improve various economic traits. However, the successful application of molecular biology to plant breeding requires extensive basic research on crop genetics, including detailed chromosome maps with DNA and protein markers (Flavell, 1985). For many agricultural crops, the information on basic Mendelian genetics is woefully inadequate.

## SUCCESSFUL USE OF EXOTIC GERMPLASM IN CROP IMPROVEMENT

Up to the early 1960s, the benefits derived from introducing foreign cultivated plants outweighed those of using exotic germplasm in the sense in which we have been using the term. However, the progress made since then has aroused widespread interest in landraces and wild taxa. Time and space limitations permit the mention of only a few outstanding successes in this area.

### Wheat

Multiple disease resistance and vigorous seedling emergence were found in a Turkish wheat, PI 178383 (Burgess, 1971). Multiple disease resistance was transferred to bread wheat from *Aegilops* and *Agropyron* species (Knott and Dvorak, 1976). Landraces from Turkey were instrumental in improving the hard red winter wheats of the USA and served as a parental source of the Japanese semidwarf wheat, Norin, and the Korean semidwarfs (Dalrymple, 1980). Wheat/rye translocation between chromosomes IB and IR are found in many European wheats (Chang, 1985c). Wide crosses are used extensively at the Centro Internacional de Mejoramiento de Maíz y Trigo (CIMMYT) to improve traits of importance in warm climates (Curtis, 1988).

### Rice

The semidwarfing gene originally found in Chinese semidwarfs, in which it probably arose from spontaneous mutations in the past, has contributed to rice yield improvements of great importance in the global context (Chang, 1984a). The 'Wild Abortive' plant in *Oryza sativa* f. *spontanea*, found on Hainan Island, provided the cytosterility for the hybrid rices (Lin and Yuan, 1980) which are now cultivated on 13 million ha in China. *O. nivara* was the sole source of resistance to the grassy stunt virus. A Chinese *spontanea* plant provided multiple pest resistance, as well as tolerance to cool temperatures and acid soils; numerous landraces found in South and South-East Asia have supplied resistance/tolerance to diseases, insects, adverse soils, drought, deep water, submergence and cool temperatures (Chang et al., 1982b). Wild species provided multiple resistance or novel sources of pest resistance (Chang, 1989) which are now being used in the breeding programmes being implemented at IRRI (IRRI, in press).

### Maize

Although the use of exotic germplasm has progressed more slowly in maize than in some other crops, rapid advances have been made in recent years in yield improvement and pest resistance (Chang, 1985c; Goodman, 1985).

### Oats

Use of genes from *Avena byzantina* and *A. sterilis* in yield improvement has produced significant results. The tetraploid, *A. barbata*, has been used to confer mildew resistance (Chang, 1985c).

### Sugarcane

Most modern varieties of sugarcane are polyglot progenies of several species. Chromosome numbers vary from 100 to 125 (Chang, 1985c).

### Potato

Many wild species of potato have provided disease and insect resistance (Harlan, 1984). More recent cytogenetical approaches to enhancing the use of wild potato species are discussed by Peloquin (1984) and Hermsen (1989).

## Triticale

Triticale is the first cereal developed artificially by plant breeders using intergeneric crosses (although wheat x rye crosses date back to 1875). It has yield advantages over wheat in marginal areas (Chang, 1985c).

## Coarse and small grains for semi-arid and arid areas

Reports on the progress made in the sorghums (Prasada Rao et al., 1989; Vidyabhushanam et al., 1989), in pearl millet (Harinarayana and Rai, 1989), and in barley, durum wheat and bread wheat (Srivastava and Damania, 1989) have recently appeared in the literature.

# CONCLUSION

The evaluation activities in exotic germplasm presented above demonstrate the value of such material in providing useful genes. However, efforts to date have not been very intensive. They have been confined largely to institutions which are well staffed and supported, and have tended to cover only the most important food crops. The main constraint for many less important crops is the lack of incentive for genetic resources specialists and evaluators to delve into exotic materials when breeders display little interest in materials with which they are unfamiliar.

A number of lessons have been learned through the collaboration between IRRI and national programmes. These are summarized here, together with some suggested improvements.

- A rich germplasm collection is crucial to finding novel sources of resistance/tolerance.
- IRRI's two-pronged approach to germplasm evaluation and utilization — interdisciplinary collaboration through the GEU programme and international collaboration through the IRTP network — has proved effective.
- Multilocational evaluation for resistance or tolerance to various biotic or eco-edaphic stresses benefits from the collaboration of national scientists because this enables national 'hot spots' to be used.
- An efficient computer database and network is essential in making information available to users.
- In-depth biological research and evaluation complement each other; the synergistic effect is heightened when increasingly diverse germplasm is used.
- Evaluation results provide useful information on the genetic diversity of conserved materials and assist further evaluation, re-collection and utilization.
- Different reactions from multilocational evaluations provide valuable information on genetic variation within insects and pathogens, as well as within plants.

- For distantly related materials, pre-breeding or germplasm enhancement efforts are necessary.
- More use should be made of core collections for evaluation work.

The enormous diversity and potential value of exotic germplasm has, for many crops, been only partly tapped. As the world faces rapidly increasing food needs in the coming years, agricultural scientists will need to make better use of exotic materials. Genetic engineering will help them to do so.

Genetic resources are the hub of all biological research activities, and specialists in this field have much to gain from successful evaluation exercises that are more than simply supplying germplasm. Genetic resources programmes can also benefit from success in evaluation in the long run. Collaboration among all participating disciplines is crucial to the meaningful use of the biological heritage in meeting the future need for increased food supply while the world's population continues to grow at an unabated pace.

## *Discussion*

S. Jana: Should one pay attention to optimal spacing in addition to optimal population (or sample) size for rejuvenating landraces and other heterogeneous accessions?

T. T. Chang: Most landraces of small grains tend to be tall, low tillering and long in growth duration, and therefore are susceptible to lodging. The cultivation practice should be relatively wide spacing, low seeding rate (to facilitate rogueing or selection) and low N application. The ideal situation is to provide an environment similar to the original habitat or to carry out seed regeneration in the home country, but that is often difficult to attain.

Wheat Genetic Resources: Meeting Diverse Needs
Edited by J. P. Srivastava and A. B. Damania

Published by John Wiley & Sons

# 3

# *Strategic Planning for Effective Evaluation of Plant Germplasm*

M. C. MACKAY

Historically, germplasm evaluation has been undertaken by plant improvement scientists on an informal basis to achieve specific research objectives. More recently, the trend has been towards evaluating larger numbers of accessions for a wider range of economically important attributes. The coordination of this work by germplasm scientists, without necessarily involving plant improvement scientists, carries a risk that the main objective — to make the best possible use of germplasm — will not be met. The primary purpose of evaluating germplasm is to enable plant improvement scientists to make more thorough use of available germplasm to improve productivity. This primary goal requires that the secondary goals of cost-effective evaluation and user access to the results be addressed.

The trend towards more systematic evaluation has resulted in the gathering of much information (the value and cost-effectiveness of which should be constantly under scrutiny) which is not always readily accessible to the plant improvement scientist. The design and implementation methods used for systematic evaluation are, therefore, open to question. In the design of an evaluation programme, consideration must be given to the selection of the economically important attributes to be evaluated, to the methods for testing accessions, and to the involvement of plant improvement scientists in collecting and using data. In other words, it is imperative that a germplasm evaluation programme is a cooperative one, involving plant improvement scientists as well as germplasm scientists.

The genebank servicing the requirements of a group of plant improvement scientists is the natural focal point for collating information on plant genetic resources, just as the plant improvement programmes are the obvious vehicles for collecting this information.

## SELECTION OF ATTRIBUTES AND GERMPLASM FOR EVALUATION

The selection of attributes for evaluation should reflect the major constraints to productivity. Nevertheless, some attributes not directly related to productivity should be also included for their future use in targeting germplasm for evaluation. Attributes such as photoperiod requirement and days to maturity can provide clues about the environment from which an accession originated. This information is useful in selecting germplasm to evaluate for specific economic attributes.

The selection of germplasm for evaluation is perhaps the most crucial part of the germplasm evaluation programme. Most countries still cannot afford to evaluate all germplasm for all attributes. In the past, genebank managers lacked the technology to systematically select germplasm for evaluation. The general approach was to evaluate all available germplasm or to randomly select germplasm for evaluation. Most genebanks now have all their data on accessions computerized, so that they can readily select discrete groups of accessions for evaluation against specific attributes.

The systematic approach has, in the past, been proposed as the most effective means of evaluation, especially with respect to non-adapted accessions. However, it is now suggested that a strategic approach should be used instead — an approach which uses the information routinely collected when material is multiplied or regenerated, so as to limit the number of accessions to be evaluated for a given attribute. This approach is based on the 'core collection' concept (Frankel and Brown, 1984). This concept of the core collection was developed in response to the massive numbers of accessions that have been conserved in genebanks over the past 15 years. These large numbers make it difficult to manage genebanks and associated activities such as evaluation. A core collection consists of a subset of accessions which is broadly representative of the genetic diversity within the whole. Passport data, together with morphological and agronomic data routinely collected by genebanks, can be used to select the 'core', thus reducing the effective size of the genebank for evaluation purposes.

Take, for example, the selection of wheat germplasm to evaluate for pre-harvest sprouting. Information about origin and maturity is useful in determining the type of environment in which an accession evolved or was selected. From such information, which is usually held as passport data, it can be deduced which accessions originated in environments where there is a likelihood of rain during the maturity stage of growth. To select accessions from other types of environments when evaluating for this attribute would be a waste of money. Thus, data on origin, which at first sight appear to have little value, can be extremely useful when combined with other attributes for selecting germplasm for future evaluation.

The value of these 'predictive' attributes will not be fully realized until more suitable tools for linking environments to accessions are developed, such as a database which cross-references soil types and climate with geographical regions. Such a tool would greatly enhance the process of germplasm utilization and plant improvement. In the meantime, predictive attributes can still be used to improve the efficiency with which germplasm is evaluated.

## MANAGEMENT AND EVALUATION OF WHEAT GENETIC RESOURCES IN AUSTRALIA

As mentioned earlier, a genebank is the obvious focal point for centralizing evaluation data, drawing the information from plant improvement scientists and collating it into a format acceptable to potential users. Evaluation data should be available on request from genebanks, ideally via direct access to the genebank computer for on-line queries. Information transfer is as vital as a strategically designed evaluation programme; in fact it should be part of the programme design.

In Australia, the focal point of cereal plant genetic activities is the Australian Winter Cereals Collection (AWCC). Most cereal introductions into Australia are made through the AWCC, as are most despatches to other countries. Australian cereal improvement scientists have found the process of obtaining genetic resources simpler and more cost-effective if the AWCC handles all requests, quarantine and seed dissemination. Of added value is the policy of the AWCC to ensure that all unique material added to the collection is made known to all scientists through a quarterly cereal introduction newsletter.

The information obtained by the AWCC is computerized and stored in files that generally follow the recommendations of the International Board for Plant Genetic Resources (IBPGR). This information, which is now available on-line to all interested parties in Australia, consists of passport data, descriptive data and evaluation data. Also included are listings of accessions held by other collections, standard cultivar abbreviations, and crosses and their parentages originating from the Centro Internacional de Mejoramiento de Maíz y Trigo (CIMMYT). The system is menu-driven, and most scientists find they can use it efficiently after a short learning period.

In addition to providing access to information, the system allows users to request seed samples, either from the collection or from other sources. Requests for seed held by the AWCC are processed and despatched within a week of being received; those for materials not held by the AWCC are ordered from the appropriate foreign institution.

The AWCC introduces almost 5000 cereal genotypes and distributes over 30 000 seed samples each year. Without its computerized management system, the AWCC would not be able to cater for this level of activity.

### Evaluation programme

Until recently, the AWCC evaluation programme was carried out in a random manner. Samples of material being cycled through the field were sent to cooperators to evaluate. This was all that could be done with the resources and technology available at the time. During the 1985-1989 period, as the AWCC expanded to include barley and oats, a new evaluation programme was planned. It consists of two parts: formal, routine evaluation; and shorter-term evaluation of specific attributes as the need arises.

The routine programme involves the selection of attributes for long-term evaluation — those of economic importance to production in Australia. Among the attributes selected are resistance to most cereal rusts, Septoria diseases, cereal cyst nematode and barley yellow dwarf virus, and tolerance to mineral toxicities and drought. This programme will be implemented in 1990 with the cooperation of scientists who routinely screen for responses to these attributes. Germplasm to be evaluated in the routine programme will be selected on the basis of ecogeographical data, to enable selection of those accessions most likely to provide a reasonable amount of genetic variation.

It is anticipated that, after the first cycle of evaluation, the data collected will enable more precise selection of germplasm for future evaluation. The objective will be to select germplasm that will provide the widest genetic variation for the attribute in question.

The specific attributes programme has now been in place for a number of years. It is used when a scientist requires germplasm to evaluate germplasm for a specific attribute, such as tolerance to boron toxicity. In this case, the AWCC and the scientist concerned make various predictive decisions about the origin of germplasm which might possess the desired attribute, such as the likely soil type, climate and days to maturity. These criteria are then used to select material for evaluation. This technique has been used with considerable success to select material with tolerance to pre-harvest sprouting, resistance to cereal cyst nematode, and tolerance to boron toxicity (Mackay, 1986).

## CONCLUSION

Eliminating the constraints to the evaluation and use of germplasm should be a high priority for genebank managers and plant improvement scientists. Responsibility for overcoming these constraints will lie largely with the genebank manager, whose task it will be to coordinate and implement carefully designed evaluation programmes and ensure that the resulting information is readily available to plant improvement scientists.

Perhaps the major constraint to evaluation is the vast number of accessions held by genebanks. Computerized databases provide the technology to manipulate large numbers of accessions. Three methods of ensuring that evaluation is made more effective are:

- use of the core collection concept to reduce the number of accessions to evaluate for particular attributes;
- selecting economically important attributes;
- making the resulting information readily available.

The technology is available and the methods are evolving to maximize the efficiency of plant genetic resource evaluation programmes.

## *Discussion*

J. P. SRIVASTAVA: I found your paper very interesting and useful. Could you please elaborate how international agricultural research centres can make use of the Australian experience in germplasm evaluation in the field as well as in information dissemination networks?

M. MACKAY: Our dissemination programme is still in the development stage, but I am sure information can be transferred/exchanged between genebanks in the form of magnetic tapes or disks perhaps on an annual basis. We need to standardize the format for this. Genebanks at international agricultural research centres could easily incorporate such information into their databases and distribute the information to the breeders and other users in the national programmes.

B. SKOVMAND: I would like to comment on Dr Srivastava's question. Dissemination of information is a very important aspect of a genebank's work. In the maize genetic resources programme at CIMMYT, they have developed a read-only-memory (ROM) disk which contains all the passport data of the local material from central America, together with the software to make this database accessible. We feel that this is an inexpensive method of transmitting information to the national programmes. They need only a personal computer to print the data.

Wheat Genetic Resources: Meeting Diverse Needs
Edited by J. P. Srivastava and A. B. Damania

Published by John Wiley & Sons

# 4

# *Phenotypic Diversity and Associations of Some Drought-Related Characters in Durum Wheat in the Mediterranean Region*

S. JANA, J. P. SRIVASTAVA, A. B. DAMANIA, J. M. CLARKE,
R. C. YANG and L. PECETTI

Parts of some 17 countries in Africa, Asia and Europe make up the Mediterranean Basin. Vavilov (1951) called this region the 'Mediterranean Centre' — one of several important world centres of diversity for plant species. According to Vavilov, 84 cultivated plant species evolved from their wild or weedy relatives in this region through the long process of domestication and conscious selection. Primitive cultivated varieties and landrace populations of several food crops in the Mediterranean region are known to abound in genetic diversity. Free-threshing tetraploid hard wheat, *Triticum turgidum* conv. *durum* (Desf.) MacKey (syn. *T. durum* Desf.), commonly known as macaroni wheat or durum wheat, is one of the most important of these food crops. It has gradually gained prominence in the Mediterranean region since the Neolithic period (Zohary and Hopf, 1988). The contemporary representative of durum wheat is widely cultivated in warmer and drier parts of the Mediterranean Basin, leaving the more favourable areas for the cultivation of bread wheat, *T. aestivum*.

Because of its long and extensive cultivation in the Mediterranean Basin, durum wheat has adapted well to Mediterranean environments. The typical Mediterranean environment may be described as having a moderately cool and humid winter, followed by a warm and dry spring (the crop-growing season), culminating in high temperature and moisture stress during the terminal grain-filling period. Considerable seasonal fluctuations occur in the amount and distribution of rainfall during the crop-growing season.

A question of obvious interest is: which plant characteristics, either singly or jointly, reflect the adaptation of durum wheat to arid and semi-arid Mediterranean environments? We sought an answer to this question by comparing the evaluation data of durum wheat germplasm collected from 15 Mediterranean countries with that of durum wheat germplasm from over 45 countries outside this region. Our study was aimed at answering the following questions:

- For some visually observable characters in durum wheat, is there more diversity within the Mediterranean region than outside it?
- For characters that are believed to confer drought tolerance in winter cereals, is there more diversity within the Mediterranean region than outside it?
- For which of the characters known to be important in drought tolerance is there a higher frequency of desirable types within the Mediterranean region than outside it?
- Are there regional differences in multicharacter associations?
- Are there regional differences in desirable multicharacter combinations?

These questions are of considerable interest in the study of the evolutionary biology of durum wheat, as well as in plant breeding. The last two questions are of particular interest in the conservation of genetic diversity and its deployment in breeding winter cereals for stressful Mediterranean environments.

## THE DATABASE

We used both characterization and evaluation data from two independent sources. For brevity, we will refer to both types of data as 'evaluation data'.

### The first source

In 1984, John Clarke and his associates conducted a comprehensive evaluation of over 4000 durum wheat accessions under rainfed conditions at Swift Current, Canada (Clarke et al., 1987). In addition to a number of morphological and phenological traits, these accessions were also evaluated for excised-leaf water retention capacity in an effort to screen germplasm for drought tolerance. The procedure and usefulness of the trait as an indicator of drought tolerance in wheats have been discussed elsewhere (Clarke and McCaig, 1982a, 1982b; Clarke, 1987; Blum, 1988). This database was particularly useful because, firstly, both annual (262 mm) and growing-season precipitation (166 mm) at the evaluation site were very low and, secondly, a large number of durum wheat accessions had been systematically evaluated for a physiological trait known to be related to drought tolerance in wheat. Furthermore, subsequent physiological studies under field conditions and controlled environments had confirmed that the methods used were sensitive enough to achieve a reliable initial classification of germplasm lines into drought-tolerant and drought-susceptible groups (Gummuluru et al., 1989).

## The second source

A large amount of data was available from an extensive multilocation evaluation of durum wheat germplasm conducted in Syria between 1984 and 1988 by ICARDA, in collaboration with the University of Tuscia in Viterbo, Italy. Nearly 8000 durum wheat accessions assembled from 60 countries had been evaluated at ICARDA's principal experimental farm at Tel Hadya and at the Center's drought research site at Breda, about 20 km south-east of Tel Hadya. At both locations, the accessions had been grown in single-row plots laid out in a modified augmented design with three common check cultivars. The accessions had been evaluated for 20 characters, nine of which were measured in each accession. For the remaining 11 characters, the accessions were evaluated on a nominal or ordinal scale, which varied from trait to trait. Details of the methods of data collection and the problems associated with such a large-scale germplasm evaluation have been described by Pecetti et al. (in prep.) and Porceddu and Srivastava (1990). Table 4.1 summarizes the information on the durum wheat accessions from 15 countries in the Mediterranean region and 35 countries outside the region that were assembled for the two-location germplasm evaluation in Syria.

**Table 4.1** Number of durum wheat accessions from the three Mediterranean subregions and other regions of the world characterized and evaluated in Canada and Syria

| Subregion or region | Number of accessions evaluated | |
|---|---|---|
| | Canada | Syria |
| Mediterranean Africa<br>(Algeria, Egypt, Libya, Morocco and Tunisia) | 270 | 550 |
| Mediterranean Asia<br>(Lebanon, Syria and Turkey) | 831 | 1516 |
| Mediterranean Europe<br>(Cyprus, France, Greece, Italy, Malta, Spain and Yugoslavia) | 445 | 1125 |
| Mediterranean region | 1546 | 3191 |
| Other regions<br>(45 countries) | 1884 | 4776 |
| Total | 3430 | 7967 |

Some meteorological data on the evaluation locations are given in Table 4.2 (*overleaf*). Altogether, 30 characters were used in the analyses reported in this paper: 19 at Swift Current, Canada; 19 at Tel Hadya, Syria; and six at Breda, Syria (*see* Table 4.3 *overleaf*). Agronomic performance, an important overall visual indicator of drought tolerance, was evaluated in relation to that of locally adapted check cultivars at Swift

Current (semi-arid) and Breda (arid), but not at Tel Hadya (semi-arid). Eight characters were evaluated at both Swift Current and Tel Hadya, and only five at both Tel Hadya and Breda. Because of the substantial location effect, the evaluation data for the same character from two locations were considered separately. Thus, there were 44 different 'character-cases' from the evaluation of 30 characters at the three locations.

**Table 4.2** Meteorological data for the three evaluation locations

| Location | Latitude | Longitude | Elevation (m) | Growing season (days) | Precipitation (mm) |
|---|---|---|---|---|---|
| Breda, Syria | 35° 56'N | 37° 10'E | 300 | 279 | 246 |
| Tel Hadya, Syria | 36° 01'N | 36° 56'E | 284 | 332 | 290 |
| Swift Current, Canada | 50° 16'N | 107° 44'W | 825 | 352 | 116 |

## DATA REDUCTION

Check entries of locally adapted cultivars in respective countries were interspersed among accessions at each location. Variation among check entries was used as a measure of soil heterogeneity and reliability of categorical data. For the 11 measurement characters listed in Table 4.3, variation of entries on no occasion exceeded that of accessions. The polymorphic index, as described by Kahler et al. (1980) and Zhang and Allard (1986), was calculated from the relative phenotypic frequencies for each of the 19 categorical traits. These estimates were zero, or not significantly different from zero, for the check entries, but were significantly greater than zero ($P = 0.05$) for the accessions.

There were over 264 200 observations, excluding those on the check entries. Two methods of data reduction were used. For the measurement data, mean and standard deviation were calculated for accessions each year. These statistics were used to classify the accessions evaluated in each year into three groups according to the following class limits:

- Group 1: less than or equal to $\bar{x}$-s;
- Group 2: greater than $\bar{x}$-s to less than $\bar{x}$+s;
- Group 3: equal to or greater than $\bar{x}$+s.

For the discrete multivariate log-linear analysis (Fienberg, 1980) of the cross-classified categorical data, this method of grouping was considered more appropriate than a classification based on the 33rd and 67th centiles. For the 10 cases of categorical classification with more than three classes, the frequency data were reduced to only three classes on the basis of usual human preference or agronomic desirability. Thus, for kernel colour, although the number of accessions in each of the four discrete

**Table 4.3** List of characters used for multilocation characterization and evaluation of durum wheat germplasm

| Type of character | Name of character | Location [a] | Type of data recorded [b] | |
|---|---|---|---|---|
| Morphological | Juvenile growth habit | TH, S | Categorical | (3) [c] |
| | Early vigour | B, TH | Categorical | (5) |
| | Growth class | S | Categorical | (3) |
| | Glaucousness | TH, S | Categorical | (3) |
| | Leaf size and shape | S | Categorical | (3) |
| | Leaf attitude | S | Categorical | (3) |
| | Chlorotic leaf spot | S | Categorical | (5) |
| | Awnedness | TH, S | Categorical | (3) |
| | Awn colour | S | Categorical | (3) |
| | Spike length | TH | Measurement | |
| | Spike density | TH | Categorical | (5) |
| | Number of seeds per spike | TH | Measurement | |
| | Kernel weight | TH | Measurement | |
| | Kernel colour | TH | Categorical | (4) |
| | Glume hairiness | S | Categorical | (3) |
| | Glume colour | S | Categorical | (3) |
| | Straw strength | TH, S | Categorical | (3, 4) |
| | Plant height | B, TH, S | Measurement | |
| Phenological | Days to flowering | B, TH, S | Measurement | |
| | Days to maturity | B, TH, S | Measurement | |
| | Grain-filling period | B, TH | Measurement | |
| Physiological | Leaf rolling | S | Categorical | (2) |
| | Agronomic performance | B, S | Categorical | (9, 5) |
| | Grain yield | TH | Measurement | |
| | Grain protein content | TH | Measurement | |
| | Tillering capacity | TH, S | Categorical | (3, 5) |
| | Rate of leaf-water loss | S | Measurement | |
| | Leaf-water retention capacity | S | Measurement | |
| | Resistance to yellow rust | TH | Categorical | (5) |
| | Resistance to common bunt | TH | Categorical | (9) |

a Letters in this column indicate evaluation sites; B = Breda, TH = Tel Hadya and S = Swift Current
b Instead of recording actual days maturity, accessions were classified into five ordinal categories at Swift Current
c Number within parentheses indicate the number of classes used for categorical classification of accessions

categories — amber, white, light red and purple — was recorded, the categories were subsequently reduced to three, from the most preferred to the least preferred, as (a) desirable (amber), (b) intermediate (white) and (c) undesirable (light red and purple).

This procedure, although somewhat arbitrary, reduced the number of classes to a manageable size. Furthermore, apart from leaf rolling, which was recorded in only two classes at Swift Current, this procedure provided three classes uniformly for each of the remaining 29 characters listed in Table 4.3. For each of these characters, the number of accessions in each class was enumerated and relative frequency estimated.

## UNIVARIATE FREQUENCIES

Global frequencies in different classes of a character were calculated by pooling data from all countries outside the Mediterranean region. In this paper we will refer to the global relative frequency in a given class of a character as the 'global frequency'. Similarly, the relative frequencies in different classes of the characters in the Mediterranean region were calculated by pooling frequency data from the 15 Mediterranean countries listed in Table 4.1. We refer to each of these class frequencies as the 'regional frequency'.

For nine of the 19 characters evaluated at Swift Current, the regional frequencies in desirable classes were significantly higher ($P < 0.05$) than the respective global frequencies (*see* Table 4.4). Seven of these traits — glaucousness, leaf size and shape, leaf attitude, leaf rolling, tillering capacity, rate of excised-leaf water loss and excised-leaf water retention capacity — are known to play important roles in enhancing drought tolerance. Agronomic performance, evaluated at plant maturity, was an overall measure of plant response to prevailing field drought conditions. As indicated later, for most of these characters the frequency of accessions in the undesirable class was also significantly lower in the Mediterranean region than in the world.

**Table 4.4** List of nine characters for which the Mediterranean region shows higher than global frequency in desirable classes evaluated at Swift Current, Canada

| **Character** | **Desirable class** | **Frequency in desirable class (x100)** | |
|---|---|---|---|
| | | **Mediterranean** | **World** [a] |
| Glaucousness | Strong | 70.2 | 52.5 |
| Leaf size and shape | Short | 16.7 | 6.5 |
| Leaf attitude | Erect | 21.2 | 8.7 |
| Chlorotic leaf spot | Green | 59.0 | 43.0 |
| Leaf rolling | Rolling | 31.2 | 27.0 |
| Agronomic performance | Excellent | 14.4 | 12.5 |
| Tillering capacity | High | 22.4 | 20.1 |
| Rate of loss | Slow | 15.3 | 7.5 |
| Leaf-water retention capacity | High | 14.2 | 10.9 |

a Excludes the Mediterranean region

Twenty characters were used for the evaluation of 7967 accessions at the two locations in Syria. Of these, five were common to Tel Hadya and Breda, giving a total of 25 character-cases. The Mediterranean region showed significantly higher frequency in the desirable class for 16 out of the 25 cases (*see* Table 4.5). All these characters are considered important under most conditions of durum wheat production. It is noteworthy, however, that for several of these characters that are particularly important in drought-prone areas (such as early vigour, glaucousness, plant height, days to flowering, grain-filling period and agronomic performance) regional frequencies in the desirable class were higher than the corresponding global frequencies.

**Table 4.5** List of 16 characters for which the Mediterranean region shows higher than global frequency in desirable categories evaluated in Syria

| **Character**[a] | **Desirable class** | **Frequency in desirable class (x100)** | |
|---|---|---|---|
| | | **Mediterranean** | **World**[b] |
| Juvenile growth habit | Erect | 54.2 | 49.1 |
| Early vigour (B) | Excellent | 17.4 | 14.9 |
| Glaucousness | Strong | 47.4 | 21.3 |
| Awnedness | Full | 94.0 | 73.7 |
| Spike density | Dense | 42.6 | 20.1 |
| Kernel weight | Heavy | 27.1 | 8.7 |
| Kernel colour | Yellow | 79.1 | 41.4 |
| Straw strength | Strong | 72.7 | 59.3 |
| Plant height | Tall | 20.8 | 13.8 |
| Plant height (B) | Tall | 20.2 | 14.0 |
| Days to flowering (B) | Early | 16.4 | 10.6 |
| Grain-filling period | Long | 14.9 | 10.5 |
| Grain-filling period (B) | Long | 21.9 | 13.1 |
| Grain yield | High | 17.1 | 14.7 |
| Agronomic performance (B) | Excellent | 20.5 | 9.2 |
| Resistance to common bunt | Resistant | 24.1 | 12.8 |

a Letter B within parentheses represents the evaluation site, Breda. All characters without this letter were evaluated at Tel Hadya
b Excludes the Mediterranean region

For about 80 per cent of the character-cases which are listed in Tables 4.4 and 4.5, the Canadian and Syrian evaluation data also revealed significantly lower regional frequencies in the undesirable classes, as shown in Table 4.6 (*overleaf*). Thus, in the Mediterranean region, shifts towards more desirable plant types were evident for most of the characters which are known to be directly or indirectly associated with drought tolerance. However, the regional frequencies were lower than the global frequencies for 11 characters, some of which are also important for increasing plant tolerance to field drought conditions (*see* Table 4.7).

On the whole, the univariate frequency data demonstrate that a relatively high proportion of durum wheats from the Mediterranean region possess attributes conducive to drought tolerance.

Two questions arise from these results. Is the Mediterranean region a world centre of diversity for drought-related traits? Are these traits synergistically organized into multicharacter combinations? We carried out further analyses of the Canadian and Syrian evaluation data to answer these two questions.

**Table 4.6** Frequencies in undesirable classes for 25 characters in the Mediterranean region and the world, estimated from the Canadian and Syrian evaluation data

| **Character** [a] | **Undesirable class** | **Frequency in undesirable class (x100)** | |
|---|---|---|---|
| | | **Mediterranean** | **World** [b] |
| Juvenile growth habit (TH) | Prostrate | 8.8 | 3.5 |
| Early vigour (B) | Poor | 25.9 | 25.8 |
| Glaucousness (TH) | Weak, absent | 9.2 | 25.1 |
| Glaucousness (S) | Weak, absent | 4.0 | 9.6 |
| Awnedness (TH) | Awnless | 0.5 | 1.2 |
| Spike density (TH) | Sparse | 7.1 | 36.8 |
| Kernel weight (TH) | Light | 11.7 | 30.0 |
| Kernel colour (TH) | Light red, purple | 12.7 | 49.8 |
| Straw strength (TH) | Weak | 8.4 | 13.1 |
| Leaf size and shape (S) | Broad, large | 30.5 | 29.1 |
| Leaf attitude (S) | Horizontal | 24.2 | 36.1 |
| Chlorotic leaf spot (S) | 100% covered | 8.6 | 15.2 |
| Plant height (TH) | Short | 9.5 | 19.4 |
| Plant height (B) | Short | 10.0 | 17.9 |
| Days to flowering (B) | Late | 16.1 | 15.5 |
| Grain-filling period (TH) | Short | 10.2 | 18.7 |
| Grain-filling period (B) | Short | 13.4 | 18.8 |
| Grain yield (TH) | Low | 12.3 | 18.0 |
| Agronomic performace (B) | Poor | 10.0 | 22.4 |
| Agronomic performance (S) | Poor | 47.4 | 56.0 |
| Leaf rolling (S) | Absent | 0.2 | 0.3 |
| Resistance to common bunt (TH) | Susceptible | 22.3 | 49.5 |
| Tillering capacity (S) | Low | 33.3 | 33.0 |
| Rate of leaf-water loss (S) | Fast | 7.1 | 18.7 |
| Leaf-water retention capacity (S) | Low | 9.4 | 12.2 |

a Letters in parentheses indicate evaluation sites; B = Breda, TH = Tel Hadya and S = Swift Current
b Excludes the Mediterranean region

**Table 4.7** List of 11 characters for which the global frequency in the desirable class is higher than the Mediterranean frequency

| **Character** [a] | **Desirable class** | **Frequency in desirable class (x100)** | |
|---|---|---|---|
| | | **Mediterranean** | **World** [b] |
| Juvenile growth habit (S) | Erect | 68.6 | 79.5 |
| Early vigour (TH) | Excellent | 43.7 | 70.5 |
| Spike length (TH) | Long | 8.6 | 24.6 |
| Glume hairiness (S) | High | 17.3 | 25.7 |
| Plant height (S) | Tall | 14.4 | 17.7 |
| Days to maturity (TH) | Early | 14.6 | 21.6 |
| Days to flowering (B) | Early | 16.4 | 18.4 |
| Days to maturity (S) | Early | 3.6 | 13.4 |
| Days to maturity (B) | Early | 6.5 | 26.6 |
| Tillering capacity (TH) | High | 50.7 | 58.8 |
| Protein content (TH) | High | 15.5 | 17.3 |

a Letters in parentheses indicate evaluation sites; B = Breda, TH = Tel Hadya and S = Swift Current
b Excludes the Mediterranean region

## MEASUREMENTS OF DIVERSITY

For each character, the accessions were classified into different nominal or ordinal categories on the basis of their respective phenotypes. Thus, a statistic describing phenotypic diversity was necessary to compare the average diversity between regions. Of the two commonly used measures of phenotypic diversity — Shannon's information statistic (Hutcheson, 1970; Bowman et al., 1971), and the phenotypic polymorphism index (Kahler et al., 1980) — the former was considered more appropriate for the present analysis because of its extensive use in estimating diversity in germplasm collections. Jain et al. (1975) used this index for the first comprehensive assessment of phenotypic diversity in a world collection of durum wheats. This statistical procedure has also been used extensively in population ecology.

We used the following formula for calculating Shannon's information statistic ($h_{s.j}$) for the 'j'th character with n character states or classes (Bowman et al., 1971):

$$h_{s.j} = -\sum p_i \ln p_i \qquad \text{for } i = 1,2,\ldots,n$$

where $p_i$ is the relative frequency in the 'i'th class of the 'j'th character. The average diversity (H') over k characters was estimated as:

$$H' = \sum h_{s.j}/k \qquad \text{for } j = 1,2,\ldots,k$$

Although the approximate variance of $h_{s.j}$ has been described (Hutcheson, 1970), the variance of H' has not been characterized. Considering the characters used for the study as a random sample of characters, we calculated an empirical variance for the average diversity: $H'_m$ for the Mediterranean region from the k estimates of $h_{s.j}$; and a similar variance for the average world diversity ($H'_w$).

First, we calculated the $h_{s.j}$ values for each of the 19 characters which had been evaluated at Swift Current, and used these values to estimate $H'_m$ and $H'_w$. Similarly, we used the 19 $h_{s.j}$ values from Tel Hadya and the six from Breda to calculate $H'_m$ and $H'_w$ statistics from the Syrian evaluation data. These estimates indicate that average diversities in the Mediterranean region and the rest of the world are essentially the same (*see* Table 4.8).

**Table 4.8** Average diversity indices (H's) estimated over all characters from the Canadian and Syrian evaluation data. Indices for Ethiopia are given in parentheses for comparison

| | Diversity index (H')[a] | |
|---|---|---|
| **Region** | **Canadian data** | **Syrian data** |
| Ethiopia | 0.697<br>(0.056) | 0.800<br>(0.019) |
| Mediterranean | 0.799<br>(0.050) | 0.823<br>(0.031) |
| World[b] | 0.825<br>(0.046) | 0.860<br>(0.019) |

a Empirical standard error is given in parentheses; k = 19 and 25 for Canada and Syria, respectively
b Excludes the Mediterranean region, but includes Ethiopia

Secondly, we selected only those characters, from both the Canadian and Syrian evaluations, which are usually considered important in conferring drought tolerance in wheat (Blum, 1988). We used $h_{s.j}$ values for these characters to recalculate the average diversity indices, $H'_m$ and $H'_w$ (*see* Table 4.9). As before, the results show that the average diversities in the Mediterranean region are nearly equal to those outside the region.

Thus the relatively high frequency of accessions with drought-related attributes in the Mediterranean region was not accompanied by a reduced level of diversity. In effect, the extent of diversity in the Mediterranean region was similar to that found elsewhere, including Ethiopia, which is the other well-known region of diversity for durum wheat.

**Table 4.9** Average diversity indices (H's) in the Mediterranean region and the world for 27 characters that are important for dryland agricultural conditions

| **Character** [a] | **Mediterranean region** | **World** [b] |
|---|---|---|
| Juvenile growth habit (TH) | 0.914 | 0.822 |
| Juvenile growth habit (S) | 0.794 | 0.596 |
| Early vigour (TH) | 1.024 | 0.765 |
| Early vigour (B) | 0.975 | 0.943 |
| Glaucousness (TH) | 0.935 | 1.010 |
| Glaucousness (S) | 0.727 | 0.931 |
| Leaf size and shape (S) | 0.999 | 0.824 |
| Leaf attitude (S) | 1.003 | 0.909 |
| Awnedness (TH) | 0.245 | 0.626 |
| Awnedness (S) | 0.248 | 0.257 |
| Plant height (TH) | 0.806 | 0.861 |
| Plant height (B) | 0.804 | 0.845 |
| Plant height (S) | 0.754 | 0.897 |
| Days to flowering (TH) | 0.882 | 0.817 |
| Days to flowering (B) | 0.857 | 0.751 |
| Days to flowering (S) | 0.863 | 0.892 |
| Days to maturity (TH) | 0.890 | 0.879 |
| Days to maturity (B) | 0.735 | 0.916 |
| Days to maturity (S) | 0.455 | 0.774 |
| Grain-filling period (B) | 0.883 | 0.841 |
| Leaf rolling (S) | 0.634 | 0.601 |
| Agronomic performance (B) | 0.809 | 0.814 |
| Agronomic performance (S) | 1.000 | 0.948 |
| Tillering capacity (TH) | 0.725 | 0.724 |
| Tillering capacity (S) | 1.062 | 1.043 |
| Rate of leaf-water loss (S) | 0.671 | 0.732 |
| Leaf-water retention capacity (S) | 0.702 | 0.703 |
| Average H's | 0.792 | 0.805 |
| Standard error of H's | 0.040 | 0.031 |
| Maximum possible H's | 1.084 | 1.084 |

a Letters in parentheses indicate evaluation sites; B = Breda, TH = Tel Hadya and S = Swift Current
b Excludes the Mediterranean region

## CORRELATIONS BETWEEN MEASUREMENT CHARACTERS

Associations between pairs of characters were examined by calculating the standard Pearson's product-moment correlation coefficients. The correlation coefficient for each character-pair was calculated separately for each year, then transformed to the z function as described by Fisher (1921). A pooled correlation coefficient was calculated from the three z values. Selected correlations of interest are given in Table 4.10.

**Table 4.10** Estimates of pooled correlation coefficients between the three phenological traits, plant height and agronomic performance from the Syrian evaluation data

| Character pair [a] | Pooled correlation coefficient | |
|---|---|---|
| | **Mediterranean region** | **World** [b] |
| Days to flowering (TH) and days to maturity (TH) | 0.769 | 0.722 |
| Days to flowering (B) and days to maturity (B) | 0.607 | 0.397 |
| Days to flowering (TH) and grain-filling period (TH) | -0.406 | -0.118 |
| Days to flowering (B) and grain-filling period (B) | -0.892 | -0.786 |
| Days to maturity (TH) and grain-filling period (TH) | 0.271 | 0.613 |
| Days to maturity (B) and grain-filling period (B) | -0.164 | 0.301 |
| Plant height (TH) and days to maturity (TH) | 0.102 | 0.393 |
| Plant height (B) and days to maturity (B) | -0.050 | 0.421 |
| Plant height (TH) and grain-filling period (TH) | -0.133 | 0.211 |
| Plant height (B) and grain-filling period (B) | 0.211 | 0.453 |
| Agronomic performance (B) and plant height (B) | 0.409 | 0.505 |
| Agronomic period (B) and grain-filling period (B) | 0.488 | 0.485 |

a Letters in parentheses indicate evaluation location; B = Breda and TH = Tel Hadya
b Excludes the Mediterranean region

As expected, days to flowering and days to maturity showed high correlation. However, the association was weaker at Breda than at Tel Hadya, probably because of the disruption of physiological harmony under extreme field drought conditions. On the other hand, under moderately favourable soil moisture conditions at Tel Hadya, the association between days to flowering and grain-filling period was substantially weaker than under the highly arid conditions at Breda. The correlation between days to maturity and grain-filling period was stronger outside the Mediterranean region. Similarly, plant height showed a stronger association with both days to maturity and grain-filling period outside the region.

It is noteworthy that the two characters of considerable importance in durum wheat production under moisture-limiting conditions — plant height and grain-filling period — showed relatively high correlations with overall agronomic performance at Breda, and the correlation coefficients in the Mediterranean region were essentially the same as outside the region. Correlations between characters were, in general, smaller at Swift Current than at Tel Hadya and Breda. For the two physiological traits of considerable importance in drought tolerance — rate of excised-leaf water loss and excised-leaf water retention capacity — there was little or no linear correlation with overall agronomic performance and other traits of interest under field drought conditions at Swift Current.

## MULTICHARACTER ASSOCIATIONS

Tests of the independence of pairs of categorical data were performed using the standard two-way contingency chi-square tests. We used the entire Syrian evaluation data, including those from the Mediterranean region for the preliminary association analyses. Out of a total of 66 two-way contingency chi-square tests involving the 12 categorical characters, only one, between early vigour and straw strength, evaluated at Breda and Tel Hadya respectively, appeared independent ($P = 0.20$). Thus bivariate associations were generally the rule rather than the exception. In an extension of the bivariate analysis to three or more variables, we adopted a discrete multivariate technique, the log-linear analysis, as described by Fienberg (1980). This technique was considered appropriate for elucidating the organization of multicharacter associations.

To reduce the complexities of data analysis and interpretation, the number of characters for the log-linear analysis was limited to five. Even so, for the large number of characters evaluated in Canada and Syria, the analyses of all possible five-character combinations and their interpretation presented a formidable challenge. For this reason, a judicious selection of characters was necessary. We carefully reviewed the large volume of literature on drought tolerance in winter cereals and decided that the group of five most important characters from the Canadian evaluation should consist of glaucousness, leaf size and shape, days to maturity, leaf rolling and the rate of excised-leaf water loss. From the Syrian evaluation, the five most important characters were considered to be early vigour, plant height and grain-filling period (evaluated at Breda) and glaucousness and juvenile growth habit (evaluated at Tel Hadya).

For the five-character association studies, because of the complexity arising from the large number of possible associations, a preliminary selection of interaction terms was performed and a basic log-linear model consisting of only significant interaction terms was constructed. Additional log-linear models were formed using a stepwise selection procedure. The likelihood ratio statistics ($G^2$) were then calculated for the respective log-linear models and tested for the goodness-of-fit of the models.

For a given combination of five characters, a parsimonious selection of the 'best-fitting' model was made by choosing the simplest model — that is, the model with a minimal number of association terms which fitted the data statistically. Both forward selection and backward elimination procedures were employed in the selection of appropriate models.

The final models for the Mediterranean region, selected from the Canadian data, included four two-factor associations, three of which were also present in the final model for the world (global model) involving six two-factor associations. Neither model included higher than two-factor associations (*see* Table 4.11 *overleaf*). In the global model, leaf size and shape was associated with each of the other four traits, followed by days to maturity with three, and by glaucousness and leaf rolling with two each. Each of these four traits showed two pairwise associations in the Mediterranean region. Excised-leaf water loss was the least interactive trait. It interacted only with leaf size and shape in the global model, and with glaucousness in the regional model.

**Table 4.11** Simplest log-linear models for the Mediterranean region and the rest of the world, derived from the Canadian and Syrian evaluation data

| | **Log-linear model** [a] | | | | | | |
|---|---|---|---|---|---|---|---|
| **Region** | **Canadian data** | | | | **Syrian data** | | |
| Mediterranean | (CD) | (CI) | (DG) | (DJ) | (AFH)<br>(BF) | (AB)<br>(EF) | (AE)<br>(EH) |
| World [b] | (CD)<br>(DG) | (CI)<br>(GI) | (CG) | (CJ) | (ABF)<br>(EHF) | (AEF)<br>(AH) | (BEH) |

a List of abbreviations for characters evaluated at Swift Current (S), Tel Hadya (TH) and Breda (B): A: juvenile growth habit (TH), B: early vigour (B), C: leaf size and shape (S), D: glaucousness (S), E: glaucousness (TH), F: plant height (B), G: days to maturity (S), H: grain-filling period (B), I: leaf rolling (S), J: rate of leaf-water loss (S)
b Excludes the Mediterranean region

The 'best-fitting' models from the Syrian data included only one three-factor association term for the Mediterranean region which did not appear in the global model. None of the four three-factor association terms in the global model was present in the region. However, of the 10 two-factor association terms in the global model, only two — early vigour x glaucousness and early vigour x grain-filling period — were absent in the regional model. Glaucousness and plant height occurred in three of the four three-factor interactions in the global model. Thus the models reveal a lower degree of association among the five characters in the Mediterranean region. No associations were detected that were unique to the region. None of the five traits appeared completely independent in either model.

Using only five traits which were presumed to be related to drought, we found that, during their long agricultural history, durum wheats in the world had developed several first-order and second-order associations. These almost certainly result from the combined effects of both human and natural selection. As far as these five traits are concerned, the environmental peculiarities of the Mediterranean Basin had not produced greater and more complex multi-character associations than had other durum wheat-growing environments in the rest of the world. Some drought-related traits, such as plant height and glaucousness, were involved in multi-character associations to a greater extent than other traits considered in the multivariate association analysis.

An extension of this approach to additional characters of importance, and to other well-defined agricultural environments, should provide a more comprehensive picture of the evolution of plant architecture in durum wheats in relation to environmental stresses.

## ESTIMATES OF JOINT FREQUENCIES

It is evident from the above results that non-random multicharacter associations occur in durum wheats both inside and outside the Mediterranean region. Two questions of practical significance are: Did these associations lead to the development of more desirable plant types for stressful Mediterranean environments? And, if so, are there regional differences in the joint frequencies of desirable combinations? For the same two five-character sets used for the multivariate association analysis, we calculated the joint frequencies in desirable classes of any two, three, four and five characters (*see* Table 4.12). The table shows that there were substantially higher proportions of desirable combinations of up to four drought-related characters in the Mediterranean region.

**Table 4.12** Joint frequencies in desirable classes of two, three, four and five characters used for the respective log-linear analyses

| **Dataset** [a] | **Number of characters** | **Frequency in desirable class (x100)** | |
|---|---|---|---|
| | | **Mediterranean** | **World** [b] |
| Canadian (S) | Two | 37.9 | 27.3 |
| | Three | 11.8 | 4.3 |
| | Four | 1.4 | 0.3 |
| | Five | 0.0 | 0.0 |
| Syrian (B, TH) | Two | 47.3 | 30.5 |
| | Three | 16.7 | 8.5 |
| | Four | 4.1 | 1.6 |
| | Five | 0.3 | 0.2 |

a The list of characters is given in the footnote of Table 4.11; S = Swift Current, B = Breda and TH = Tel Hadya
b Excludes the Mediterranean region

Whether or not plant species tend to evolve towards a 'model' plant architecture for any particular environmental situation is an open question (Smith, 1987). However, the high joint frequencies in stressful Mediterranean environments suggest a comparative advantage of these environments in the selection and preservation of some drought-related traits. The simple empirical analysis indicates that, in the never-ending search for a 'model' architecture for moisture-limiting conditions, there exists a higher probability of finding such a model in the Mediterranean durum wheat germplasm than elsewhere.

## CONCLUSION

In most areas, including the Mediterranean Basin, durum wheat is usually cultivated under rainfed conditions in which it is exposed to drought. Plant characteristics conducive to development and survival under drought stress are therefore likely be common in the crop's primitive cultivars and landrace populations. Genetic improvement of durum wheat for drought-prone areas may be achieved by incorporating these favourable attributes into advanced breeders' lines and modern cultivars.

For nearly 8000 durum wheat accessions used in our evaluation, it was possible to determine the countries of origin from the available passport information. However, it was not possible to determine the exact biological status of the accessions. From the incomplete passport information, it was evident that most of the accessions from regions outside North America (only 145) were primitive varieties and landraces (old germplasm). The proportion of old germplasm was certainly very high (estimated at over 95 per cent) from the countries bordering the Mediterranean Sea and from other parts of Africa and Asia. It is reasonable to assume that these accessions reflect the combined result of hundreds of generations of natural and human selection. It is not surprising, therefore, that non-random desirable associations of several important drought-related characters were observed in these accessions.

Plant breeding efforts, particularly for harsh Mediterranean environments, should take into account these favourable gene combinations. Furthermore, conservation strategies should be devised that not only preserve but also promote the development of the various synergistically organized multilocus structures that enhance crop productivity in these environments. That is, whereas the short-term goal of plant breeders should be to exploit genetic diversity for multivariate organizations of characters, the long-term objective of genetic conservationists must be to enhance genetic diversity for such adaptively organized multivariate characters in breeding populations.

For several important traits, Mediterranean environments maintained a higher frequency of desirable plant types than did all other regions of the world pooled together. However, these higher frequencies did not result in the reduction of overall phenotypic diversity in the Mediterranean region.

For the two sets of five characters that were considered important for drought tolerance in wheats, desirable combinations of two or more characters appeared in substantially higher frequency in the Mediterranean region than in the rest of the world. Evaluation under arid Mediterranean conditions in Syria revealed that plant height and glaucousness were the most interactive traits. Evaluation under semi-arid temperate conditions in Canada revealed that leaf size and shape appeared most frequently in pairwise associations with other traits, followed by glaucousness, suggesting the relatively high interactive nature of the two traits.

The Mediterranean Basin is a region that harbours high genetic diversity in durum wheat. It is also a region in which relatively high frequencies of desirable multi-character combinations useful in arid conditions have been maintained. There is probably no single durum wheat architecture that is ideal for a stressful environment. However, the evidence presented here strongly supports the view that a progressive increase in the frequencies of several desirable character combinations is possible. These combinations will be extremely useful in breeding durum wheat for stressful Mediterranean environments.

## *Discussion*

M. TAHIR: In view of the great differences in the days to maturity, growth habit and photoperiod sensitivity of wheat genotypes which are subjected to highly variable temperatures and moisture regimes, what is the reliability of data on grain-filling period?

S. JANA: Unreplicated single-row observations are not very reliable, although checks planted at frequent intervals were used to increase reliability of the measurements. Nevertheless, the grain-filling period data provide very useful information on an overall basis.

C. DE PACE: The Shannon-Weaver diversity index and log-linear model allows the analysis of categorical and frequency data between populations from different geographical areas. According to your experience, which method should be adopted for obtaining meaningful information on characters and populations that show the highest phenotypic diversity?

S. JANA: The Shannon-Weaver procedure is used to measure diversity within and between accessions. The log-linear analysis is used to assess associations between three or more characters among accessions.

Wheat Genetic Resources: Meeting Diverse Needs
Edited by J. P. Srivastava and A. B. Damania

Published by John Wiley & Sons

# 5

# *Evaluation of Greek and Turkish Durum Wheat Landraces*

A. BIESANTZ, P. LIMBERG and N. KYZERIDIS

In a report to the Food and Agriculture Organization (FAO) of the United Nations, Frankel (1973) discussed the worldwide loss of valuable durum wheat genetic resources. He specified the Mediterranean and the West Asia regions as having suffered considerably from this depletion, and cited the introduction of modern high-yielding varieties as a major cause for the displacement of existing landraces. Zamanis (1989) reported that only about 3 per cent of the area cultivated to durum wheat in Greece still consisted of landraces.

The exhaustion of these genetic resources may lead to the disappearance of such valuable properties as drought resistance, yield stability, and a variety of quality characteristics. This concern led to efforts to collect remaining durum wheat germplasm in the East Aegean islands of Greece (Zamanis, 1983), and in all the regions of Turkey. The research outlined in this paper was based on the evaluation of these landraces in field and pot experiments conducted under different ecological conditions (Biesantz et al., 1985-87).

## MATERIALS AND METHODS

Over a 2-year period, 32 Greek durum wheat landraces and five released varieties were tested at Salonica and Serres in Greece and at the main station of the International Center for Agricultural Research in the Dry Areas (ICARDA) at Tel Hadya, near Aleppo in Syria. Field experiments using a double lattice design with four replications were employed to test different yield and quality parameters. The same variables were examined using 19 Turkish and five Greek landraces and 12 released varieties at sites near Adana and Ankara, in Turkey. The field design in Turkey was the same as the one used in Greece. The origins of the released varieties tested were:

| | |
|---|---|
| Greek material : | Santa, Sapho and Vergina (bread wheat) |
| French material: | Mondur |
| Italian material: | Japiga and Mandrone |
| Turkish material: | Balcali, Bintepe, Cakmak, Diyarbakir, Gediz, Genc (Aest), Gökgöl and Kunduru |
| Syrian material: | Haurani and Sham 1 |

Different levels of mineral fertilizer were applied at each location. The following list indicates basic and additional rates at the different locations.

| | |
|---|---|
| *Basic fertilization at planting* | |
| Salonica and Serres, 1985 and 1986: | 40 kg N/ha + 30 kg P/ha |
| Aleppo, 1985 and 1986: | 70 kg N/ha + 60 kg P/ha |
| Adana, 1987: | 30 kg N/ha + 70 kg P/ha |
| Ankara, 1987: | 30 kg N/ha + 70 kg P/ha |
| *Additional fertilization at the beginning of tillering* | |
| Salonica and Serres, 1985 and 1986: | 30 kg N/ha |
| Aleppo, 1985 and 1986: | 25 kg N/ha |
| Adana, 1987: | 40 kg N/ha |
| Ankara, 1987: | 35 kg N/ha |
| *Additional fertilization at booting stage* | |
| Salonica and Serres, 1985 and 1986: | 30 kg N/ha |
| Aleppo, 1985 and 1986: | 25 kg N/ha |
| Adana, 1987: | 30 kg N/ha |
| Ankara, 1987: | 35 kg N/ha |

In a pot experiment conducted in Salonica, different levels of water supply were correlated with protein content and vitreous kernel count in three released varieties and eight landraces. During the growing period, the plants were given different levels of water (40, 60 or 80 per cent of the water retention capacity).

## RESULTS

### Comparative performance of released varieties and Greek landraces

The main purpose of the research conducted in Greece was to evaluate the performance of Greek landraces when cropped in the same way as released varieties. The 2-year period of the study was characterized by inconsistent climatic conditions. In 1985, drought and unusually high temperatures were experienced (annual precipitation was 167 mm at Salonica, 199 mm at Aleppo and 218 mm at Serres). In 1986, rainfall was adequate and temperatures were lower, favouring higher yields (annual precipitation was 594 mm at Salonica, 316 mm at Aleppo and 457 mm at Serres; crops at Aleppo received an additional 40 mm of irrigation at planting).

The grain yields of Greek landraces and of Greek and foreign released varieties in 1985 are illustrated in Figure 5.1 (*overleaf*). The soils at Serres, Salonica and Aleppo vary significantly. Serres has silty clay, with a field capacity two to three times greater than at Salonica. Salonica is characterized by poor soils and low field capacity. In Aleppo, the high percentage of clay in the soil results in a high field capacity, but water availability is extremely limited. As shown in Figure 5.1, the released varieties were superior under the good soil and climatic conditions in Serres. However, under the poor conditions of Aleppo and Salonica, the landraces outyielded the released varieties. These results substantiate the superior performance of landraces in terms of drought resistance and their ability to adapt to unfavourable environments.

The mean yield of all field experiments (2 years, three locations) indicated that representatives of landraces could be found in each yield class (*see* Table 5.1).

**Table 5.1** Grain yield classes of Greek landraces and released varieties. Mean values of six field experiments in Salonica, Serres and Tel Hadya, 1985 and 1986

| | **Variety** | **Grain yield (dt/ha)** |
|---|---|---|
| Yield classes 1 [a] | Atsiki 4 | 28.1 |
| | Santa released varieties | 27.9 |
| | Sapho released varieties | 27.7 |
| | Atsiki 1 | 27.3 |
| Yield classes 2 | Atsiki 3 | 26.1 |
| | Moundros | 26.0 |
| | Romanou | 25.8 |
| | Vergina released varieties (*Aestivum*) | 24.7 |
| | Atsiki 5 | 24.2 |
| Yield classes 3 | Kaminia | 23.9 |
| | Kontopouli 3 | 23.3 |
| | Moundros 3 | 23,2 |
| | Local Heraklion | 22.2 |
| Yield classes 4 | Haurani released varieties | 21.4 |
| | Myrina | 20.7 |
| Yield classes 5 | Mavratheri Chion | 19.5 |
| | Asprostaro Chion | 19.4 |
| | Limnos | 18.9 |
| | Deves | 17.1 |
| | Chania | 17.0 |
| | Mavragani Arkados | 15.8 |
| x varieties | | 22.9 |

a Yield classes according to Schuster and von Lochow, 1979

**Figure 5.1** Grain yield of Greek landraces and released varieties in three locations, 1985

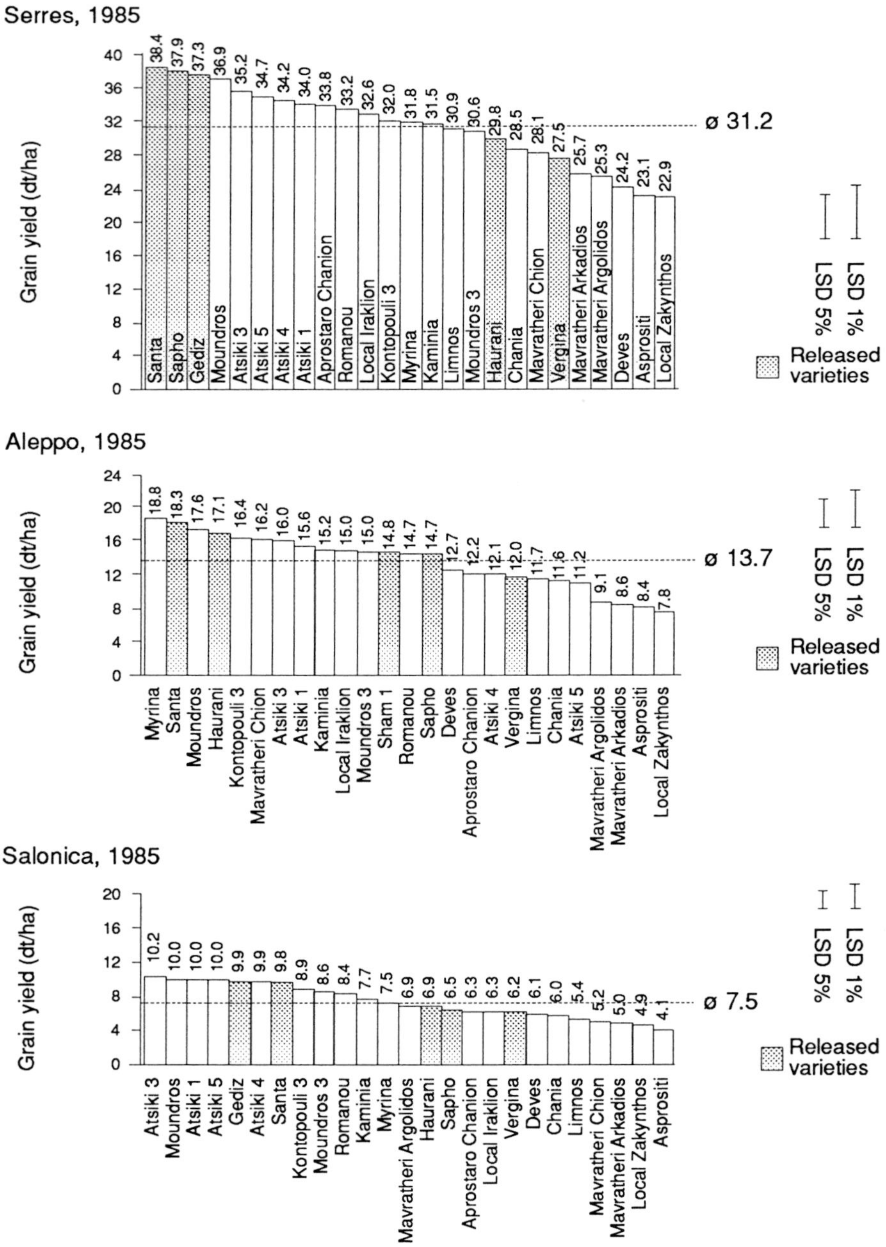

Noteworthy is the fact that two landraces (Atsiki types) were included in the highest yield class. The excellent yields of four Atsiki types and of Moundros and Romanou were attributable to their short stems and high harvest indices. The harvest indices of these landraces were higher than those of some released varieties. However, the long-stemmed landraces were found in the lowest yield class. Yield losses in this class were attributable mainly to lodging.

It should be mentioned here that ICARDA's electrophoretical comparisons of the Atsiki landraces with the Italian cultivar Capeiti indicated similar storage protein banding patterns (Damania and Somaraoo, 1988). A common parental stock can therefore be assumed, raising the question as to whether the Atsiki types can be considered as true landraces. Despite the similarity of storage protein patterns, significant differences in yields and protein content were measured between the two types for all the field trials conducted in 1986. It was concluded that the Atsiki landraces differ considerably from released varieties in Greece, and are in fact a valuable genetic resource in their own right.

Figure 5.2 (*overleaf*) shows the influence of different percentages of water retention capacity on vitreousness and protein content. At a level of 40 per cent all varieties exhibited a high percentage of vitreous kernels. When water availability was increased to 80 per cent of capacity, lower percentages of vitreousness were observed in all but one variety. Gediz, a released variety, maintained a high level of vitreousness, but the high-yielding variety Santa gave unsatisfactory results. In comparison, four landraces preserved their high levels of vitreousness, proving to be genetically more stable than the released varieties.

Similar results were found under natural conditions in the field. In Serres, all the released varieties achieved over 90 per cent vitreousness under dry conditions in 1985 (218 mm precipitation), but in 1986 (457 mm precipitation) 10 landraces performed better than the best released varieties. Most released varieties failed to reach the required minimum of 75 per cent vitreousness in 1986 (*see* Figure 5. 3 *overleaf*).

## Comparative performance of released varieties and Turkish landraces

Figure 5.4 (*page 52*) shows the yields and grain quality of Turkish landraces at locations near Ankara (400 mm precipitation) and Adana (670 mm precipitation) in 1987. The results can be compared with those observed in Greece: Ankara, with poor climatic and soil conditions, resembles Salonica; while Adana, with more favourable conditions, is similar to Serres.

At Ankara, five landraces had yields which were similar to the four best released varieties, achieving levels of approximately 3.5 t/ha. Because the germination density ($\bar{x}$ varieties = 237 plants per m$^2$) and the spikes per m$^2$ ($\bar{x}$ varieties = 264) were extremely low, the correlation coefficient between yield and spikes per m$^2$ was very high ($r = 0.76^{**}$). High coefficients were also found between yield and productive tillering ($r = 0.44^{**}$), germination density ($r = 0.30^{**}$) and the number of grains per ear

**Figure 5.2** Pot experiment 1986 in Salonica — the relation of grain yield, protein content and vitreousness of *Triticum durum* to water capacity

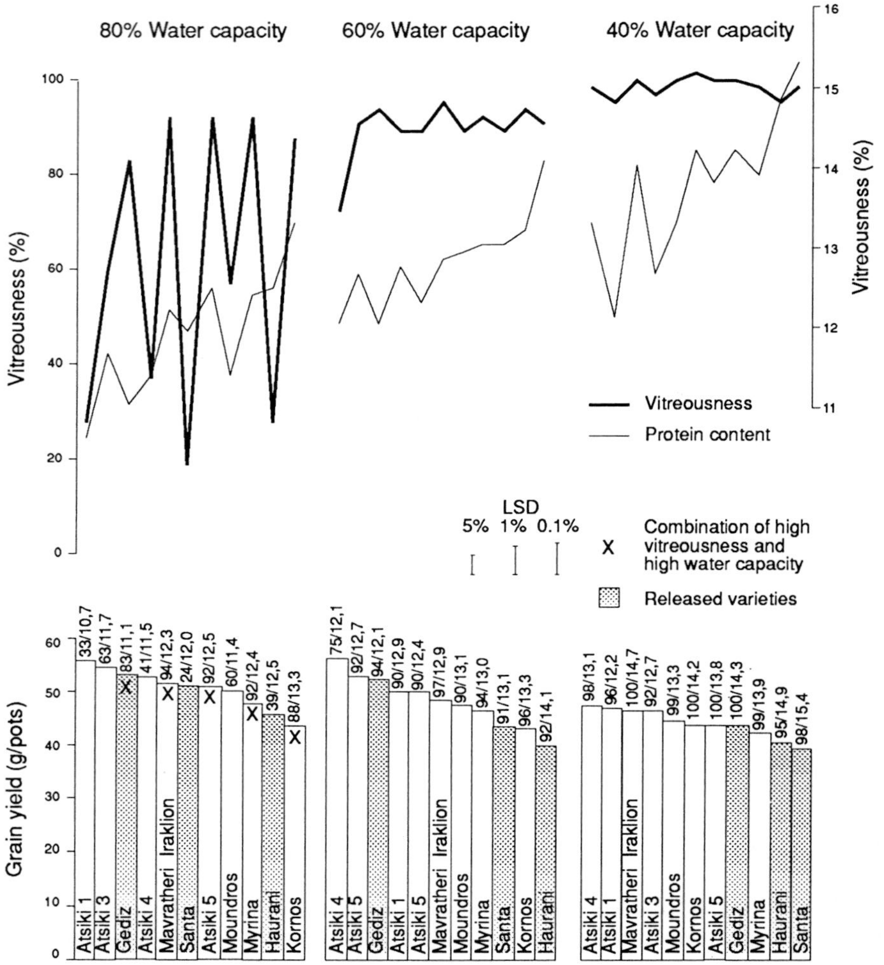


(r = 0.25**). However, there was no significant correlation between yield and number of spikelets per spike, the percentage of sterile spikelets, the height of plants, days to heading, single ear weight and thousand-kernel-weight (TKW) (*see* Table 5.2 *overleaf*).

In Adana, under the very favourable climatic conditions and high soil fertility of the Cukurova plain, the Turkish landraces were clearly outyielded by the released varieties. Even the promising Greek landraces (Atsiki and Moundros) were signifi-

**Figure 5.3** Vitreousness and protein content of Greek landraces and released varieties in Serres, 1985 and 1986

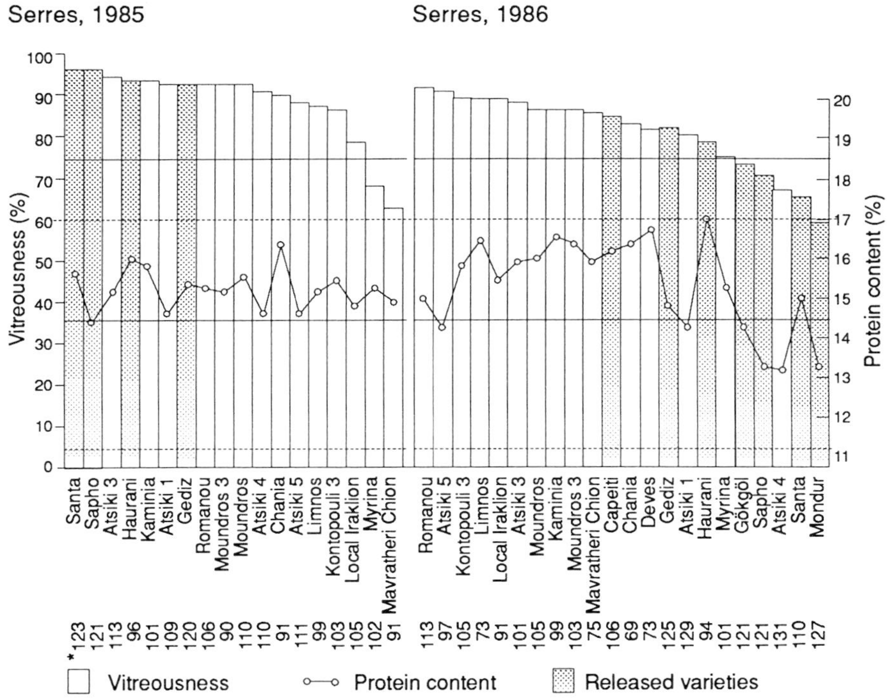


* The numbers indicate the relative yield in comparison to x varieties = 100%

cantly inferior in terms of yield. Their yield potential appears not to exceed 5 t/ha, whereas the best released varieties achieved approximately 6.6 t/ha. All the Turkish landraces produced long stems: their plant height ranged from 125 to 153 cm, compared with 100 to 118 cm for the high-yielding varieties. Heavy lodging, late heading, a low number of spikes per $m^2$ and a single ear weight were the most prominent causes of serious yield losses by the landraces at this location. In contrast to the results observed at Ankara, significant negative correlations were found between grain yield, on the one hand, and height of plants, days to heading and TKW on the other (*see* Table 5.2 *overleaf*).

Christiansen-Weniger (1970) analysed Turkish wheat yields from 1928 to 1961. During this period, yields were 'poor to disastrous' in 15 years and 'medium to excellent' yields in the other 19 years. These results, combined with those obtained at Ankara in 1987, demonstrate the considerable climatic variability of rainfed regions in Turkey and reinforce the importance of introducing germplasm with high yield stability.

**Figure 5.4** Grain yield, vitreousness and protein content of Turkish landraces and released varieties in Ankara and Adana, 1987

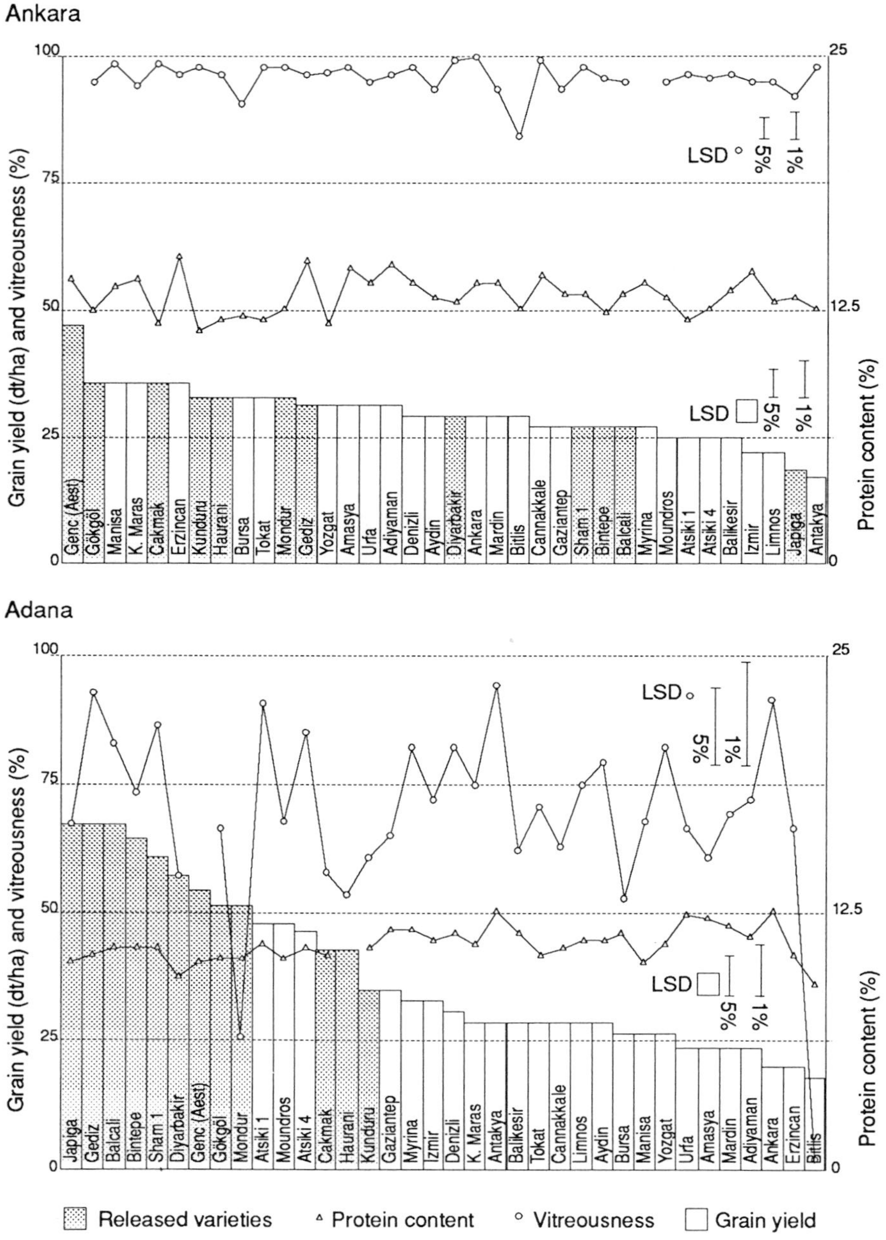

**Table 5.2** Correlation coefficients of grain yield and yield components in Ankara and Adana 1987

| **ANKARA** | | 1 | 2 | 3 | 4 | 5 | 6 | 7 | 8 | 9 | 10 | 11 | 12 |
|---|---|---|---|---|---|---|---|---|---|---|---|---|---|
| Grain yield | 1 | 1.00 | | | | | | | | | | | |
| Grains per ear | 2 | 0.25 | 1.00 | | | LSD 5% = 0.16 | | | LSD 1% = 0.21 | | | | |
| Spikelets per spike | 3 | 0.14 | 0.46 | 1.00 | | | | | n = 144 | | | | |
| % sterile spikelets | 4 | 0.00 | -0.67 | -0.28 | 1.00 | | | | | | | | |
| Germination density | 5 | 0.30 | -0.11 | -0.03 | 0.19 | 1.00 | | | | | | | |
| Height of plant | 6 | 0.11 | -0.39 | 0.28 | 0.52 | 0.27 | 1.00 | | | | | | |
| Days to heading | 7 | 0.06 | -0.21 | 0.21 | 0.18 | 0.27 | 0.39 | 1.00 | | | | | |
| Harvest index % | 8 | 0.16 | 0.38 | -0.13 | -0.33 | -0.18 | -0.48 | -0.31 | 1.00 | | | | |
| Spikes per m$^2$ | 9 | 0.76 | -0.30 | -0.23 | 0.33 | 0.32 | 0.18 | 0.15 | -0.05 | 1.00 | | | |
| Single ear weight | 10 | 0.13 | 0.80 | 0.51 | -0.49 | -0.11 | -0.11 | -0.13 | 0.35 | -0.50 | 1.00 | | |
| Thousand kernel weight | 11 | -0.15 | -0.47 | -0.06 | 0.38 | 0.06 | 0.50 | 0.18 | -0.10 | -0.14 | 0.05 | 1.00 | |
| Productive tillering | 12 | 0.44 | -0.18 | -0.19 | 0.13 | -0.50 | -0.03 | -0.09 | 0.10 | 0.61 | -0.33 | -0.12 | 1.00 |
| | | | | | | | | | | | | | |
| **ADANA** | | 1 | 2 | 3 | 4 | 5 | 6 | 7 | 8 | 9 | 10 | 11 | 12 |
| Grain yield | 1 | 1.00 | | | | | | | | | | | |
| Grains per ear | 2 | 0.70 | 1.00 | | | LSD 5% = 0.16 | | | LSD 1% = 0.21 | | | | |
| Spikelets per spike | 3 | -0.08 | 0.39 | 1.00 | | | | | n = 144 | | | | |
| % sterile spikelets | 4 | 0.16 | 0.04 | 0.22 | 1.00 | | | | | | | | |
| Germination density | 5 | 0.09 | 0.11 | -0.02 | 0.21 | 1.00 | | | | | | | |
| Height of plant | 6 | -0.73 | -0.40 | 0.24 | -0.24 | -0.15 | 1.00 | | | | | | |
| Days to heading | 7 | -0.59 | -0.39 | 0.46 | 0.06 | -0.09 | 0.55 | 1.00 | | | | | |
| Spikes per m$^2$ | 8 | 0.82 | 0.29 | -0.33 | 0.27 | 0.14 | -0.74 | -0.49 | 1.00 | | | | |
| Single ear weight | 9 | 0.54 | 0.85 | 0.39 | -0.11 | -0.01 | -0.18 | -0.36 | -0.01 | 1.00 | | | |
| Thousand kernel weight | 10 | -0.24 | -0.23 | 0.03 | -0.27 | -0.24 | 0.38 | 0.02 | -0.47 | 0.27 | 1.00 | | |
| Productive tillering | 11 | 0.78 | 0.25 | -0.33 | 0.16 | -0.24 | -0.66 | -0.45 | 0.92 | 0.02 | -0.36 | 1.00 | |

According to Kling (1985), high vitreousness in durum wheat grain depends on hot dry climatic conditions during the ripening phase, provided that the protein content is sufficient. Thus, in Ankara, where poor rainfall conditions and relatively low air humidity (57 per cent) were recorded, all varieties exhibited vitreousness levels of over 90 per cent. The protein content ranged from 12 to 15 per cent, indicating excellent quality.

However, in Adana, the high precipitation (670 mm) and higher air humidity (67 per cent) caused heavy yellow berry and low protein content (9 to 12.5 per cent). Only three released varieties (Gediz, Balcali and Sham 1) had vitreous kernel counts of over 75 per cent, whereas five Turkish and four Greek landraces were represented at this quality level.

The generally low levels of vitreousness found in released varieties in Adana confirms the need to improve the relationship between yield and quality in future durum wheat breeding programmes in Mediterranean countries. Genetic material from the Salonica and Menemen/Izmir genebanks might help to meet this need, and would certainly broaden the existing genetic base.

## *Discussion*

C. Josephides: I am in agreement with your emphasis on landraces. Can you tell us when and where the Atsiki landraces were collected in Greece?

A. Biesantz: I am glad you have mentioned these Atsiki landraces, as there is some controversy about them. They were collected in the East Aegean islands. The peculiar thing about them was that they were of short stature and resembled cultivars, while the other landraces from the same mission were tall. An electrophoretic study of storage protein was made at ICARDA and it was found that the Atsiki landraces were very similar to the Italian variety Capeiti in terms of numbers and patterns of the biotypes. However, in experiments at various sites we had significant differences in harvest index, grain yield and quality among the Atsiki landraces. If they are descendants from Capeiti, how do we explain these differences? Also, if they are not landraces, genebank personnel should make a note of it.

A. Abou-Zeid: You mentioned that in Turkey greater attention is given to introduced high-yielding material than to local landraces. I wish to confirm this observation. But this is not only in Turkey but in several developing countries, such as Kenya. We received 200 landrace accessions from ICARDA under the Durum Germplasm Evaluation Network at the genebank in Nairobi, but we could not get any of the breeders interested in the material and it was eventually lost because of neglect. I would like to see more emphasis given to the evaluation and utilization of landraces by ICARDA in order to broaden the genetic base. I think this should be discussed.

T.T. Chang: A few words about the International Rice Testing Program (IRTP) at the International Rice Research Institute (IRRI) would be relevant at this stage. We send out two types of nurseries — the yield nurseries, which consist of improved material, and the special nurseries. We invite other countries to contribute their best lines to the yield nursery, and therefore it is not all IRRI material; this helps spread or increase diversity. The special nurseries include landraces reported to be associated with desirable traits such as cold tolerance and disease resistance. So I think the international agricultural research centres do give emphasis to the evaluation and utilization of landraces.

Wheat Genetic Resources: Meeting Diverse Needs
Edited by J. P. Srivastava and A. B. Damania

Published by John Wiley & Sons

# 6

# *Evaluation for Useful Genetic Traits in Primitive and Wild Wheats*

A. B. DAMANIA, L. PECETTI and S. JANA

West Asia is the primary centre of diversity for wheat and barley, and almost certainly the region in which these food crops were first cultivated about 10 000 years ago (Zohary, 1969). Crop evolutionists believe that domesticated cereals were developed when farmers began sowing wild cereals and harvesting their crop by reaping entire inflorescences rather than beating the grain off the plants. The success of the plants no longer depended on their ability to release grain but on their ability to contribute to the harvest. Significantly, the shift from a brittle rachis (in wild emmer, *Triticum dicoccoides*) to a non-brittle one (in cultivated emmer, *T. dicoccum*) is governed by a single major gene (Zohary and Hopf, 1988). Eventually, primitive forms gave way first to landraces and then to relatively 'modern' wheats, which, besides producing higher yields under moderate- to high-input technology, were more suitable for pasta and bread making. Pantanelli (1944) stated that with the spread of modern varieties of durum and bread wheats, primitive forms such as *T. dicoccum* fell into disuse in Italy, and presumably all over Europe and the Mediterranean region.

The key to success in crop improvement for dry areas lies, to a considerable extent, in tapping the variability of existing plant genetic resources for use in breeding adapted germplasm. Being located close to a major centre of diversity for winter cereals, the International Center for Agricultural Research in the Dry Areas (ICARDA) is in a unique position to enrich its research programmes with useful genetic material and to share this valuable resource with scientists all over the world.

Given the steadily increasing population pressures on natural resources, the primary challenge to agricultural research for the drylands will be to generate technologies which can increase yields through the judicious use of inputs, without risking instability of production. The deployment of genetic resources to improve and stabilize crop production in the face of biotic and abiotic stresses is a key component in ICARDA's strategy to develop appropriate technology for difficult environments (Damania and Srivastava, 1990).

In addition to landraces and primitive forms, wild relatives can also provide valuable genes for disease resistance, high protein content, tillering, drought tolerance and other economically desirable attributes (Srivastava and Damania, 1989). Thus, progenitors of cultivated species should be considered as an important source of variability for broadening the genetic bases of cultivated crops (Lange and Balkema-Boomstra, 1988). However, relatively little attention has been given to the study of variability in primitive and wild wheats in the past few decades, although these are known to possess higher protein and lysine contents (Sharma et al., 1981) as well as genes for tolerance to the stresses in unfavourable environments. Their exploitation has been insufficient for three reasons. First, germplasm collections in the past have been fragmentary as well as scanty, with the result that the material available is not representative. Second, work on wild forms has concentrated primarily on evolutionary and taxonomic studies. And third, variability within populations of wild species has not been studied in adequate detail, and utilization has hardly begun (Srivastava et al., 1988).

*Aegilops* germplasm is generally used in wheat breeding as a source of genes for disease resistance. However, for the efficient utilization of genetic resources, other desired traits must be identified and transferred. Despite the enormous pool of genetic variability in wild species and the sophisticated techniques currently available for transferring this to cultivated wheat, there are still problems in identifying desirable characteristics in alien species for stressed environments.

In recent years variability in *Aegilops* species has been studied by several workers (Dhaliwal et al., 1986; Waines et al., 1987), whereas in the past more emphasis was placed on *Triticum turgidum* L. var. *dicoccoides* Koern, the wild progenitor of all modern wheats (Lawrence et al., 1958; Avivi, 1979). There is still a lack of information on various characteristics of *Aegilops* and, to a lesser extent, on the primitive forms.

The present study was undertaken to explore the variability of wild and primitive forms of wheat for agronomic traits of economic importance. The reaction to naturally occurring diseases was also observed. The genotypes with positive attributes can be recommended to breeders for further tests and subsequent utilization in their crossing blocks as parental material.

## MATERIALS AND METHODS

The experiment consisted of 86 accessions of primitive wheats from ICARDA's genetic resources collection (*see* Table 6.1). The *Aegilops* species evaluated by ICARDA in northern Syria is given in Table 6.2. They were space-planted in single rows at two sites. One site was at ICARDA's main farm at Tel Hadya, situated 37 km south of Aleppo at 36°05'N and 36°35'E, with an annual average rainfall of 342 mm (504 mm in the 1987-88 season) and an elevation of 284 m. The soils here are luvisols, transitional to vertisols (cracking soils), and also of the terra rosa or red Mediterranean type. The other site was at Breda, 40 km south-east of Tel Hadya, at 35°56'N and 37°11'E, with an annual average rainfall of 280 mm (414 mm in the 1987-88 season) and an elevation of 300 m. The soils are calcic xerosols or semi-desert brown type. The incidence of yellow rust was recorded at both sites for each accession.

**Table 6.1** Primitive forms evaluated at two sites in northern Syria during two seasons

| **Species** | **No. of accessions** |
|---|---|
| *Triticum carthlicum* | 1 |
| *T. dicoccum* | 78 |
| *T. ispahanicum* | 1 |
| *T. monococcum* | 1 |
| *T. polonicum* | 2 |
| *T. spelta* | 1 |
| *T. sphaerococcum* | 1 |
| *T. turgidum* | 1 |
| Total | 86 |

**Table 6.2** *Aegilops* species and number of accessions planted in 1987-88 at two sites in northern Syria

| **Species** | **No. of accessions** | **Species** | **No. of accessions** |
|---|---|---|---|
| *Aegilops bicornis* | 3 | *Ae. searsii* | 2 |
| *Ae. biuncialis* [a] | 25 | *Ae. sharonensis* | 3 |
| *Ae. caudata* | 4 | *Ae. speltoides* [a] | 30 |
| *Ae. columnaris* [a] | 15 | *Ae. squarrosa* [a] | 131 |
| *Ae. comosa* | 4 | *Ae. triaristata* | 4 |
| *Ae. crassa* | 4 | *Ae. triuncialis* [a] | 125 |
| *Ae. cylindrica* [a] | 48 | *Ae. umbellulata* [a] | 28 |
| *Ae. kotschyi* | 18 | *Ae. uniaristata* | 4 |
| *Ae. ligustica* | 1 | *Ae. variabilis* | 4 |
| *Ae. longissima* | 2 | *Ae. variabilis* ssp. *cylindrostachys* | 1 |
| *Ae. lorentii* [a] | 18 | *Ae. vavilovii* | 6 |
| *Ae. mutica* | 4 | *Ae. ventricosa* | 1 |
| *Ae. neglecta* | 8 | *Aegilops* species [b] | 29 |
| *Ae. ovata* | 86 | | |
| Total | | | 629 |

a Species analysed statistically
b Unclassified intermediate forms

## RESULTS

The collection was screened for two of the most common wheat diseases in the region, yellow rust and common bunt. Resistance to at least one of these diseases was found in several accessions of primitive forms. The list of these accessions, with their ICARDA genebank numbers, is given in Table 6.3. Three accessions of *T. dicoccum* were resistant to both diseases.

**Table 6.3** List of primitive forms evaluated at two sites in northern Syria

| **Resistant to yellow rust** | | **Resistant to common bunt** | |
|---|---|---|---|
| *Triticum dicoccum* | IC 9840 | *T. dicoccum* | IC 12432 [a] |
| | IC 12424 | | IC 12440 |
| | IC 12432 | | IC 12446 [a] |
| | IC 12435 | | IC 12454 [a] |
| | IC 12437 | | IC 12468 |
| | IC 12443 | | IC 12469 |
| | IC 12445 | | IC 12499 |
| | IC 12446 | | IC 12508 |
| | IC 12449 | | IC 12526 |
| | IC 12453 | | IC 12530 |
| | IC 12454 | | IC 12536 |
| | IC 12460 | | IC 12546 |
| | IC 12516 | | IC 12558 |
| | IC 12523 | *T. ispahanicum* | IC 9807 |
| | IC 12524 | *T. spelta* | IC 9810 |
| *T. carthlicum* | IC 12180 | *T. polonicum* | IC 2194 |
| *T. sphaerococcum* | IC 12274 | | IC 12196 |

a Resistant to both diseases

The most susceptible *Aegilops* species were *Ae. columnaris*, *Ae. cylindrica*, *Ae. kotschyi*, *Ae. squarrosa* and *Ae. triaristata*. Others, such as *Ae. biuncialis*, *Ae. lorentii*, *Ae. ovata*, *Ae. speltoides* and *Ae. triuncialis*, were relatively resistant. The results, together with ploidy level and gene symbols, are given in Table 6.4.

Germination at Breda was poorer than at Tel Hadya. Several factors were responsible for this, the major ones being moisture and cold stress. Nevertheless, there was considerable variability for characters observed at both sites. The minimum and maximum values of observations on *Aegilops* species are given in Table 6.5. At both sites *Ae. kotschyi* was early in heading and maturity and *Ae. speltoides* was late. Although *Ae. speltoides* was the tallest of all species, there was significant variability in the mean plant height recorded at Tel Hadya (67.2 cm) and at Breda (57.1 cm).

**Table 6.4** Yellow rust infection on *Aegilops* species at two sites

| Species | Ploidy | Genome | % of infected accessions | |
|---|---|---|---|---|
| | | | Tel Hadya | Breda |
| *Aegilops biuncialis* | 4x | UM | 8.0 | 8.3 |
| *Ae. columnaris* | 4x | UM | 73.3 | 69.2 |
| *Ae. cylindrica* | 4x | CD | 97.9 | 100.0 |
| *Ae. kotschyi* | 4x | US | 61.1 | 42.8 |
| *Ae. lorentii* | 4x | UM | 5.5 | 16.6 |
| *Ae. ovata* | 4x | UM | 3.5 | 3.8 |
| *Ae. speltoides* | 2x | S | 7.1 | 8.0 |
| *Ae. squarrosa* | 2x | D | 78.6 | 68.8 |
| *Ae. triaristata* | 4x, 6x | UM, UMUn | 41.7 | 50.0 |
| *Ae. triuncialis* | 4x | UC | 16.8 | 16.2 |
| *Ae. umbellulata* | 2x | U | 42.8 | 31.0 |

**Table 6.5** Minimum and maximum mean values of *Aegilops* species for four characters at two sites

| | | Tel Hadya | | No. | Breda | | No. |
|---|---|---|---|---|---|---|---|
| Days to heading | Min | 140.9 | *(Ae. kotschyi)* | 18 | 146.1 | *(Ae. kotschyi)* | 14 |
| | Max | 167.8 | *(Ae. speltoides)* | 28 | 167.5 | *(Ae. speltoides)* | 25 |
| Days to maturity | Min | 169.3 | *(Ae. kotschyi)* | 18 | 172.8 | *(Ae. kotschyi)* | 14 |
| | Max | 67.2 | *(Ae. speltoides)* | 28 | 57.1 | *(Ae. speltoides)* | 25 |
| No. tillers per plant | Min | 78.6 | *(Ae. squarrosa)* | 130 | 30.7 | *(Ae. squarrosa)* | 109 |
| | Max | 223.2 | *(Ae. triuncialis)* | 125 | 110.2 | *(Ae. biuncialis)* | 24 |

Similarly, the number of tillers per plant for *Ae. triuncialis* and *biuncialis* was highest at Tel Hadya and Breda respectively, but at Breda tillering was reduced by 50 per cent for both species because of stress.

An analysis of variance for four characters at both sites showed significant variability between species. The analysis for observations recorded at Tel Hadya is given in Table 6.6 (*overleaf*). The large variability observed for desirable traits in primitive and wild forms would be of practical use in breeding. In wild species such as *Aegilops* it is possible to select single plants for specific traits which can be maintained as pure lines and used as parental material in crosses.

During the 1988-1989 season, 323 selected accessions from 22 species which had shown resistance to yellow rust in the previous season were space-planted at Tel Hadya and Breda, where the rainfall was 234 and 193 mm respectively (considerably less than

**Table 6.6** Analysis of variance for four characters in *Aegilops* species at Tel Hadya

| Source of variation | DF [a] | SS | MS | F-Value | Probability |
|---|---|---|---|---|---|
| Days to heading | | | | | |
| Treatment (species) | 11 | 13402.8926 | 1218.4448 | 24.2477 | 0.0000 |
| Error | 467 | 23466.6953 | 50.2499 | | |
| Days to maturity | | | | | |
| Treatment (species) | 11 | 10473.2744 | 952.1158 | 22.7697 | 0.0000 |
| Error | 467 | 19527.6074 | 41.9150 | | |
| Plant height | | | | | |
| Treatment (species) | 11 | 27368.9785 | 2488.0889 | 37.2271 | 0.0000 |
| Error | 467 | 31212.1230 | 66.9354 | | |
| No. tillers per plant | | | | | |
| Treatment (species) | 11 | 1428038.50 | 129821.6797 | 33.7496 | 0.0000 |
| Error | 467 | 1796369.75 | 3846.6162 | | |

a D.F. = degrees of freedom; S.S. = sum of squares; M.S. = mean of squares; F-value = frequency value

the long-term average, and less than half that for the previous season). There was a period of over 40 days in January-February when the night temperature fell below zero. Thus, screening for drought tolerance and frost damage in the vegetative phase was carried out. There were 153 accessions from the following species which survived these abiotic stresses: *Ae. biuncialis* (16 accessions), *Ae. columnaris* (2), *Ae. ovata* (52), *Ae. kotschyi* (3), *Ae. lorentii* (14), *Ae. neglecta* (7), *Ae. squarrosa* (13), *Ae. triuncialis* (45) and *Ae. variabilis* (1).

## CONCLUSION

One of the major constraints to increased use of non-conventional germplasm by plant breeders is the amount of time and effort needed to identify and isolate a required gene from the wide range of variability existing in wild progenitors and primitive forms. Not all wild forms are useful in plant breeding. Some are susceptible to diseases or possess other undesirable traits. The considerable variability present in wild and primitive forms allows identification of desirable genetic stocks, provided these are properly evaluated and documented.

The data assembled after conserving, evaluating and documenting the genetic resources of primitive and wild wheat germplasm have been systematically analyzed with the aim of deciphering variability within populations, between populations, and between countries and regions as a whole. Wherever variability is found to be intense, further missions will be planned by the Genetic Resources Unit at ICARDA in collaboration with the national programmes, so that collections can be made before genetic erosion occurs.

## *Discussion*

A. K. ATTARY: What is the difference between *Triticum polonicum* and *T. ispahanicum* and who first reported the latter species? I am asking this because *ispahanicum* has been designated as a subspecies of *T. aestivum* in the Iranian collection at Karadj.

A. B. DAMANIA: The spikes of *T. ispahanicum* are longer and more slender than the normal *polonicum* spikes. This species was discovered by Heslot near Isfahan in Iran, hence the name. However, most taxonomists consider it a different morphological form of *polonicum*. It is definitely a tetraploid and should not be considered as a subspecies of *T. aestivum*.

H. MEKBIB: Have you observed cultivation of *polonicum*, *dicoccum* or other primitive wheats in the Mediterranean countries?

A. B. DAMANIA: Yes, *polonicum* is still cultivated in Algeria. *T. turgidum* var. *turgidum* or poulard wheat is, I believe, still cultivated in Morocco and *T. dicoccum* has been reported from the Appennine mountains of southern Italy.

J. P. SRIVASTAVA: I have seen pure fields of *polonicum* being grown in Syria also.

J. G. WAINES: What characters were transferred from *T. turgidum* var. *polonicum* to the new cultivar Sebou at ICARDA?

A. B. DAMANIA: Earliness, long vitreous kernels and tolerance to high temperatures during grain-filling period were the main characters which were transferred to this durum cultivar from *T. polonicum*.

M. N. SANKARY: Have you looked at disease resistance in *Aegilops* from a taxonomical point of view in the way Vavilov did?

A. B. DAMANIA: No. I would be hesistant to associate resistance to various diseases with certain species, as I have found that among a single species and within a population sample there can be individuals with varying response to diseases. Recently, Dr Mamluk has reported flag smut on *Aegilops*. During the last season (1987-88) the rainfall was much higher than the long-term average, which encouraged the infestation of yellow rust. Observations were therefore made on presence or absence of infection. The resistant plants have been separated and are being maintained as pure lines. But we have yet to look at the species and their genomes to see if there is any correlation between resistance and taxonomical classification.

M. N. Sankary: You have assembled a very good collection of *Aegilops* species and some of your lines are highly productive as far as tillers are concerned. Some of these species could be used as forage plants. Can you tell the origin of this material?

A. B. Damania: These accessions form part of the genetic resources collection at ICARDA and they originate from various sources. Some come from earlier collections made in the region by Dr J. Witcombe. Others were collected in Algeria, Morocco and Pakistan by ICARDA scientists. Syria and Lebanon have also been explored. Professor Jana and Dr Jaradat have collected accessions from Turkey and Jordan. So we have a whole range of environments from where this material originates. *Aegilops* is commonly referred to as 'goat grass' and so it could be used as a forage if we can grow it in sufficient quantity.

D. R. Knott: You mentioned some accessions that had striking tolerance to salinity and high temperatures. What was this material?

A.B. Damania: These four accessions were *T. durum*. Three were from Afghanistan and one from Turkey. Out of the three from Afghanistan, two were originally collected by Wilbur Harlan and one by Koelz from the bazaar in Kabul. Dr Jack Harlan, in a personal communication, commented that the Irano-Afghan border would be a likely place to collect genes for salt and high temperature tolerance, and we are planning to make some explorations once the situation in the area is peaceful.

Wheat Genetic Resources: Meeting Diverse Needs
Edited by J. P. Srivastava and A. B. Damania

Published by John Wiley & Sons

# 7

# *The Use of Restriction Fragment Length Polymorphisms in the Evaluation of Wheat Germplasm*

R. D'OVIDIO, O. A. TANZARELLA, D. LAFIANDRA and E. PORCEDDU

The dramatic achievements in recombinant DNA and cell biology research have stimulated very high, probably excessive, expectations of the applications of such techniques to plant genetics and breeding. The results obtained in the past few years have indeed been impressive but, at least for the most advanced fields (including somatic hybridization through protoplast fusion and direct transfer of cloned genes), they have been limited to model species, without practical interest.

The only technique derived from the progress in molecular biology research whose application in crop genetics and breeding has already produced very interesting results is the analysis of restriction fragment length polymorphisms (RFLPs). The use of these molecular markers is extremely useful, since one of the main limiting factors to genetic analysis in most crop species is the lack of markers. RFLPs have furnished a very powerful method, which can be used in virtually any plant species, to obtain detailed maps of genetic linkage. Furthermore, as genetic markers RFLPs have some very convenient peculiarities. They are codominantly expressed, do not have pleiotropic effects on agronomic traits, and the number of possible markers they provide is virtually infinite.

Another advantage of RFLPs over biochemical markers such as isozymes, which require different staining and electrophoretic techniques, is that the same laboratory method can be used to detect the hybridization patterns of all the available probes. In addition, RFLPs should detect much more polymorphism than biochemical markers, because many of the probes are non-coding, less conserved sequences. The possibility of obtaining very dense, saturated genetic maps could allow the exploitation of RFLPs for gene tagging and the dissection of quantitative trait loci. These applications could prove particularly valuable for screening wheat germplasm and for monitoring the transfer of useful genes.

The main drawbacks of RFLPs are that they are more expensive than biochemical markers (Beckmann and Soller, 1983; Gale and Sharp, 1988) and that they require the manipulation of radioactive isotopes. However, the development of cheaper techniques of plant DNA extraction and of non-radioactive detection methods is reducing the costs and simplifying the procedures.

In spite of these exciting prospects, RFLPs have so far been applied to the genetic analysis of only a few plant species. Among these, maize (Neuffer and Coe, 1974) and tomato (Rick, 1974) were already well characterized, with detailed genetic maps of morphological and biochemical markers, which have been substantially enriched by many RFLP loci. Even more promising are the results obtained in crop species whose genetics were almost unknown, such as lettuce (Landry et al., 1987), or which had few localized morphological markers, such as *Brassica* species (Figdore et al., 1988). These results suggest that, through the application of RFLP methods, substantial progress in genetic analysis can be obtained in almost any plant species.

In wheat, however, despite its importance as a food crop, little research has so far been done on the isolation and characterization of clones to be used for RFLP analysis (Chao et al., 1988; Gale and Sharp, 1988). The use of such clones for genetic analysis and breeding applications in wheat, notwithstanding the polyploidy of its genome, should be facilitated by the availability of aneuploids, such as nullitetrasomic and ditelosomic lines, and of addition lines that greatly ease the chromosomal assignment and arm localization of DNA probes. This paper describes some preliminary results of a programme to isolate and characterize durum wheat clones to be used for RFLP analysis. We are planning to use the clones of sequences mainly to assess the variability of durum wheat germplasm and the potential for its exploitation in wheat breeding.

## MATERIALS AND METHODS

The wheat genotypes examined were: *Triticum monococcum* (2x); *T. boeoticum* (2x); *T. urartu* (2x); *T. durum* (4x) cv. Langdon, cv. Berillo, cv. Creso, cv. Lira, cv. Quadruro, cv. Trinakria, cv. Valforte, cv. Valnova; and *T. aestivum* (6x) cv. Chinese Spring and cv. Raeder. The wheat plants were grown from seeds in moist vermiculite in dark conditions at 24°C for 15 days.

### DNA extraction and cloning

High-molecular-weight (HMW) genomic DNA was purified following the procedure reported by Dvorak et al. (1988). DNA from *Triticum durum* cv. Langdon was further purified on a CsCl gradient in a VTi 65. Then 1 μg of Langdon DNA and 1 μg of pUC19 plasmid (Bethesda Research Laboratories) were mixed and digested with Pst I and Bam HI restriction endonucleases.

The resulting DNA fragments and linearized pUC19 were ligated and one tenth of the ligation mixture was used to transform *Escherichia coli* strain JM83. The transformed bacteria were selected on LB plates which contained 100 μg/ml ampicillin, 160 μg/ml isopropylthio-B-galactoside and 400 μg/ml 5-bromo-4-chloro-3-indolyl-B-D-galactoside. White clones were individually transferred into LB plates containing 100 μg/ml ampicillin. Recombinant clones containing non-repetitive DNA inserts were selected by clone hybridization and dot blot (Mason and Williams, 1985).

### DNA restriction and gel electrophoresis

DNA was digested by incubating at a ratio of 1 μg DNA to 4 units of restriction endonucleases for 5-6 hours in the buffer under conditions specified by the manufacturer. In a horizontal gel apparatus, 1 per cent agarose gels were run at 40 V for 15 hours in a buffer containing 89 mM Tris, 89 mM boric acid and 2 mM EDTA pH 8.0.

### DNA blot and labelling

Capillary transfer of the gel-fractionated DNA to nylon membrane (Bio-Rad) was carried out using the procedure reported by Southern (1975) after HC1 depurination (Wahl et al., 1979). For DNA labelling, 1 μg of genomic DNA (dot blots and colony hybridization) or 0.5 μg of purified insert (Southern blots) were labelled by the random primed method with digoxigenin-dUTP (Boehringer).

### Blot hybridization

The filter was first soaked in 2X SSC (0.3 M NaCl and 0.03 M Na-citrate) and placed in a plastic bag with 20 ml of hybridization solution per 100 $cm^2$ of membrane. The hybridization solution contained 5X SSC, 0.5 per cent blocking reagent (Boehringer), 0.1 per cent N-lauroylsarcosine and 0.02 per cent sodium dodecyl sulphate (SDS). The pre-hybridization was carried out in a water bath, with agitation, at 68°C for at least 3 hours. The hybridization was done under the same conditions, but using 10 ml of hybridization solution per 100 $cm^2$ of membrane and 0.5-1 μg of digoxigenin-dUTP labelled DNA. The detection of DNA hybrids was done according to the instructions of the manufacturer (Boehringer).

### Rehybridization of DNA on nylon membrane

Rehybridization of DNA on nylon membrane was done after removing the colour precipitate with N-N-dimethylformamide (Gebeyehu et al., 1987).

The membrane was washed in N-N-dimethylformamide at 65°C in a glass tray for about 1 hour by changing the solution every 15 minutes. The membrane was then washed in 2X SSC, 0.1 per cent SDS, and treated with 0.5 mg/ml proteinase K (Boehringer) in 2X SSC, 0.1 per cent SDS, at 65°C for 1 hour. The membrane was washed again in 2X SSC, 0.1 per cent SDS, and the probe was removed by incubating the membrane in 50 per cent formamide, 10 mM sodium phosphate pH 6.5 at 65°C for 1 hour. The membrane was finally washed in 2X SSC, 0.1 per cent SDS, and pre-hybridized and hybridized as described above.

## RESULTS

We decided to follow two different approaches for cloning wheat sequences: from genomic DNA; and through the isolation of a cDNA library.

**Figure 7.1** Colony hybridization of durum wheat clones

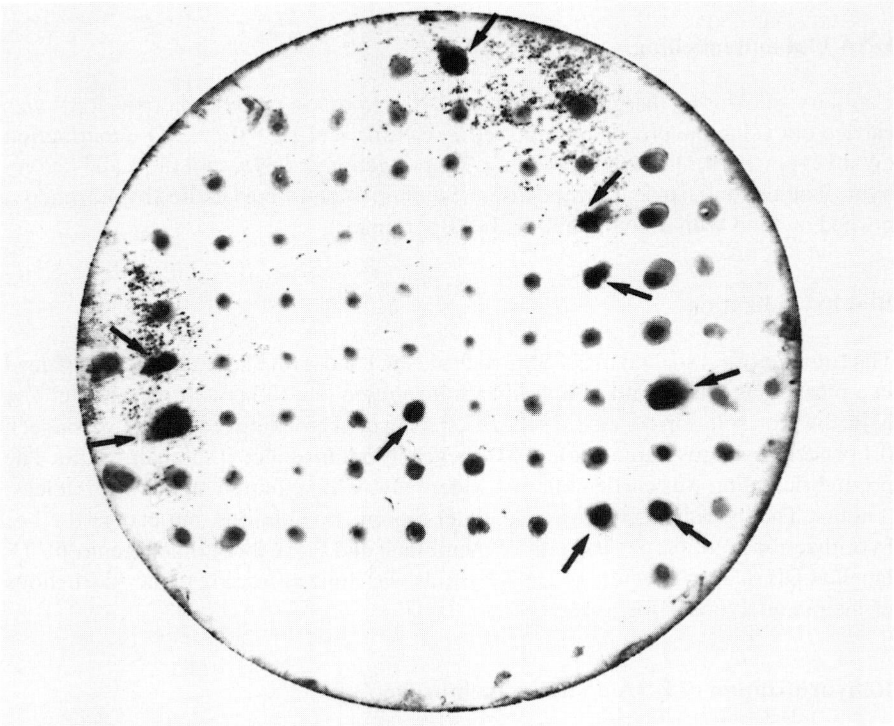

The arrows show the clones containing repetitive DNA; these clones were not used for dot blot analysis

The first approach consisted of the digestion of genomic DNA with the restriction endonuclease Pst I, which allows preferential cutting of non-repetitive DNA and enrichment in cloned sequences, and the insertion of these sequences into the plasmid pUC19. In this first phase of the programme we used only one fifth of the ligation mixture, which, when plated in a selective medium, produced 500 positive white colonies; their colony hybridization with genomic DNA allowed evaluation of the level of repetition (*see* Figure 7.1). About 5 per cent of the colonies gave a very strong signal, while the remaining ones showed different, but much lower, intensities of hybridization.

A further screening of 50 clones, carried out by dot blot hybridization (*see* Figure 7.2), showed that 20 per cent of these clones displayed a very weak signal, 30 per cent gave intermediate hybridization, and 50 per cent displayed quite strong intensity. The use of three of these clones (*see* Figure 7.3 *overleaf*) for probing Southern blots of genomic DNA digests demonstrated the efficiency of dot blot hybridizations for discriminating between clones of repetitive and non-repetitive DNA sequences. Clone pTDL1, which displayed a very weak signal in the dot blot, gave faint bands also in Southern blot hybridization of restricted genomic DNA digests (*see* Figure 7.4 *overleaf*), whereas clone pTDL2, which hybridized more completely in the dot blot, showed more intense bands in the Southern blots (*see* Figure 7.5 *overleaf*).

**Figure 7.2** Dot blot analysis of the clones selected by colony hybridization

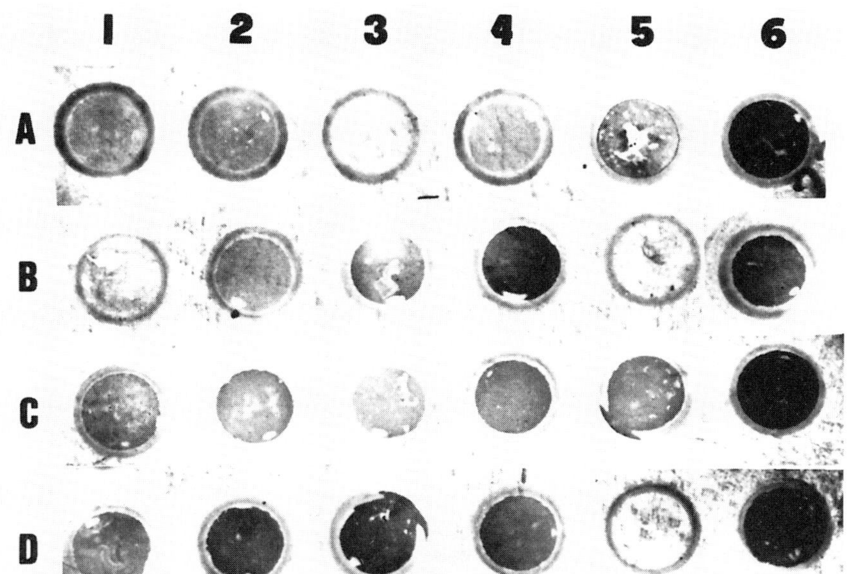


The clones A3 and C3 were respectively called pTDL1 and pTDL2; they are not representative of the clone analyzed by Southern blot

**Figure 7.3** One per cent agarose gel of three clones selected by dot blot

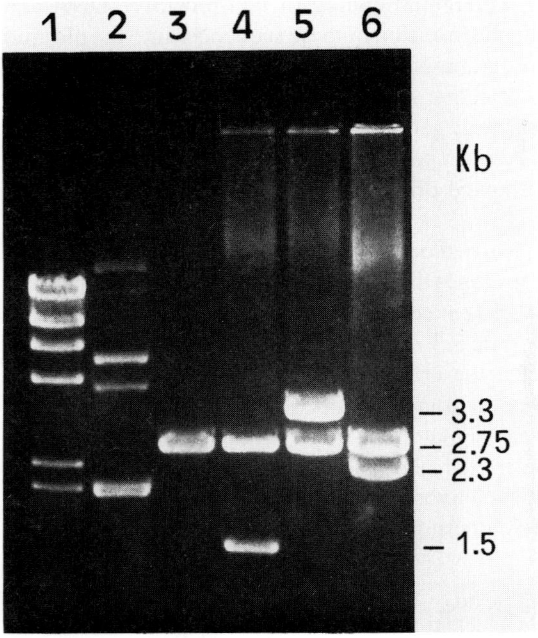


Legend:

1. Molecular weigh marker II (Boehringer)
2. Uncut pUC19 plasmid
3. Pst I-Bam HI digested pUC19 plasmid
4. pTDL3 clone
5. pTDL1 clone
6. pTDL2 clone

The insert contained in each clone was cut out by Pst 1-Bam HI digestion

**Figure 7.4** Southern blot of Pst I-Bam HI digested wheat DNA hybridized with pTDL1 clone

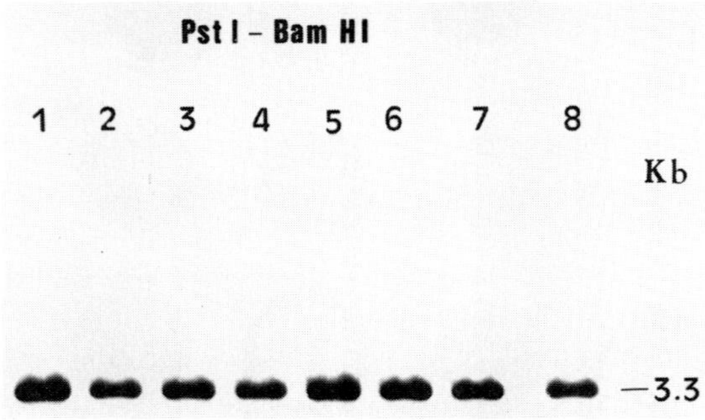


Legend:

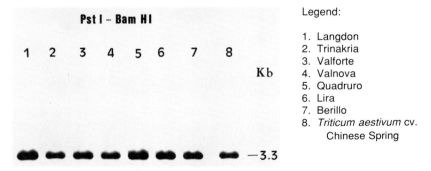

1. Langdon
2. Trinakria
3. Valforte
4. Valnova
5. Quadruro
6. Lira
7. Berillo
8. *Triticum aestivum* cv. Chinese Spring

The samples 1-7 represent *Triticum durum* varieties

**Figure 7.5** Southern blot of Pst I-Bam HI digested wheat DNA hybridized with pTDL2 clone.

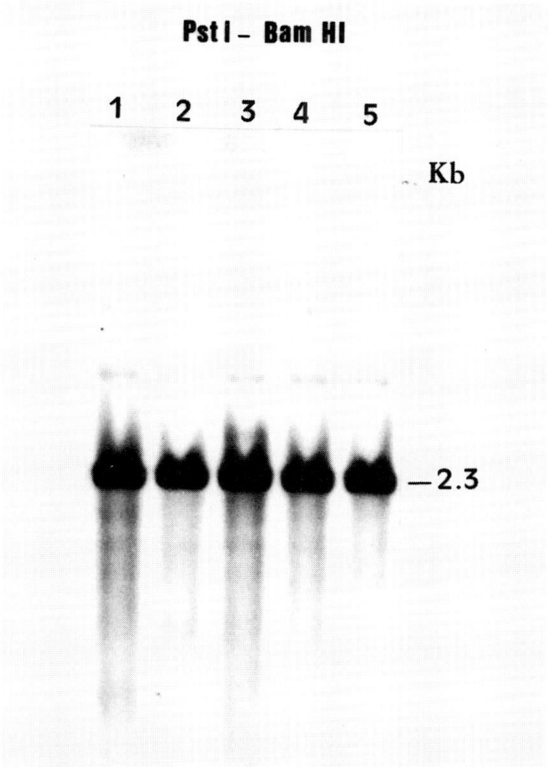


Legend:

1. Langdon
2. Quadruro
3. Valforte
4. Valnova
5. *Triticum durum* cv. Chinese Spring

The samples 1-4 represent *Triticum durum* varieties

The next stage in this approach will be the characterization of other clones which gave weak or quite weak signals in the dot blot, and the chromosomal localization using Southern hybridization with genomic digests of wheat aneuploid lines. At the same time, the clones will be used for screening the germplasm of wheat and related species, looking for polymorphisms. We believe that the extent of variability detected in wheat germplasm, even with biochemical markers, will allow the identification of a number of polymorphic loci.

The second approach, the isolation of a cDNA library, will enable us to obtain more probes. We have already obtained the poly $A^+$ fraction of mRNA by extraction of total RNA through a column of oligo dT cellulose. This poly $A^+$ mRNA will be used for cDNA synthesis, which will be cloned into a Lambda gt10 vector. We are following this alternative strategy because cDNA should be more efficient for cloning single-copy sequences. The drawback of cDNA clones in RFLP analysis is that they are coding sequences, which should be better preserved and therefore less polymorphic than random genomic sequences. It will be interesting to compare the efficiency of the clones obtained by the two methods in detecting RFLPs.

The long-term goal of our programme is to obtain a saturated chromosomal map of wheat based on RFLP markers. This will allow the tagging of useful genes, which will simplify the screening of germplasm collections and the monitoring of their transfer to wheat cultivars.

## CONCLUSION

Although we have not yet detected any RFLPs, we consider these preliminary results very important because they show that our technical approach is effective. A further reason for continued interest in this method is to demonstrate that RFLP analysis can be carried out by using a non-radioactive detection system. This has two advantages over the use of radioactive isotopes: it is safer and simpler to use in the laboratory, and it is less expensive, partly because the labelled probe can be stored for very long periods and the same probe can be used several times with different Southern blots.

Wheat Genetic Resources: Meeting Diverse Needs
Edited by J. P. Srivastava and A. B. Damania

Published by John Wiley & Sons

# 8

# *Seed Storage Proteins and Wheat Genetic Resources*

D. LAFIANDRA, S. BENEDETTELLI, B. MARGIOTTA,
P. L. SPAGNOLETTI-ZEULI and E. PORCEDDU

The storage proteins of the wheat endosperm are usually divided into two main groups, gliadins and glutenins, on the basis of solubility criteria in different solvents. A peculiar aspect of both groups is their high level of heterogeneity, determined strictly by the genotype. This characteristic has been used in varietal identification, detecting off-types in pure seed production, and related problems. The importance of storage proteins in determining nutritional and technological properties in wheat is also well established. Poor nutritional value is associated with both groups of proteins, particularly with the gliadin fraction, whose lysine content is very low. Also, the viscoelastic properties of the gluten are determined by the composition and interactions of both groups, with gliadins imparting extensibility and glutenins strength and elasticity to the dough. Both groups are lodged in protein bodies in the developing endosperm. They have no other function than to supply amino acids for the developing seedling.

Storage protein genes also provide molecular biologists with an excellent model for studying developmentally regulated genes. Biochemical, genetical and molecular studies have received great impetus in the past. More recently, storage proteins have been used in studies on genetic resources in genebanks, on topics such as the migration of species from centres of diversity and the genetic variation in collected materials.

## CLASSIFICATION OF WHEAT STORAGE PROTEINS

Detailed reviews of the biochemistry and genetics of wheat storage proteins have recently been published (Shewry and Miflin, 1985; Bietz, 1987; Shepherd, 1988; Wrigley and Bietz, 1988).

Gluten proteins were fractionated by Osborne (1907) on the basis of their solubility in different solvents. Gliadins are readily soluble in alcohol/water mixtures, while glutenins are soluble in alkaline or acidic solutions. Gliadins are composed of monomeric polypeptides stabilized by intrachain disulphide bridges. Glutenins, on the other hand, are more complex mixtures, made up of several polypeptidic subunits joined by intermolecular disulphide bridges. Upon reduction, glutenins give rise to several polypeptide chains, which have been subdivided into high-molecular-weight (HMW) glutenin subunits and low-molecular-weight (LMW) glutenin subunits.

Shewry et al. (1986) proposed a new classification of gluten proteins, based on their biological characteristics as well as on their chemical and genetic relationships, whereby the proteins are divided into: the HMW group (corresponding to the HMW glutenin subunits), the sulphur-poor prolamin group (corresponding to the ω gliadins), and the sulphur-rich prolamin group (corresponding to the α, β and ψ gliadins and the LMW glutenin subunits). These authors also suggested that, as individual monomers or reduced subunits of each group conformed with Osborne's definition of being rich in proline and amide nitrogen, these should all be called prolamins.

## ANALYZING WHEAT STORAGE PROTEINS

### Electrophoretic techniques

In the early 1960s, Woychick et al. (1961) separated wheat gliadins by using starch gel electrophoresis at acidic pH in aluminum lactate buffer. Separation in these conditions is based mainly on charge differences, although a molecular sieving effect also occurs. Gliadins were subdivided into α, β, ψ and ω fractions according to their decreasing mobility. Improving the condition of separation, along with the use of thinner polyacrylamide gels, revealed that each of these groups was composed of several polypeptides, and that wheat cultivars could be identified on the basis of differences in number, mobility and staining intensity of the gliadin components. Computerized procedures have also been developed to compare and analyze gliadin electrophoretic patterns for wheat cultivar identification (Sapirstein and Bushuk, 1986).

The use of two-dimensional techniques greatly improved resolution and revealed further gliadin heterogeneity. Several different techniques have been used for analyzing wheat gliadins. The first, employed by Wrigley and Shepherd (1973), combined isoelectric focusing with a second dimensional separation in starch gel under acidic pH. Subsequently, Du Cros et al. (1983) modified this technique, employing polyacrylamide gels for both first and second dimensional separation.

The second technique, employed by Mecham et al. (1978), used a combination of two different pH levels for first and second dimensional separation (pH 3.2, followed by pH 9.2). Further improvements to the two-pH procedure were made by Lafiandra and Kasarda (1985). Computerized procedures for analyzing and storing information obtained by using this two-dimensional technique have been published (Tomassini et al., 1989). A third technique was used by Payne et al. (1982), who employed a separation at acidic pH followed by a second dimensional separation in sodium dodecyl sulphate (SDS)-polyacrylamide gel electrophoresis (PAGE). Another two-dimensional technique is to combine isoelectro focusing with SDS-PAGE. The first two procedures separate proteins mainly on charge differences, while the second two are accomplished by combining charge differences in the first dimension with differences in molecular weight in the second dimension (*see* Figure 8.1).

**Figure 8.1** Comparison of three different two-dimensional separations of gliadins extracted from the durum wheat cultivar Creso

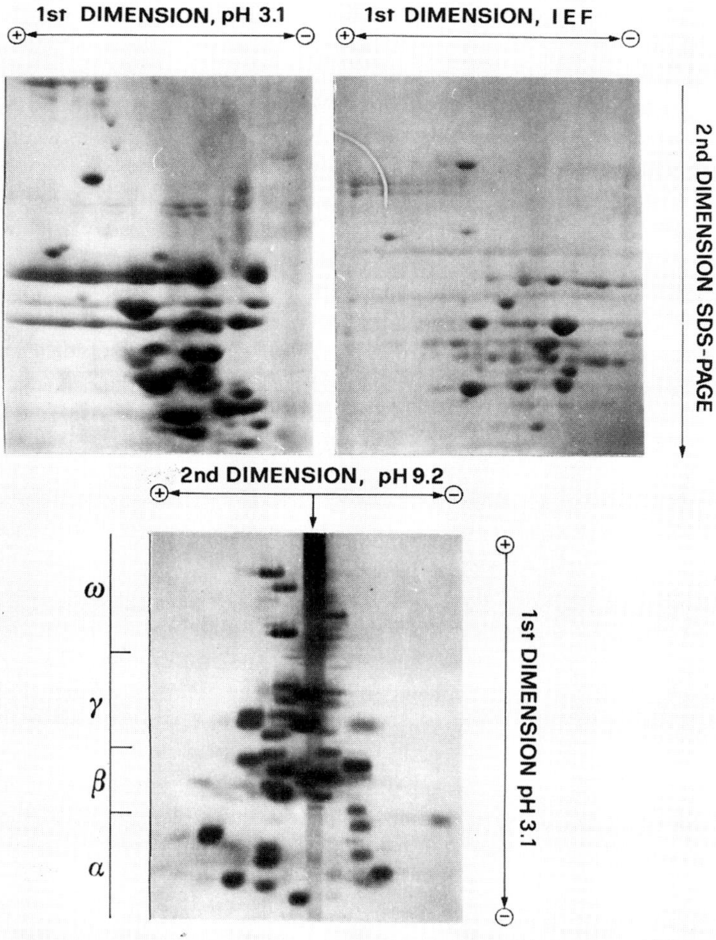

Combining reducing agents with the detergent SDS made it possible to analyze glutenin subunits (Bietz and Wall, 1972). Two-dimensional procedures have also been used to study glutenin subunits (Brown et al., 1979) and particularly the LMW glutenin subunits, which have proved difficult to analyze in one-dimensional SDS-PAGE because, on reduction, they overlap with some gliadins (Payne et al., 1984a).

### Chromatographic techniques

Until recently, these techniques were not being used for routine analysis because they were time-consuming and required a large amount of sample material. Their use was limited to the purification of components for further studies. The recent introduction of high-performance liquid chromatography (HPLC) in the analysis of cereal proteins (Bietz, 1983, 1985) has provided new and powerful separation methods. Reversed-phase high-performance liquid chromatography (RP-HPLC), in which proteins are separated on the basis of different surface hydrophobicities, is the method most often used.

The technique is rapid and requires small amounts of proteins, and thus analyses on half kernel are easily accomplished. RP-HPLC has been used to characterize wheat varieties (Bietz et al., 1984), and is useful in predicting pasta-cooking and bread-making qualities and in analyzing diploid and tetraploid wheat ancestors and related species (Burnouf and Bietz, 1987). Its use in genetic studies of wheat proteins has also been reported (Burnouf and Bietz, 1987; Lafiandra, 1988). Chromatographic data are easily stored and can be processed by computer. Different chromatograms can be compared by replotting stored data. RP-HPLC may therefore be considered an invaluable tool in handling wheat genetic resources.

## CHROMOSOMAL LOCATION AND GENETICS OF WHEAT STORAGE PROTEINS

The availability of aneuploid stocks in both durum and bread wheats has made it possible to assign gliadin and glutenin components to specific chromosomes and chromosome arms. This has been achieved by using one- and two-dimensional electrophoretical techniques and more recently by using RP-HPLC (Bietz and Burnouf, 1985; Burnouf and Bietz, 1985).

Early studies located gliadin genes on the short arm of chromosomes of the homoeologous groups 1 and 6, and all subsequent studies have confirmed these results, both in durum wheat (Du Cros et al.,1983; Lafiandra et al., 1983) and in bread wheats (Wrigley and Shepherd, 1973; Payne et al., 1982; Lafiandra et al., 1984), as well as in related species (Lawrence and Shepherd, 1981; Dvorak et al., 1986). Genes for HMW glutenin subunits were located on the long arms of chromosomes of homoeologous group 1 (Bietz et al., 1975; Payne et al., 1980). Loci for gliadins, designated *Gli-A1*, *Gli-B1*, *Gli-D1*, *Gli-A2*, *Gli-B2*, and *Gli-D2*, were found near the end of the short arm of the chromosomes of homoeologous groups 1 and 6 (Payne et al., 1984a).

The *Gli-1* loci correspond to a family of associated genes which code for three structurally distinct protein families, namely $\omega$ and $\psi$ gliadins and the LMW glutenin subunits (Payne et al., 1984c). Singh and Shepherd (1988) observed no recombination between genes controlling some major LMW glutenin subunits, which they designated *Glu-3*, and those controlling gliadin components localized on chromosome arms 1AS and 1DS, while 1.7 per cent recombination was found for those present on chromosome 1B, indicating that these proteins are coded by separated genes. The *Gli-2* loci correspond to $\alpha$ and $\beta$ gliadins. The loci *Glu-A1*, *Glu-B1* and *Glu-D1* correspond to the genes for HMW glutenin subunits. The variation in HMW glutenin subunits is because there are genes at five loci: two genes on chromosome 1D, controlling two different subunits — 1Dx, with slower electrophoretic mobility on SDS-PAGE, and 1Dy, with faster mobility; one or two on chromosome 1B, similarly indicated 1Bx and 1By; and one or none on chromosome 1A (Payne et al., 1980, 1981).

Genetic analyses have indicated that both gliadin and glutenin genes are tightly associated, inherited as blocks in crossing experiments; and several allelic forms have been described at each complex locus (Sozinov and Poperelya, 1980; Payne and Lawrence, 1983). Additional loci coding for gliadin-type proteins, located on the short arms of chromosomes 1A and 1B, equidistant between *Glu-1* and *Gli-1* and designated *Gli-A3* and *Gli-B3* respectively, have also been described (Payne et al., 1988).

## USE OF SEED STORAGE PROTEINS IN GENEBANK ACTIVITIES

### Use of wild species in exploiting genetic resources

Recently, several studies have focused on the variability of storage proteins in wild wheat relatives. Waines and Payne (1987) analyzed the HMW glutenin subunit composition of 497 accessions of diploid A genome species, while accessions of *Triticum turgidum* var. *dicoccum* have been analyzed by Vallega and Waines (1987). Collections of *T. turgidum* var. *dicoccoides* have been analyzed by Mansur-Vergara et al. (1984), Nevo and Payne (1987), Levy and Feldman (1988) and Levy et al. (1988).

Studies conducted by Nevo and Payne (1987) on samples of *T. dicoccoides* from well-defined ecogeographical sites showed the presence of extensive allelic variation at both *Glu-A1* and *Glu-B1* loci, in contrast with lower levels of allozyme allele diversity per locus. Correlations were also observed between the diversity of HMW glutenin subunits and the frequencies of specific glutenin alleles with physical and biotic factors. Studies by Levy and Feldman (1988) showed that intrapopulation variability could be predicted from geographical distribution, marginal populations being more uniform than those at the centre of distribution. Significant correlations were also found between molecular weight, size of subunits and ecological conditions.

Geographical variation in 212 accessions of *Aegilops squarrosa* from different countries was assessed using gliadin analyses by Cox and Harrell (1988). Polymorphic indices for *Gli-D1* and *Gli-D2* were 0.85 or higher in five regions; in material from Turkey, the values were 0.00 for both loci, giving precise information on where to undertake further collections.

## Use of cultivated species in exploiting genetic resources

Gliadins were used by Damania et al. (1983) to analyze *T. turgidum* and *T. aestivum* landraces from Yemen and Nepal and to assess variation in germplasm samples. Variation for HMW glutenin subunits in landraces of hexaploid wheat from Afghanistan and Nepal has been studied (Lagudah et al., 1987; Margiotta et al., 1988). The variation appeared independent of the altitude and geographical location of collection sites in the material analyzed by Lagudah et al. (1987). The germplasm from Nepal, from four different river valleys, has recently been analyzed for both gliadin and glutenin components (Lafiandra et al., 1987c).

Gliadin analyses showed a high degree of variation at each of the *Gli-1* and *Gli-2* loci. The consistent presence of seeds lacking the entire clusters of gliadin components controlled by the 1D chromosome was observed in some samples. These genotypes, termed 'null types', were found mainly in mixtures with seeds having normal gliadin patterns. Some samples included only seeds without 1D-controlled components. Lines without the 1D- controlled gliadin components were found throughout the area of collection. Electrophoretic analysis indicated that, in most cases, null 1D types differed also at the remaining *Gli-1* and *Gli-2* loci. Moreover, when plants were grown from these null type seeds, considerable variation for morphological traits was also evident, indicating that different materials from the same region possessed this characteristic. Positive correlation between altitude and incidence of null types at the *Gli-D1* locus was found, indicating a possible association with the stress conditions prevalent at higher altitudes.

Allelic variation at the three loci coding for HMW glutenin subunits was also assessed, and seeds which lacked HMW glutenin subunits, particularly 1Dx, 1Dy or 1By subunits, were detected (Lafiandra et al., 1987c; Margiotta et al., 1988). No correlation was found in this case between altitude and seeds which lacked HMW glutenin subunits. The distribution of populations in different valleys was deduced from correspondence analysis which was carried out on the absolute frequencies observed for the different *Glu-1* alleles. Variation among valleys was evaluated by discriminant analysis. The distribution of populations analyzed is shown in Figure 8.2. The first discriminant function, which accounts for 71 per cent of the variation, separates the material collected in one river valley from the material collected in the other three valleys. This function is closely associated with a particular allele at the *Glu-D1* locus, not reported in modern cultivars and present only in this particular population.

**Figure 8.2** Distribution of different bread wheat populations collected in four different Nepalese river valleys as deduced by their variation at the different *Glu-1* loci

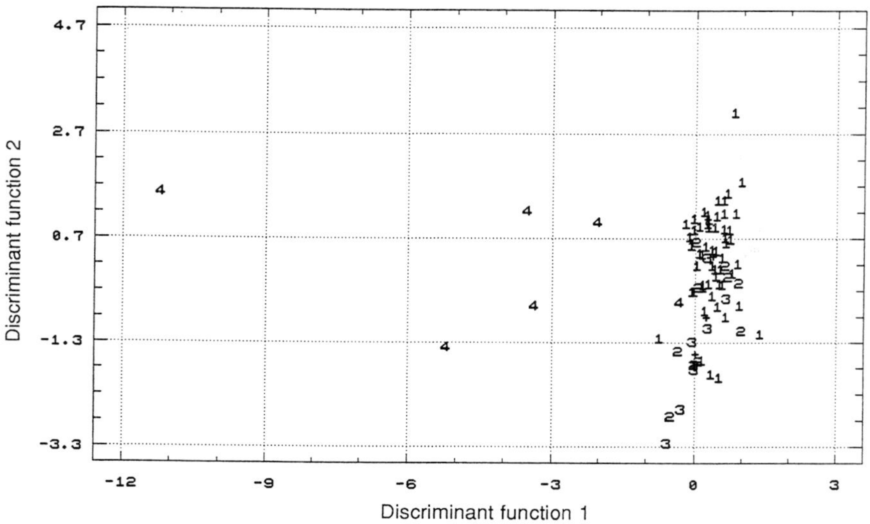


## Studies of the genetic structure of germplasm

As seed viability decreases during storage, seed samples must be periodically rejuvenated. As a result of natural selection, hybridization and other factors, the genetic structure of the accession can be modified. To keep costs at a reasonable level, it is necessary to define the minimum population size suitable to prevent strong dispersive effects on genotypic frequencies. If necessary, each accession can be split into less heterogeneous subpopulations, but whenever these effects are negligible it would be advantageous to store a small number of entries by pooling seed lots from different accessions.

The electrophoretic patterns of storage proteins were used by Sergio et al. (1988) to discriminate among genotypes when assessing the changes in the genetic structure of two durum wheat 'landraces' from Sicily during seed multiplication. The two populations were grown for 7 years at different planting densities — 100 kernels/$m^2$ and 550 kernels/$m^2$ — and the genotypic frequencies were assessed on the basis of the electrophoretic patterns of the gliadin components. An analysis of gliadin electrophoretic patterns allowed the identification of several genotypes in the original population. The relative frequency of durum wheat genotypes did change over time and some genotypes were lost, especially at the higher planting density (*see* Figure 8.3 *overleaf*).

**Figure 8.3** Genotypic frequencies assessed by means of acid-PAGE of gliadin proteins in a durum wheat landrace (MG 8001) grown at two planting densities after one (left) and seven cycles (right) of seed rejuvenation

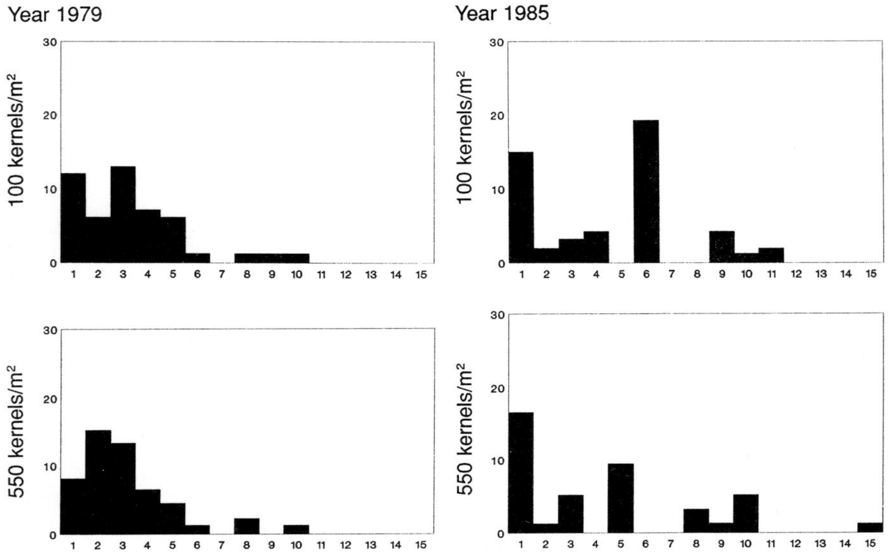


## Storage proteins and wheat evolutionary relationships

Because of their large intraspecific variation, gliadins and glutenins have not been used extensively for tracing wheat evolution. The high rate of mutations affecting these proteins and their quick fixation in a genome, probably because of the lack of severe constraints, make these proteins highly variable. Several authors (Johnson and Hall, 1965; Johnson, 1972; Ladizinsky and Johnson, 1972), in their extensive electrophoretic studies of different wheat species, used the non-gliadin fraction of the 70 per cent ethanol extract of seed proteins, since these proteins, essentially albumins and globulins, showed little intraspecific variation, were genome-specific, and could discriminate various degrees of affinity among genomes. Nevertheless, valuable information can also be obtained when analyzing the electrophoretic patterns of storage proteins, and the use of two-dimensional electrophoretic separation shows unrevealed genome specificity for these proteins.

Kasarda (1980) noted that wheat cultivars in general may be divided into Type 1 and Type 2 according to their α gliadin electrophoretic patterns. The two types of patterns result from two different allelic blocks of the 6A locus. In view of the similarities between α gliadins in Type 1 wheats and those in some *T. urartu* accessions, Kasarda speculated that *T. urartu* contributed the A genome to Type 1 wheats and that the A genome of Type 2 wheats must have been contributed by a different type of *T. urartu*. Alternatively, it could have been contributed by *T. boeoticum*.

Studies on the presence of certain allelic gliadin bands, 42 and 45, associated with technological properties in durum wheat and in the wild diploid and tetraploid ancestors of cultivated wheat, were conducted by Kushnir et al. (1984). The occurrence of gliadin 45, which correlates with strong gluten and high pasta quality in durum wheat, in lines of *Ae. sharonensis* and *T. dicoccoides* led them to postulate that band 45 is most likely of earlier origin. The absence of band 42 in both *Ae. sharonensis* and *T. dicoccoides* implies that this component is of more recent origin, probably the consequence of a mutation in cultivated tetraploid wheats.

Kasarda et al. (1984) used two-dimensional patterns of gliadins controlled by the extracted A and B genomes of hexaploid wheats and found that the A genome extract pattern of hexaploid wheats was similar to some accessions of *T. monococcum* and *T. urartu*. The extracted B genome pattern was found to be in accordance with the patterns of *Ae. searsii*, *Ae. longissima* and *Ae. speltoides*, but the composite pattern of different *Ae. searsii* accessions was closer to the extracted B genome pattern of hexaploid wheats. Also, the SDS-PAGE of the HMW glutenin subunits showed that patterns of *Ae. searsii* are more similar to the B genome-coded HMW glutenin subunits present in bread wheats (Cole et al., 1981).

The D genome patterns of Chinese Spring and Cheyenne, as deduced by chromosomal assignment of gliadin components in their intervarietal substitution lines (Lafiandra et al., 1984), were found to be similar to accessions of *Ae. squarrosa* ssp. *strangulata*. Johnson (1972) demonstrated that all the subspecies included in the hexaploid wheat of genomic formula AABBDD (namely *spelta, macha, vavilovii, compactum, sphaerococcum* and *vulgare*) have a very uniform electrophoretic pattern, simulated by the profile which is produced by mixing proteins from *T. dicoccum* and *Ae. squarrosa*, favouring the monophiletic origin of all hexaploid wheats.

Analyses of aneuploids, intervarietal substitution lines and segregating material from hybrid combinations have shown a moderate degree of variation in D genome-encoded gliadin and glutenin proteins in the subspecies *vulgare* (Lawrence and Shepherd, 1980; Payne and Lawrence, 1983; Metakovsky et al., 1984). Similar limited variation was found in other subspecies of the *aestivum* group. On the contrary, large variation has been shown to be present in collections of *Ae. squarrosa* at the *Gli-D2* and *Glu-D1* loci (Porceddu and Lafiandra, 1986; Cox and Harrell, 1988; Lagudah and Halloran, 1988), indicating that only a small amount of this variation is present in bread wheats, probably because of the very few crosses among tetraploids and *Ae. squarrosa* strains. However, the possibility that some of the variation initially introduced into hexaploid wheats from *Ae. squarrosa* has subsequently been lost cannot be excluded (Lawrence and Shepherd, 1980).

Genetic processes such as diploidization and gene dosage compensation, implicated in the reduction of the number or activity of redundant genes, may have contributed to the evolutionary processes of allopolyploid wheats both in the wild and under cultivation (Feldman et al., 1986). The influence of these processes on storage proteins were studied by Galili and Feldman (1983), Feldman et al. (1986) and Galili et al. (1988). Their results indicated that the process of diploidization for HMW glutenin subunits genes non-randomly affects mainly the A genome genes. Wild tetraploid wheat underwent relatively little diploidization, while the cultivated forms exhibit a massive diploidization for these genes, probably brought about by gene repression, deletion or mutation. Recent evidence that DNA sequences of HMW glutenins exist on chromosome 1A of the cultivar Chinese Spring of hexaploid wheat, which lacks HMW glutenin bands of this chromosome (Thompson et al., 1983), supports the conclusion of inactivation through mutation.

## MUTANTS FOR WHEAT STORAGE PROTEINS

Extensive electrophoretic analyses in tetraploid and hexaploid wheat collections have revealed the presence of seeds lacking certain groups of gliadin and glutenin components controlled by genes at a given complex locus (Lafiandra et al., 1987a; 1987c). This lack concerns either an entire cluster of tightly associated components (*see* Figure 8.4), or only a few polypeptides. To date, mutants at eight out of the nine complex loci coding for wheat storage proteins have been identified in materials from Nepal, China, Ethiopia and Algeria (Lafiandra et al., 1988).

## USE OF STORAGE PROTEIN VARIABILITY IN IMPROVING TECHNOLOGICAL PROPERTIES OF DURUM AND BREAD WHEATS

### Durum wheats

In durum wheat, the chromosome 1B is considered to be responsible for differences in technological quality. Damidaux et al. (1978) first reported that durum wheat cultivars possessing the $\psi$ gliadin 45 were superior in cooking quality to cultivars which possessed $\psi$ gliadin 42. The two gliadins were found to be coded by two co-dominant alleles of a single gene on chromosome 1B (Damidaux et al., 1980). It has also been shown that each of these two components belongs to a more complex group of proteins, whose genes are tightly linked, comprising $\omega$ gliadins and LMW glutenin subunits; usually the $\omega$ 45 component is associated with the $\omega$ gliadin component 35 and LMW-2 glutenin subunits, while $\psi$ 42 is associated with the $\omega$ components 33, 35 and 38 and LMW-1 glutenin subunits (Payne et al., 1984b). Payne et al. (1984b) postulated that LMW glutenin subunits might be responsible for qualitative differences, and that bands 42 and 45 are only genetic markers. That LMW glutenin subunits are the actual cause of qualitative differences in durum wheat was recently demonstrated by Pogna et al. (1988), using the Italian durum wheat cultivar Berillo, in which recombination occurring at the complex *Gli-B1* locus was already known (Margiotta et al., 1987).

**Figure 8.4** Two-dimensional electrophoretic patterns of gliadin proteins from normal seeds and seeds lacking components controlled by 6A (upper right) and 1D (lower right) chromosomes

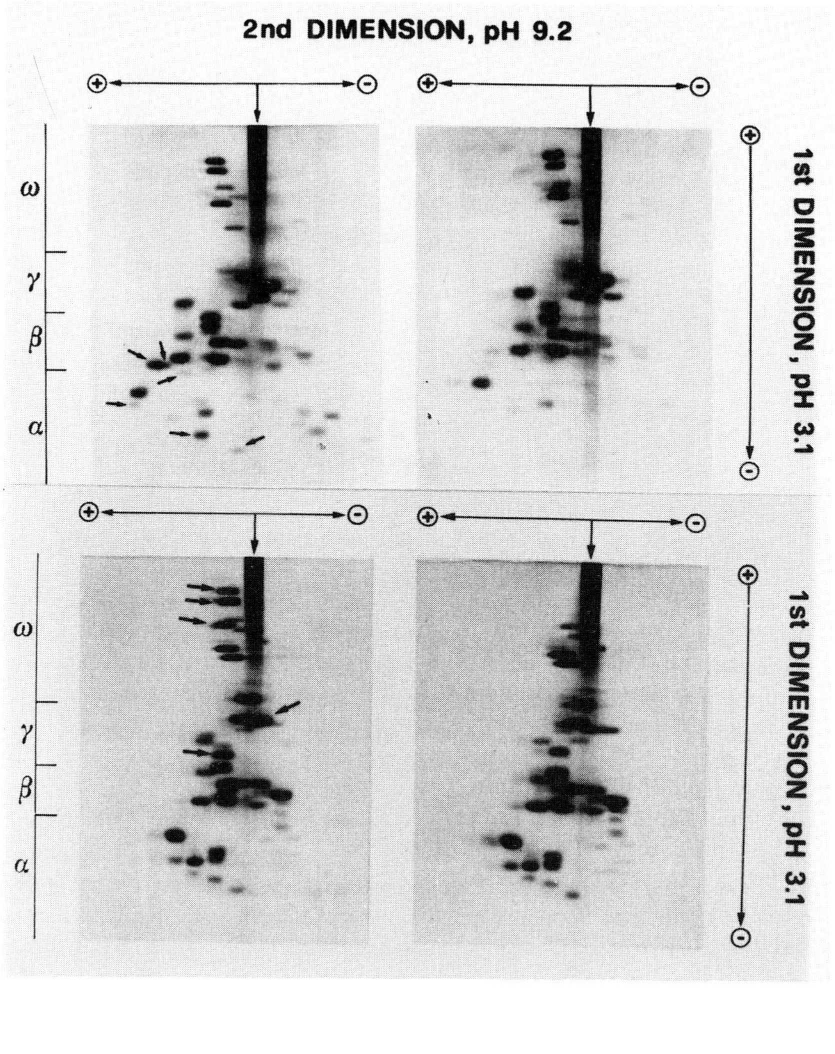


Analyses of world collections of durum wheats have shown that either band 42 or 45 are always present. Recent analyses of landraces and old cultivars from different countries have revealed the presence of alternative forms to bands 42 and 45 (Margiotta et al., 1987); a large number of these different forms is present, for example, in tetraploid collections from Ethiopia. Tests of SDS-sedimentation, which correlates with gluten strength, indicated the existence of wide variation in 42 and 45 types (*see* Figure 8.5).

**Figure 8.5** Frequencies of SDS-sedimentation value for durum wheats possessing γ component 42, γ 45 and components of different mobilities found in Ethiopian material

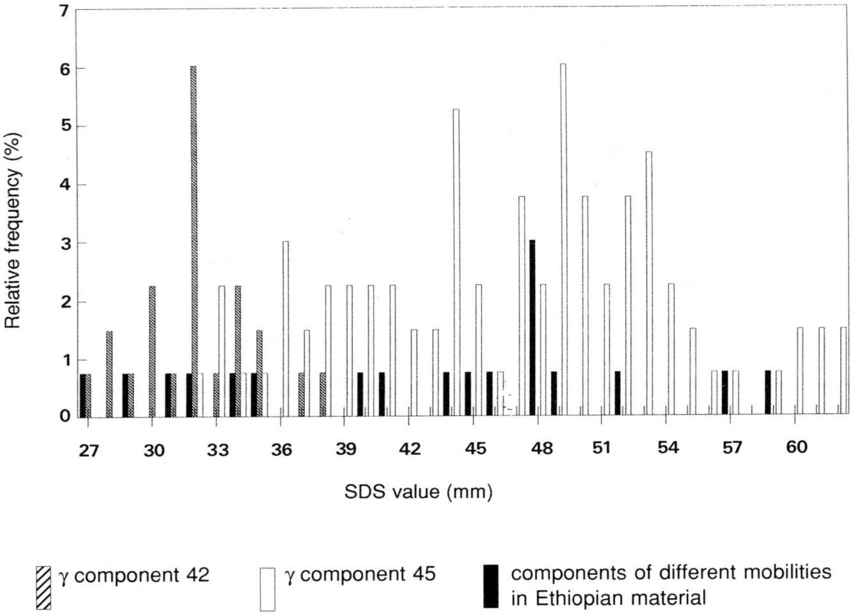


## Bread wheats

Allelic variation at each of the three complex loci *Glu-1* is the main reason for quality differences in bread wheat, and the different alleles at each locus have been ranked according to their influence on bread making (Payne et al., 1984a). The possibility of identifying new sources of variation in bread wheat landraces has been stressed, and extensive electrophoretic analyses have indicated this possibility (Payne et al., 1984a; Lafiandra et al., 1987b).

Allelic variation at gliadin loci has also been shown to be responsible for differences in baking quality and dough strength (Sozinov and Poperelya, 1980; Wrigley, 1980). The following ranking order among glutenin and gliadins has been reported: *Glu-1* > *Gli-1* > *Gli-2* (Payne et al., 1984a). Since gliadins coded by genes on Group 1 chromosomes are tightly associated to LMW glutenin subunit genes, it is very likely that LMW glutenin subunits are directly responsible for qualitative differences (Gupta and Shepherd, 1988). Payne et al. (1984b) suggested that the ψ gliadin 45 block of genes, responsible for good technological quality in durum wheat, could be transferred to bread wheat to test its effect on bread-making quality. Analyses of bread wheat landraces indicated that this block is already present in bread wheats, facilitating its transfer to breeding lines without the need for interspecific crosses (*see* Figure 8.6).

**Figure 8.6** Comparison of gliadin components from bread wheat (left) and durum wheat (right) lines, both possessing γ component and ω 35

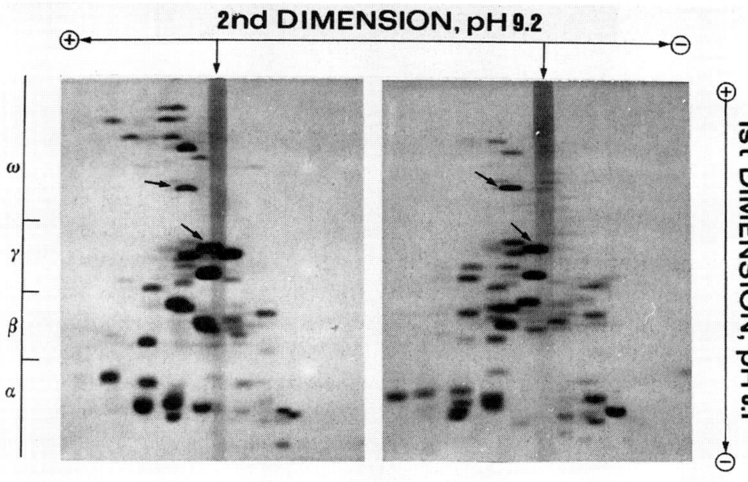


The existing variation for components associated with technological properties can be further enlarged by making use of the variation present in wild wheat relatives, as stressed by Waines and Payne (1987), and by Nevo and Payne (1987), among others. Limited variation is present for the D genome-controlled protein components in bread wheat, and thus the large collections of *Ae. squarrosa* already available can provide a rich and valuable source of new genes for the improvement of bread-making quality.

## STORAGE PROTEINS AND NUTRITIONAL VALUE

As already mentioned, lysine is the first amino acid limiting the nutritional value of wheat. Large-scale screening of world durum and bread wheat collections have revealed only a few lines with promising levels of lysine content (Johnson et al., 1978), and none was detected with genetic effects as large as those found in maize, barley and sorghum. Since gliadins are responsible for the low nutritional value of wheats, one way of improving quality would be to modify the coding sequence of their structural genes in order to create proteins containing more lysine residues. However, Croy and Gatehouse (1985) questioned this approach, as it would be limited by the presence of constraints on structures and sequences of storage proteins, and only certain changes would be allowed. The detection of mutants lacking entire clusters of gliadin components offers a different perspective to the improvement of wheat nutritional value. Through crosses, it has in fact been possible to assemble lines that lack more than one cluster of gliadin components (Lafiandra et al., 1987c). The effect on nutritional quality might be significant when more than two different clusters are eliminated.

## CONCLUSION

Storage proteins cannot supply the range of genetic markers provided by isozymes for measuring genetic diversity, studying population structure and producing chromosome maps, but in some instances their use in genebank activities can be just as valuable as isozyme markers and a useful complement to them. Their use has so far been rather limited compared to isozymes, but in some cases they have been helpful in tracing the distribution patterns of species from their centres of origin. Also, the lack of severe constraints to mutations can be useful when studying the evolution of cultivated plants.

Storage proteins from species other than wheat have been used to describe *Pisum* genetic resources (Przybylska, 1988), assess polymorphism in wild barley (Nevo et al., 1983) and trace bean dissemination patterns (Gepts, 1988). Improved electrophoretic and chromatographic techniques, combined with computing procedures, can be very helpful in handling the considerable amount of data which is produced when evaluating genetic resources. The electrophoretic analysis of seed storage proteins can be effectively used to study the genetic structure of wheat entries. This is particularly useful when the viability of seed lots is very low, making it impossible to observe other characters.

Plant improvement will continue to depend on the genetic variation contained in existing genepools, which must continue to be collected, preserved and evaluated. New allelic variants for storage proteins have been screened in landraces, and their possible incorporation into advanced material for developing new varieties with improved bread- and pasta-making qualities is being evaluated. Even greater variation is represented by the wild diploid and tetraploid ancestors, allowing the rapid assessment of effects on quality.

Identification of mutants for storage proteins not only has practical implications for breeding wheat with a more balanced amino acid composition, but is also important in establishing technological relationships and in shedding light on molecular aspects of complex gene families, such as those of storage proteins.

## *Discussion*

A. B. DAMANIA: Did you find any correlation between the deletion on 1D chromosome and altitude of the collection site?

D. LAFIANDRA: The only correlation found between altitude and the frequency or lack of gliadin bands in the omega region was a positive one. That is, material from higher altitudes was very rich indeed, lacking the 1D controlled gliadin components. Some of the landraces collected from above 3000m altitude were completely without 1D gliadins.

A. B. DAMANIA: Could this result from mutation caused by the high ultra-violet rays, with the deterioration of the ionosphere cover — the so-called 'greenhouse effect'?

D. LAFIANDRA: This is a possibility. But in some cases the entire cluster was not lost, indicating that there may be other causes besides mutation.

J.G.WAINES: Dr Johnson used protein patterns in his studies at Riverside, California. Not much has been said about them over the years. Do you have any information on them?

D. LAFIANDRA: His work concentrated mainly on the main components of the storage protein gliadins and glutenins, rather than on albumins and globulins, which are in the endosperm.

J. G.WAINES: Some of the recent work on flour quality involves certain chromosomes of wheat which Johnson had looked at in his studies on wheat evolution.

D. LAFIANDRA: In a paper that was presented at the last gluten workshop, scientists at the Plant Breeding Institute, Cambridge showed that there is a relationship between these proteins and flour quality. For some of these bands, the relationship is negative.

C. JOSEPHIDES: The presence of band 45 and absence of band 42 is correlated with good pasta-making quality in durum wheat by distinguishing between strong and weak gluten. How great is the variability for gluten strength in the durum cultivars in the group with band 45?

D. LAFIANDRA: There is good variation. Some cultivars with band 45 have a very high SDS-sedimentation value but others have a low value. I think that in most cases this is directly related to protein content.

C. JOSEPHIDES: Are you suggesting that there are genes on other chromosomes that suppress expression of the genes on the short arm of 1B?

D. LAFIANDRA: No, I am suggesting that there is variation for SDS-sedimentation values in those cultivars with band 45. It could result from other factors. But protein content is very much involved, because low protein cannot be associated with strong gluten.

M. OBANNI: You mentioned that in the Ethiopian durums you found other bands in the region where bands 42 and 45 are usually present. Do you know the genetic basis for this?

D. LAFIANDRA: No, but from their positions we could ascertain the chromosomes and genomes which were the same for the bands 42 and 45, so they must be allelic forms.

Wheat Genetic Resources: Meeting Diverse Needs
Edited by J. P. Srivastava and A. B. Damania

Published by John Wiley & Sons

# 9

# *One-Dimensional Electrophoretic Separation of Gliadins in a Durum Wheat Collection from Ethiopia*

S. BENEDETTELLI, M. CIAFFI, C. TOMASSINI, D. LAFIANDRA and E. PORCEDDU

Modern plant breeding has reduced the genetic variability in many plant species, including wheat. The major hope for future crop improvements lies in using the rich genepool of populations, landraces and wild relatives (Feldman and Sears, 1981; Nevo and Payne, 1987; Porceddu et al., 1988). Programmes to identify variation should maximize the efficiency of sampling strategies by using ecological-genetic factors and biochemical markers as predictors. Recently, electrophoretically discernible seed storage proteins have been used to assess variation in cereal populations, landraces and cultivars. A special characteristic of storage proteins is their high level of heterogeneity which is strictly genetically determined. As these proteins are direct gene products, relatively free of environmental effects, they can provide, either independently or in addition to other analyses, a reasonably accurate measure of genetic diversity.

Wheat landraces and populations of wild relatives collected from various countries in the primary and secondary centres of diversity have been evaluated for their variation in seed storage proteins (Lafiandra et al., 1987a, 1987b; Lagudah et al., 1987; Margiotta et al., 1988). Although these studies have indicated the importance of variation in protein components in breeding, no attempt has been made to understand the overall pattern of genetic variation of populations. Analyses of *Triticum dicoccoides* samples have shown that variation in high-molecular-weight (HMW) glutenin subunits could be accounted for by environmental factors and could be used to determine the genetic structure of populations (Nevo and Payne, 1987).

This paper looks at some aspects of geneticvariation at the biochemical level, using the electrophoretic patterns of gliadin proteins in a tetraploid wheat collection from Ethiopia.

## MATERIALS AND METHODS

Eight populations of tetraploid wheat, *Triticum turgidum* var. *durum*, collected from Ethiopia were used for this study. The latitude, longitude and altitude of collection sites and the accession numbers of each population are given in Table 9.1. The populations are identified by their respective Mediterranean Germplasm (MG) numbers. Figure 9.1 shows the geographical distribution of the collection sites in the country.

**Table 9.1** Geographical coordinates and number of accessions of the studied populations

| Population | Latitude N | Longitude E | Altitude (m) | No. of accessions |
|---|---|---|---|---|
| MG 7764 | 9°45’ | 39°06’ | 2900 | 29 |
| MG 7766 | 9°54’ | 39°06’ | 2900 | 51 |
| MG 7768 | 9°54’ | 39°06’ | 2900 | 53 |
| MG 7769 | 9°24’ | 38°47’ | 2780 | 57 |
| MG 7770 | 9°34’ | 38°52’ | 2800 | 47 |
| MG 7771 | 9°42’ | 38°50’ | 2800 | 49 |
| MG 7780 | 9°56’ | 38°48’ | 2100 | 48 |
| MG 7781 | 9°56’ | 38°47’ | 2200 | 48 |

Electrophoretic analyses were carried out on single seeds collected from a randomly selected spike in each accession; 382 seeds were analyzed, representing 382 samples.

Gliadin proteins were extracted with 1.5 M dimethylformamide and fractionated by gel electrophoresis in an aluminium lactate buffer using the method described by Bushuk and Zillman (1978), with minor modifications. The migration distances of individual bands were measured from the photographs of electrophoregrams. The reference protein component used to standardize relative mobility (Rm) was band 50 (Bushuk and Zillman, 1978), present in the Italian durum wheat cultivar Karel. Errors of electrophoretic separation or observation were corrected according to the procedure proposed by Benedettelli et al. (in prep.). This procedure permitted reduction of the 72 Rm values observed to 33 separate protein components.

### Data matrix

The electrophoretic pattern of each accession was considered as a vector where the presence or absence and mobility of each component (group Rm value) was recorded. The matrix resulting from all these vectors was considered as a two-dimensional contingency table (individuals x protein components) in which each accession was described by a set of 0 and 1, according to the presence or absence of each component group previously described.

**Figure 9.1** Geographical distribution of collection sites in Shewa Provice, Ethiopia

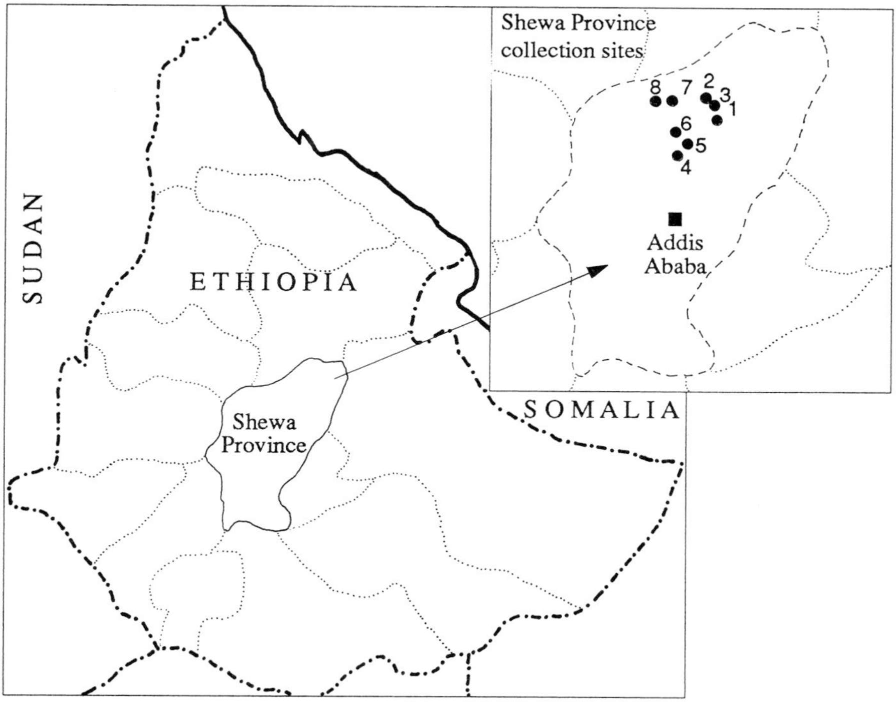


## Correspondence analysis

Multivariate correspondence analysis was originally propounded by Benzecry (1973) and later described by Hill (1974) and Barrai (1986). It allows a swarm of original data (discontinuous or continuous) to be described as dispersions from a common centroid and gives the relative position of each original observation.

The swarm of the original observations contained in the two-dimensional contingency table can be described by a chi-square i*j distance matrix computed for each observation from a common centroid. This matrix can be analyzed either by means of the principal component (PC) analysis (Benzecry, 1973; Hill, 1974) or by means of the discriminant analysis proposed by She et al. (1987).

In this study, discriminant analysis was carried out to examine the distribution of different accessions between and within populations. All statistical analyses were performed on a personal computer using the Numerical Taxonomy and Multivariate System (NTSYS) and Stratgraphic STSC software. After performing correspondence analysis, a similarity matrix (Euclidean distance) and cluster UPGMA algorithm were used to detect the association between different protein components.

## RESULTS

Table 9.2 gives the absolute frequencies of different bands in all eight populations. The chi-square value calculated from the contingency table was highly significant.

**Table 9.2** Cumulative frequencies of 33 principal components among the eight populations

| Principal component | Populations (MG) | | | | | | | | |
|---|---|---|---|---|---|---|---|---|---|
| | 7764 | 7766 | 7768 | 7769 | 7770 | 7771 | 7780 | 7781 | Total |
| 15 | 2 | 1 | 0 | 0 | 0 | 0 | 4 | 6 | 13 |
| 18 | 6 | 3 | 0 | 0 | 0 | 3 | 7 | 8 | 27 |
| 23 | 27 | 48 | 51 | 55 | 41 | 48 | 46 | 47 | 363 |
| 29 | 18 | 44 | 42 | 46 | 39 | 37 | 10 | 8 | 244 |
| 33 | 16 | 41 | 42 | 43 | 31 | 41 | 10 | 20 | 244 |
| 34 | 9 | 31 | 39 | 44 | 28 | 39 | 13 | 7 | 210 |
| 37 | 3 | 1 | 1 | 1 | 1 | 1 | 26 | 8 | 42 |
| 42 | 11 | 15 | 11 | 4 | 17 | 10 | 26 | 8 | 129 |
| 44 | 9 | 16 | 10 | 9 | 14 | 0 | 17 | 11 | 86 |
| 45 | 24 | 41 | 53 | 54 | 42 | 49 | 47 | 40 | 350 |
| 47 | 23 | 48 | 50 | 53 | 41 | 45 | 45 | 46 | 351 |
| 50 | 25 | 48 | 50 | 54 | 44 | 45 | 42 | 35 | 343 |
| 51 | 24 | 35 | 47 | 50 | 41 | 37 | 36 | 35 | 305 |
| 54 | 14 | 35 | 32 | 40 | 18 | 31 | 35 | 39 | 244 |
| 57 | 19 | 41 | 35 | 24 | 41 | 23 | 30 | 29 | 242 |
| 58 | 12 | 21 | 23 | 30 | 20 | 19 | 15 | 14 | 154 |
| 59 | 16 | 31 | 39 | 38 | 27 | 33 | 18 | 27 | 229 |
| 61 | 19 | 38 | 30 | 42 | 35 | 34 | 40 | 33 | 271 |
| 62 | 13 | 35 | 37 | 42 | 33 | 33 | 24 | 25 | 242 |
| 63 | 16 | 30 | 31 | 25 | 19 | 26 | 30 | 19 | 196 |
| 64 | 16 | 27 | 36 | 39 | 28 | 31 | 26 | 34 | 237 |
| 65 | 13 | 16 | 33 | 26 | 22 | 24 | 20 | 12 | 166 |
| 67 | 25 | 29 | 21 | 22 | 26 | 22 | 39 | 33 | 217 |
| 71 | 12 | 32 | 37 | 15 | 18 | 6 | 14 | 17 | 151 |
| 73 | 18 | 43 | 46 | 48 | 39 | 43 | 39 | 35 | 311 |
| 76 | 13 | 27 | 21 | 30 | 15 | 27 | 18 | 23 | 174 |
| 77 | 14 | 23 | 23 | 24 | 21 | 15 | 15 | 25 | 160 |
| 78 | 10 | 23 | 22 | 22 | 22 | 21 | 16 | 9 | 145 |
| 79 | 14 | 21 | 32 | 32 | 20 | 19 | 17 | 15 | 170 |
| 80 | 10 | 15 | 15 | 19 | 18 | 16 | 16 | 9 | 118 |
| 81 | 7 | 19 | 34 | 28 | 22 | 30 | 13 | 17 | 170 |
| 83 | 20 | 25 | 16 | 23 | 21 | 13 | 35 | 34 | 187 |
| 85 | 12 | 19 | 7 | 13 | 14 | 21 | 13 | 15 | 114 |

$X^2 = 527.114$; Degrees of Freedom = 224

Eigenvalues and percentage estimates of total variance expressed by each value, computed from the correspondence analysis (*see* Table 9.3), indicated that because of the relatively small contribution of each eigenvalue, the first 25 values were required to describe 95 per cent of the total variability. The relative frequencies and distances from the centroid of each variable indicated that the variables with low relative

**Table 9.3** Eigenvalues calculated from the dispersion matrix of the correspondence distances and relative frequencies, distances from centroid of the considered variables (bands)

| Eigenvalues | Percent | Cumulative | Variable | Relative frequency | Distance$^2$ |
|---|---|---|---|---|---|
| 0.12018 | 13.35 | 13.35 | 15 | 0.00197 | 24.99559 |
| 0.09737 | 10.81 | 24.16 | 18 | 0.00409 | 12.57904 |
| 0.07697 | 8.55 | 32.71 | 23 | 0.05496 | 0.05936 |
| 0.06695 | 7.44 | 40.14 | 29 | 0.03694 | 0.52827 |
| 0.05559 | 6.17 | 46.32 | 33 | 0.03694 | 0.52441 |
| 0.05464 | 6.07 | 52.39 | 34 | 0.03179 | 0.74438 |
| 0.04466 | 4.96 | 57.35 | 37 | 0.00636 | 8.38578 |
| 0.04010 | 4.45 | 61.80 | 42 | 0.01953 | 2.06622 |
| 0.03596 | 3.99 | 65.79 | 44 | 0.01302 | 3.57500 |
| 0.03134 | 3.48 | 69.27 | 45 | 0.05299 | 0.10354 |
| 0.02998 | 3.33 | 72.60 | 47 | 0.05314 | 0.09846 |
| 0.02534 | 2.81 | 75.42 | 50 | 0.05193 | 0.11156 |
| 0.02418 | 2.69 | 78.10 | 51 | 0.04618 | 0.24526 |
| 0.02229 | 2.47 | 80.58 | 54 | 0.03694 | 0.58927 |
| 0.02173 | 2.41 | 82.99 | 57 | 0.03664 | 0.58435 |
| 0.01940 | 2.15 | 85.15 | 58 | 0.02332 | 1.43806 |
| 0.01757 | 1.95 | 87.10 | 59 | 0.03467 | 0.65662 |
| 0.01597 | 1.77 | 88.87 | 61 | 0.04103 | 0.43378 |
| 0.01540 | 1.71 | 90.58 | 62 | 0.03664 | 0.56085 |
| 0.01164 | 1.29 | 91.87 | 63 | 0.02967 | 0.93938 |
| 0.01083 | 1.20 | 93.08 | 64 | 0.35880 | 0.63751 |
| 0.00902 | 1.00 | 94.08 | 65 | 0.02513 | 1.29129 |
| 0.00828 | 0.92 | 95.07 | 67 | 0.03285 | 0.78880 |
| 0.00754 | 0.84 | 95.83 | 71 | 0.02286 | 1.47532 |
| 0.00716 | 0.79 | 96.63 | 73 | 0.04709 | 0.24653 |
| 0.00641 | 0.71 | 97.34 | 76 | 0.02634 | 1.20268 |
| 0.00527 | 0.59 | 97.93 | 77 | 0.02422 | 1.37768 |
| 0.00514 | 0.57 | 98.50 | 78 | 0.02195 | 1.61664 |
| 0.00440 | 0.49 | 98.99 | 79 | 0.02574 | 1.24265 |
| 0.00377 | 0.42 | 99.41 | 80 | 0.01787 | 2.17215 |
| 0.00309 | 0.34 | 99.75 | 81 | 0.02574 | 1.23279 |
| 0.00226 | 0.25 | 100.00 | 83 | 0.02831 | 1.10920 |
| | | | 85 | 0.01726 | 2.27058 |

frequencies are those with the greatest distance. The relative frequencies are very important for discriminating among the accessions. The absolute contributions of these variables to the PCs are also relatively high.

Only eight PCs were used for discriminating among accessions dispersed by columns (*see* Table 9.4). Eigenvalues with relative percentages and canonical correlation values of the seven discriminant functions, and their statistical tests of significance, are given in Table 9.5. It can be seen that the first three functions describe more than 90 per cent of the total variability considered.

**Table 9.4** Absolute contribution (ac) of the original variable (comp) to the principal component variable utilized for discriminant analysis

| Variable 1 | | Variable 2 | | Variable 3 | | Variable 5 | |
|---|---|---|---|---|---|---|---|
| comp | ac | comp | ac | comp | ac | comp | ac |
| 15 | .156225 | 76 | .135896 | 15 | .136908 | 37 | .262784 |
| 18 | .163202 | 77 | .134535 | 18 | .116700 | 44 | .138415 |
| 42 | .116131 | 78 | .106011 | 80 | .176211 | 71 | .132046 |
| | | 79 | .120409 | 81 | .129395 | | |
| | | 83 | .105232 | | | | |
| | | 85 | .110968 | | | | |
| **Variable 6** | | **Variable 7** | | **Variable 10** | | **Variable 11** | |
| **comp** | **ac** | **comp** | **ac** | **comp** | **ac** | **comp** | **ac** |
| 42 | .119976 | 44 | .114153 | 71 | .282496 | 54 | .146060 |
| 44 | .152974 | 58 | .155121 | | | 57 | .181898 |
| 64 | .119154 | | | | | 67 | .181982 |

The variation expressed by these three functions was highly significant. The standardized discriminant function coefficient loading of protein components used in the discriminant analysis to determine the value of the canonical variate (*see* Table 9.6) indicate that the first canonical discriminant function described 68.07 per cent of the total variation. The highest contributions came from PC 1 (0.936) (protein components 15, 18, 37 and 42 in Table 9.4) and from PC 5 (0.410) (protein components 37, 44 and 71 in Table 9.4).

The second canonical variable (14.81 per cent of the total variability) was negatively correlated with PC 5 and PC 14 (protein components 54, 57 and 67 in Table 9.4), and positively correlated with PC 7 (protein components 44 and 58 in Table 9.4). The third discriminant function (9.97 per cent of total variability) was positively correlated with PC 6 (protein components 42, 44 and 64 in Table 9.4) and negatively correlated with PC 5. The total distribution of 382 accessions, shown in Figure 9.2 (*overleaf*), formed

**Table 9.5** Discriminant analysis of 382 accessions by eight populations

| Discriminant function | Eigenvalue | Relative percent | Canonical correlation |
|---|---|---|---|
| 1 | .8164856 | 68.07 | .67044 |
| 2 | .1776583 | 14.81 | .38840 |
| 3 | .1196354 | 9.97 | .32688 |
| 4 | .0348286 | 2.90 | .18346 |
| 5 | .0284353 | 2.37 | .16628 |
| 6 | .0130592 | 1.09 | .11354 |
| 7 | .0093564 | .78 | .09628 |

| Functions derived | Wilks lambda | Chi-square | DF[a] | Significance level |
|---|---|---|---|---|
| 0 | .3836607 | 357.33274 | 56 | .00000 |
| 1 | .6969142 | 134.68768 | 42 | .00000 |
| 2 | .8207268 | 73.69175 | 30 | .00002 |
| 3 | .9189147 | 31.54160 | 20 | .04843 |
| 4 | .9509193 | 18.77163 | 12 | .09419 |
| 5 | .9779589 | 8.313228 | 6 | .21604 |
| 6 | .9907303 | 3.47373 | 2 | .17607 |

a DF = Degrees of Freedom

**Table 9.6** Contribution of principal components to discriminant function

| Variables | 1 | 2 | 3 | 4 | 5 | 6 | 7 |
|---|---|---|---|---|---|---|---|
| 1 | 0.93645 | 0.19653 | 0.23035 | 0.01427 | -0.04293 | -0.17492 | 0.07676 |
| 2 | 0.29420 | 0.07135 | -0.03190 | -0.03776 | 0.05085 | 0.90366 | 0.25727 |
| 3 | -0.22530 | -0.30102 | 0.34787 | 0.28316 | 0.03684 | -0.02953 | 0.70194 |
| 5 | 0.41032 | -0.63053 | -0.47511 | 0.19794 | 0.37865 | -0.12086 | 0.06597 |
| 6 | -0.04788 | -0.11008 | 0.48862 | 0.21828 | 0.58279 | 0.20958 | -0.52899 |
| 7 | -0.10561 | 0.45719 | 0.17240 | 0.39029 | 0.33849 | -0.20750 | 0.26414 |
| 10 | -0.00344 | -0.30239 | 0.39571 | -0.76175 | 0.23406 | -0.12981 | 0.15782 |
| 14 | 0.08945 | -0.47624 | 0.42186 | 0.29263 | -0.54992 | 0.02685 | -0.19044 |

two distinct groups along the first axis. The accessions in the group on the left are better separated than those in the group on the right.

Population numbers MG 7766 (2), MG 7768 (3), MG 7769 (4), MG 7770 (5) and MG 7771 (6) can be grouped together by the position of their centroids (*see* Figure 9.3 (*overleaf*). The centroids of populations MG 7764 (1), MG 7780 (7) and MG 7781 (8) are distinct, and far from the main group. The dispersions of these populations are similar, except for MG 7770 (5) (*see* Figure 9.4), which appears more dispersed around its centroid than the others. The dispersion of the accessions is very high in MG 7780 (7) in the first and second canonical variables (*see* Figure 9.5).

**Figure 9.2** Distribution of the 382 accessions analyzed

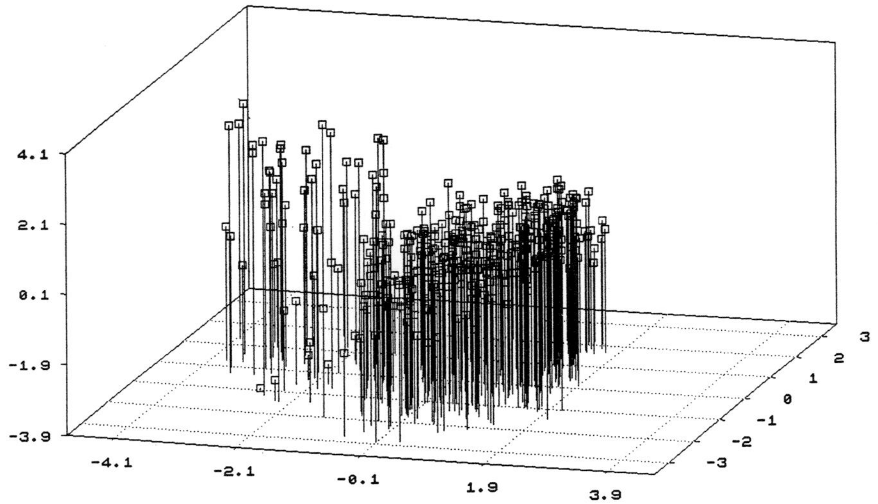


**Figure 9.3** Position of centroids of the eight populations studied

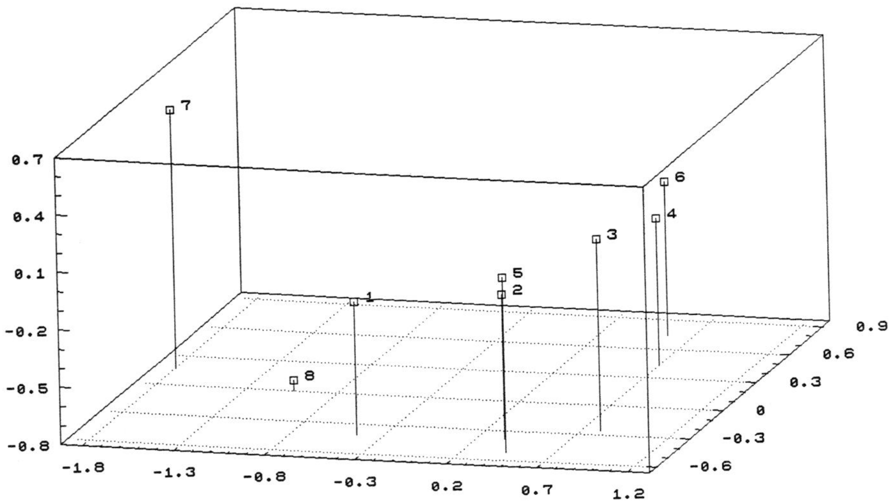

**Figure 9.4** Dispersion of population five around its centroid

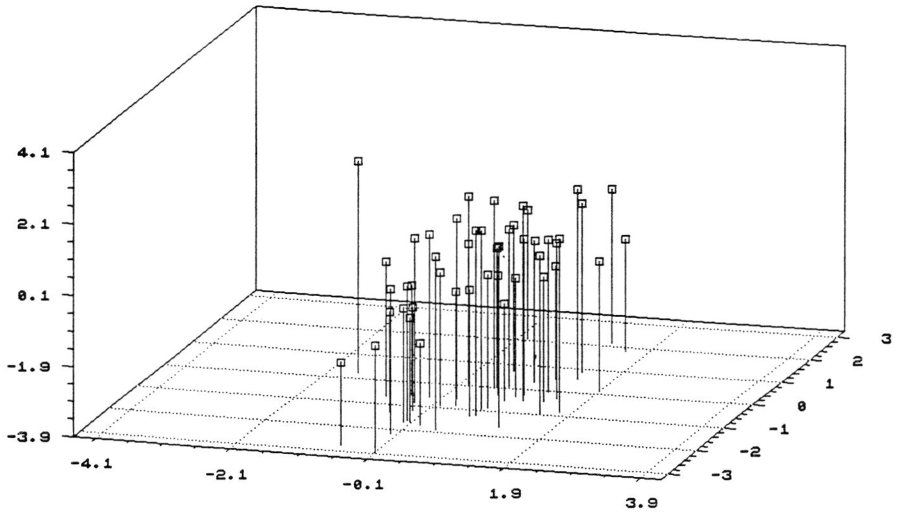


**Figure 9.5** Dispersion of population seven around its centroid

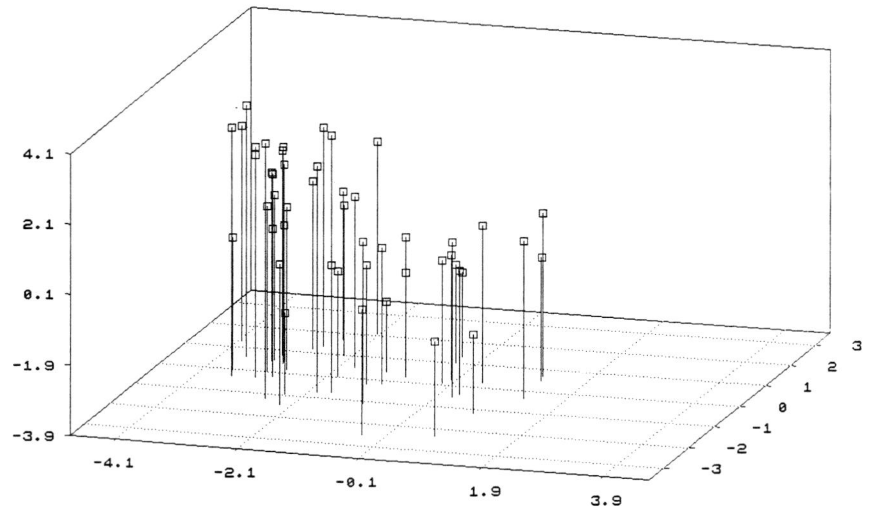

When the representative electrophoretic patterns of populations plotted far from each other, as in MG 7771 (6) and MG 7780 (7), it was possible to detect a fairly large difference in protein bands with low electrophoretic mobility (*see* Figure 9.6). In fact,

**Figure 9.6** Comparison of population MG 7771 (6) and MG 7780 (7) showing differences in protein components with lower mobility

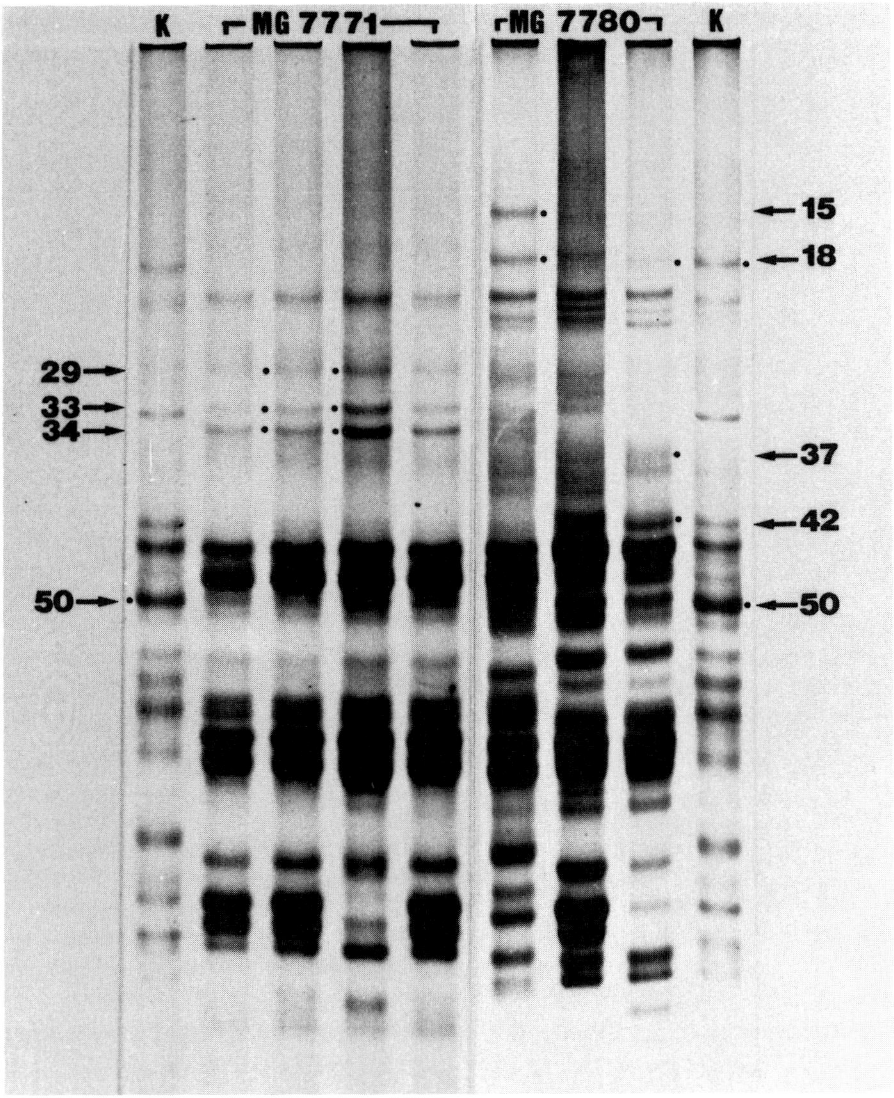

the population MG 7771 (6) was characterized by components (29, 33, 34, 45, 47 and 50) that were not present in population MG 7780 (7), which had components 15, 18, 37 and 42. These results are confirmed by the frequencies of these components, which differ between populations (*see* Table 9.2).

The separation of populations obtained by this method partially reflects their geographic distribution, as shown in Figure 9.1 and Table 9.1. The populations can be classified by the altitude of the collecting site.

The correspondence analysis for the dispersion of the columns by rows, carried out to examine the association of the different protein components, showed some association groups which include the original variables: components 23, 47 and 50 are in fact clustered together. Studies on genetic control (Lafiandra et al., 1983) suggest that these proteins are encoded by genes located on the short arm of chromosome 1A.

Another group of associations is the one which is composed of bands 29, 33 and 34 (*see* Figure 9.7), and is probably controlled by genes which are located on the short arm of chromosome 1B (Lafiandra et al., 1983). The third cluster is composed of components 77 and 79, and is probably controlled by genes located on the short arm of chromosome 6A.

**Figure 9.7** Dendrogram and electrophoretic patterns showing group associations

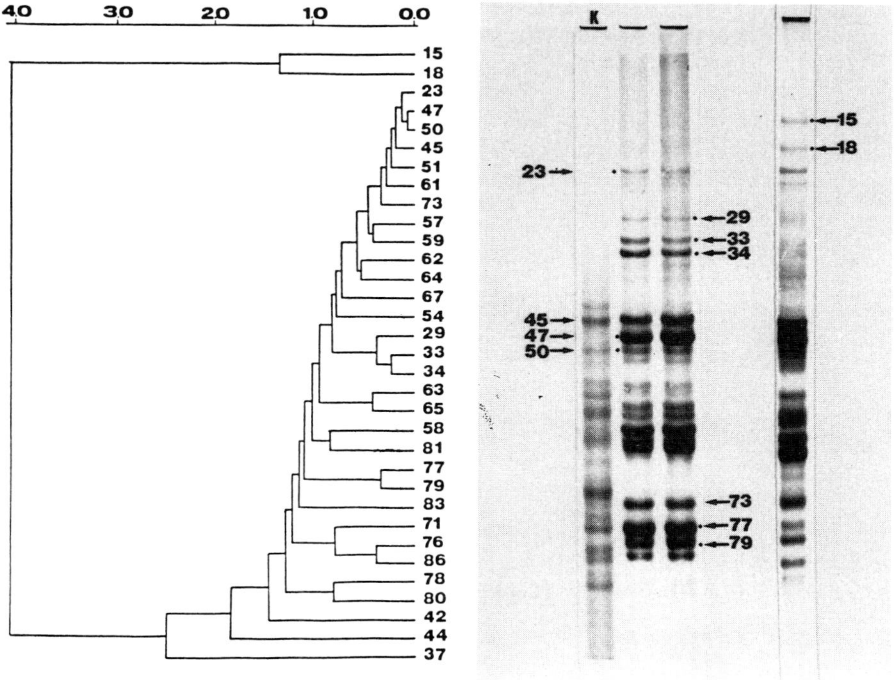

It is also worth mentioning the group constituted by bands 15 and 18, which frequently appear together in the electrophoretic gel, but are shown by very few accessions (*see* Table 9.2). The cluster method assigns them to the same group, although the relative distance, shown by the dendrogram, is not as close as expected. These two protein components are probably controlled by genes on chromosome 1A.

## CONCLUSION

The approach described in this paper is a good way of analyzing one-dimensional electrophoretic gels of storage proteins, even if the evaluation of variability between and within populations cannot be done. The procedure reveals some genetically meaningful associations of different protein bands, without identifying all possible alleles at each locus. However, the closeness of the protein pattern to the effective genetic constitution of the organism holds considerable promise for assessing the genetic diversity in germplasm collections by this method.

The correspondence analysis, together with PC analysis, reduces the parameters needed to describe a particular distribution of points in a multidimensional space and simplifies the study and the comprehension of a multivariate distribution. As described by Burt and Banks (1947), factor analysis is a statistical technique for reducing a large number of correlated variables to a small number of uncorrelated variables.

In this study, the main objective was to discriminate between populations. The description of variability within populations was made mainly on the basis of the variables that describe heterogeneity between groups. This approach can also be applied to the analysis of gliadin proteins together with other molecular markers such as isozymes, restriction fragment length polymorphisms (RFLPs) and HMW glutenins, for the purpose of studying genetic distribution and associating the variability with traits of breeding interest.

# PART 2

# Evaluation Constraints and Germplasm Networks

Wheat Genetic Resources: Meeting Diverse Needs
Edited by J. P. Srivastava and A. B. Damania

Published by John Wiley & Sons

# 10

# *Constraints to Germplasm Evaluation*

J. G. WAINES and D. BARNHART

This paper discusses the procedural and genetic constraints to evaluating and using the wild relatives of wheat. The constraints include biosystematic and nomenclatural constraints, physical constraints arising from the difficulty of handling wild germplasm in field experiments, genetic characters that hinder field or greenhouse evaluation, and characters that affect use in breeding programmes.

## BIOSYSTEMATIC CONSTRAINTS

One of the major problems confronting workers involved in the evaluation of wild wheat germplasm is that authors of the classification systems used for wild material do not agree on the biosystematic validity of the various taxa. We are referring here to: the validity of *Triticum urartu* Tum. as a species biosystematically separate from the other wild diploid wheat, *T. monococcum* L. ssp. *boeoticum* Boiss.; the validity of *Aegilops sharonensis* Eig as a species biosystematically separate from *Ae. longissima* Schweinf. and Muschli.; and the validity of *Ae. peregrina* (Hackel) Maire and Weiller (syn. *Ae. variabilis* Eig) as a species biosystematically separate from *Ae. kotschyi* Boiss.

These questions are important because *T. urartu* was recently proposed as the donor of the A genome of durum and bread wheat on the grounds of similarity of restriction fragment length polymorphisms (RFLPs) of repeated nucleotide sequences (Dvorak et al., 1988). *Ae. sharonensis* was proposed as the donor of the B genome of durum and bread wheat (Kushnir and Halloran, 1981), and its seed storage proteins are distinct (Waines and Johnson, 1972). It is important, for the purposes of germplasm collection, preservation and evaluation, for us to be clear on the biosystematic speciation in the wheat group.

The confusion arises because a group of cytogeneticists presently working in Missouri and Palestine base species boundaries mainly on genome analysis — that is, on chromosome pairing at meiosis in $F_1$ hybrids (Kimber and Feldman, 1987; Kimber and Sears, 1987). If chromosomes pair in these intertaxa $F_1$ hybrids, then the two taxa have the same genome, and are considered the same species. This has the effect of lumping together species that are reproductively isolated. Genome analysis is not a very finely tuned experimental technique. There are even differences in the treatment of species between Kimber and Feldman (1987) and Kimber and Sears (1987), which are not explained and hence are confusing to the average breeder. On the other hand, a group of geneticists and biosystematists working mostly in California and France base species boundaries on either genetic or ecological reproductive isolation. If there are substantial barriers to gene exchange between two taxa in the field, irrespective of whether they share the same genome, then they are considered separate species (Johnson and Dhaliwal, 1976; Yaghoobi-Saray, 1979; Sharma and Waines, 1981; Waines et al., 1982; Lucas and Jahier, 1987; Waines and Payne, 1987; Dvorak et al., 1988; Smith-Huerta et al., 1989). If there is reproductive isolation between wild wheat taxa in natural populations, there is the possibility of genetic differentiation.

We believe this has happened sufficiently between *T. urartu* and *T. monococcum* ssp. *boeoticum*, and between *Ae. sharonensis* and *Ae. longissima*, for them all to be considered separate biosystematic species with different genes useful to the plant breeder. We are not sure about *Ae. peregrina* and *Ae. kotschyi*. There is a need for further field studies involving natural populations of these three pairs of species to clarify the situation. To use wild germplasm effectively, we must have a useful system of species classification.

## PROBLEMS WITH GENOME FORMULAE

Another problem in the wheat group arises from the fact that Kimber and Sears (1987) are still using an incorrect method of writing the genome formula. That the B genome donor contributed the cytoplasm of durum wheat, and was presumably the female parent, was demonstrated by Suemoto (1968, 1973). As plant breeders, we have to write the female parent first in cross notation. Hence we should write the genome formula of durum wheat as BBAA, and that of bread wheat as BBAADD. This is important when amphiploids are made of *Ae. speltoides* or *Ae. searsii* with *T. urartu*. If the *Aegilops* species is the female parent, then the genome formula is SSAA, and we shall attempt to equate SSAA with BBAA, not with AABB.

## NOMENCLATURAL PROBLEMS

The nomenclatural problem in wild wheats and their relatives derives from the group being placed historically in two genera, *Triticum* L. and *Aegilops* L., by morphological taxonomists. This system was followed by Hammer (1980), in the Flora of Turkey (Davis, 1985), and in other descriptions of regional floras where *Triticum* and *Aegilops* grow wild. The impetus to merge the two genera derives mostly from Morris and Sears (1967), who followed Bowden (1959) and the International Code of Nomenclature for Cultivated Plants (Fletcher et al., 1958).

The basic premise of these authors is that the B genome (maternal genome) of tetraploid wheat is from an *Aegilops* species and, hence, following the code, *Aegilops* must be merged with *Triticum*. However, geneticists and cytogeneticists do not yet agree which *Aegilops* species is the donor of the B genome of the BBAA tetraploid wheats (we have followed plant breeding practice and listed the maternal genome first). Perhaps we should wait until the source of the B genome is agreed upon before we change the names of genera. If the International Code of Botanical Nomenclature (Lanjouw et al., 1956) is followed, we may have to merge the genus *Dasypyrum* (*Haynaldia*) with *Triticum*, because *D. villosum* may form spontaneous amphiploids with durum and bread wheat in Sardinia and southern Italy, and a supposed amphiploid has been called *X Haynaldo-triticum sardoum* (Meletti et al., 1977). This claim of spontaneous amphiploid formation needs to be verified. Will we also have to merge *Secale* with *Triticum*?

These biosystematic and nomenclatural problems are a great constraint to germplasm evaluation. A workshop is needed to consider these problems in our major cereal group. This workshop should be held in southern Europe or West Asia, so that a majority of the people who work with *Triticum* and *Aegilops* can attend.

## PHYSICAL CONSTRAINTS

Physical constraints include the problems encountered in sowing, growing and harvesting wild wheats in field or glasshouse trials, and then in threshing the kernels.

### Sowing

The dissemination unit in wild *Triticum* species is a spikelet containing one or two kernels surrounded by pales and glumes, one or more of which are often awned. In *Aegilops* species, the inflorescence spike sometimes breaks up into individual spikelets, or more often it stays intact and separates from the culm at an abscission zone at the base of the spike. In *Ae. speltoides* Tauch, the population is dimorphic for the inflorescence spike: var. *speltoides* (syn. *aucheri*) has an inflorescence axis which persists through grain maturity and the lateral spikelets lack awns; var. *ligustica* has an inflorescence axis which breaks up when the grains are mature into individual spikelets, and the lateral spikelets have lemmas with awns. In *Aegilops*, the dissemination unit is either the entire spike, made up of several spikelets, or individual spikelets, as in wild *Triticum*.

The awns can be removed from a spikelet either by hand, or in a threshing box, or with a machine that will beat them off without damaging the grains. When the awns are removed, it is not difficult to sow spikelets by hand in rows in a field. It is difficult, however, to sow complete spikes by hand. In species with intact inflorescences, the awns are often more difficult to remove, and the inflorescence has to be broken by hand into individual spikelets, or the kernels threshed out by hand. To avoid threshing, we prefer to sow spikelets. Moreover, threshing may damage thin kernels.

## Emergence

Many of the diploid *Aegilops* and *Triticum* species have slender coleoptiles which, at Riverside, California (the site of our study), have difficulty breaking though the surface soils. Our soils are formed from decomposed granite and have very little humus. The surface sets like concrete after wetting. To overcome seedling emergence and obtain an even stand of wild diploid wheats, we usually sow spikelets in a University of California soil mix in flats and then transplant 1- to 2-month-old seedlings to plots in the field. Sowing spikelets in flats in the cold glasshouse also prevents bird damage to germinating seedlings. Migrating birds in autumn and winter can decimate small plots of wild wheats. Domesticated diploid and polyploid wheats have more robust coleoptiles and seedlings have less difficulty emerging in our soils. However, they are just as susceptible to bird damage as wild wheats at this stage.

## Population biology

Since most wild wheat and *Aegilops* spikelets contain two kernels, one somewhat larger than the other, sowing 100 spikelets in a plot will result in up to 200 seedlings. We sow 100 spikelets to avoid genetic drift. If the parental species tends to outcross, the two seedlings that develop from a spikelet can be half-sibs, or they can be full-sibs if the parent tends to self. In a species with many spikelets in a spike, the resulting seedlings can all have the same female parent, but they can have different male parents and show multiple paternity.

It is important to remember that many wild wheat and *Aegilops* accessions can and do outcross more than domesticated wheats. As pollen is wind blown, the extent of outcrossing will depend greatly on climatic factors. In cool, humid conditions, we would expect more outcrossing than in hot dry conditions. Estimates of outcrossing in wild diploid *Triticum* and *Aegilops* species can be found in Yaghoobi-Saray (1979) and Smith-Huerta et al. (1989). Only if grains are threshed out of spikelets and sown separately will individual plants be of a single genotype. If a two-grained spikelet is sown, the resultant plant-clump can contain two genotypes. Further study is needed of the reproductive and population biology of diploid wheat and *Aegilops* species in West Asia.

## Germplasm increase

For the past 3 years, with assistance from the International Board for Plant Genetic Resources (IBPGR), we have increased the wild wheat collection at the University of California (Riverside) and duplicated it at the United States Department of Agriculture (USDA) Small Grains Collection, which is now at Aberdeen, Idaho, and at the Kyoto Germplasm Collection in Japan.

We increased germplasm in the field in the following way. Four rows each on raised beds 76 cm apart were sown with spikelets by hand, or germinated seedlings were transplanted into the rows. The row length was 5 m. Each 15 $m^2$ plot contained 125 spikelets (thus, potentially, over 200 genotypes) and was surrounded by a 1 m-wide border (two rows) of domesticated oats. Irrigation was by sprinkler or furrow, in addition to natural winter rainfall. Weed control was mechanical between the beds and by hand hoe between plants within the row. Broad-leaved weeds were also sprayed with a selective herbicide. Such a patchwork of wheats and oats allows each accession to develop independently of others. Because we experience damage by jack rabbits and other herbivores, the whole germplasm increase area is surrounded by wire netting.

## Harvesting

If there are not too many plots, when wild wheats are mature, spikelets can be harvested by hand each morning when dew is still present. If there are many plots, spikelets can be allowed to drop to the soil surface, which has been kept weed free; they can be raked up, or picked up by hand or by a vacuum machine. Later, soil and leaf material will be separated from the spikelets. It is possible to obtain replicated plots and components of yield data of wild *Triticum* and *Aegilops* accessions in the field, with the help of patient graduate students and field workers. In this way, 70 accessions of *Aegilops* species with three replications were harvested in summer 1988. We are now analyzing the material for yield components, grain protein and lysine content.

Other techniques used in harvesting fragile, rachised wheats include taping the awns of the spike with scotch tape in two places, near the spike base and at the spike apex. Alternatively, it is possible to place all the effective spikes on a plant in a plastic bag which has holes cut into it with a paper punch. Cheesecloth bags or bags made of a fine net can also be made and used to enclose the spikes. However, all these methods are time-consuming and can be applied only to plants in a small number of plots. Furthermore, under our conditions, plastic bags tend to collect condensed moisture, with the result that fungal saprophytes may grow on the glumes. For ease of handling, we prefer to let the spikelets drop naturally and to rake them up from the soil surface. In this way, we have increased the seed of more than 500 accessions in a given year. Spikelets are put into large brown paper bags and taken to the field house for cleaning, packaging and shipping. If the *Aegilops* species under increase has a spike that falls entire, without breaking into individual spikelets, then it is easy to pick up spikes from the soil surface. The border of oats that surrounds each plot prevents spike movement between plots, but it does not prevent cross-pollination.

### Location of increase

Wild wheat accessions grow better at a coastal field station, such as Irvine, California (elevation 123 m) or a mountain station such as Idyllwild (elevation 1615 m) than at intermediate elevations such as Riverside (297 m) or Moreno (463 m) (Guzy et al., 1989). This is because Riverside and Moreno have a shorter growing season and the temperature becomes too hot too quickly. High temperatures at flowering tend to kill the pollen of wild *Triticum* species, but not necessarily of *Ae. longissima* or *Ae. searsii* (Ehdaie and Waines, 1989). A small project that investigated phenotypic plasticity (Bradshaw, 1965) in diploid and polyploid wild wheats grown at these four sites in southern California indicated that diploid wheats have greater plasticities for yield components than other taxa (Guzy et al., 1989). *T. monococcum* exhibited higher spikelet numbers relative to wild tetraploids, although seed size was reduced.

### Threshing

The tough glumes of wild wheats protect grains from birds, but they can be a problem at threshing time. Many wild wheats and primitive domesticates, such as *T. dicoccum* and *T. monococcum*, can be threshed in a threshing box lined with ribbed rubber. A wooden brick is used to rub the glumes against the rubber, releasing the grains. However, this can be time-consuming. It may be better to weigh a sample of spikelets, take a subsample, weigh it, and then thresh the subsample by hand with a pair of forceps. The weight of threshed grains can be determined and hence the weight of grains in the original sample.

## GENETIC CONSTRAINTS

### Evaluation

Only small samples of grain are necessary for micro protein and lysine determination. Large quantities of threshed grain (up to 200 g) are necessary for micro loaf-baking tests and cookie diameter tests. Fortunately, many electrophoretic and seed protein techniques can be performed on single grains of wild wheats or on single leaves (Waines and Payne, 1987; Smith-Huerta et al., 1989), and the genetic control of protein bands studied. The flag leaves of some wild *Triticum* and *Aegilops* species are small and thin, and this may hinder water retention experiments. Austin et al. (1986) reported that a *T. urartu* accession with a high photosynthetic rate had wide flag leaves, which is encouraging.

## Using the evaluated germplasm

Most crosses of wild *Triticum* and *Aegilops* species as males with tetraploid and hexaploid wheat cultivars as females are successful if sufficient care is paid to setting up the cross. The male parent can affect the development of hybrid seed (Gill and Waines, 1978), and this is thought to be under simple genetic control. The genes that control intergeneric hybrid formation in the wheat group are known, and are thought to be similar to the rye-wheat crossability system (Thomas et al., 1981). We have found that some accessions of *T. urartu* give semi-lethal or lethal seedlings with other species of *Triticum* and *Aegilops*. There are accessions that hybridize more easily. We have not investigated the genetics of this characteristic.

Accessions of diploid species with desirable characteristics should be crossed with tetraploid and hexaploid wheats to incorporate the character into material that can be crossed with modern cultivars. Roberts et al. (1982) reported that a collection of *Ae. squarrosa* L. from Afghanistan was resistant to the root knot nematodes *Meloidogyne incognita* and *M. javanica*. The resistance is dominant and is inherited as a single Mendelian trait (Kaloshian, 1988). The resistance is expressed in an amphiploid, Prosquare, formed from *T. turgidum* var. *durum* Produra and *Ae. squarrosa* G3489 (Kaloshian et al., 1989a). This amphiploid is also resistant, in glasshouse tests, to the Columbia Basin root nematode, *M. chitwoodi* (Kaloshian et al., 1989b). Lacking evidence to the contrary, the simplest explanation is that a single locus controls resistance to these three *Meloidogyne* species. Further tests may indicate that more than one locus is involved. The 'gene' conferring resistance to *M. javanica* is located on chromosome 6D of *Ae. squarrosa* (Kaloshian, 1988). There is reason to suspect that Produra may confer hybrid necrosis genes to the amphiploid that will be expressed in crosses with other commercial wheats. To overcome this possible problem, we are attempting to cross the durum wheat Cocorit with *Ae. squarrosa* G3489. Field tests are being conducted this summer in southern and northern California to see if the amphiploid showing resistance to *M. javanica* and other root knot nematodes will express the resistance under field conditions. If so, the resistance in the Prosquare-based and Cocorit-based amphiploids will be transferred to commercial wheats.

A more direct way to transfer a desirable trait in *Ae. squarrosa* into bread wheat would be to cross it directly to the bread wheat cultivar and backcross the tetraploid hybrid to the bread wheat cultivar. This technique was successful when bread wheat was the female parent (Merkle and Starks, 1985) and when *Ae. squarrosa* was the female parent (Gill and Raupp, 1987). *T. monococcum* ssp. *boeoticum* was successfully crossed with bread wheat (The and Baker, 1975), as was *Ae. speltoides* (Dvorak, 1977), and characters were transferred. To what extent this is possible with other *Aegilops* and related species needs to be investigated.

If *T. urartu* is the A genome donor of commercial tetraploid and hexaploid cultivars, as the work of Dvorak et al. (1988) suggests, this will allow further possibilities for the use of wild germplasm. *T. urartu* contains accessions with high concentrations of protein and lysine (Waines et al., 1987). The protein concentration of six accessions ranged from 22.5 to 27.8 per cent; lysine content ranged from 4.49 to 6.50 μg/mg. The adjusted lysine/protein ratio of one accession, G1754, was 27.05 μg/mg. We would like to investigate the possibility of improving the nutritional quality of bread wheat by substituting *urartu* chromosomes for their homoeologous chromosomes in the A genome of a commercial bread wheat. We are attempting to cross the Glenson monosomic series to *urartu* G1754 to see if this line of research is feasible. The Glenson monosomic series was developed by the Institute for Plant Science Research, Cambridge, UK on behalf of the Centro Internacional de Mejoramiento de Maíz y Trigo (CIMMYT).

## CONCLUSION

Although working with wild *Triticum* and *Aegilops* species is not as easy as working with domesticated wheats, it can be just as rewarding. It is worth bearing in mind that, some 10 000 years ago, hunter gatherers and early farmers in West Asia harvested and evaluated wild wheats. If they could do it then, we should certainly be able to harvest, evaluate and use the same species today!

## *Discussion*

L. TANASCH: Genome analysis has shown that the donor of B genome in bread wheat is not yet certain. Some researchers think that *T. urartu* may be the donor. Is there some work done using the RFLP technique to solve this problem?

J. G. WAINES: B. L. Johnson said *T. urartu* was the B genome donor in bread wheat. But this was disproved by Miller and Chapman in Cambridge and by Dvorak who was in Canada at that time, with chromosome pairing studies using the telocentrics of Chinese Spring. Johnson assumed that *T. monococcum* was the A genome donor, as did everyone else. Work by Professor Konarev in Leningrad suggested that on immunochemical grounds *T. urartu* was a parent of *T. turgidum*. However, recent work conducted by Dvorak and others (1988) suggests that *urartu* donated the A genome on the basis of RFLP analysis. The RFLP pattern of *T. turgidum* is identical to that of *urartu*, but differs from that of *monococcum*.

S. JANA: What is the basis for selecting the figure '120' as a population size that would prevent genetic drift in wild accessions in rejuvenation nurseries?

J. G.WAINES: It is based on textbooks of population genetics and recommendations of the International Board for Plant Genetic Resources (IBPGR).

S. JANA: Why do wild wheats maintain seed viability longer in storage if preserved as spikes rather than as cleaned seeds?

J. G. WAINES: Threshing wild wheats and *Aegilops* species which have hard glumes requires a great deal of pressure. This is usually applied from a block of wood or brick. The block can damage either the embryo or the endosperm of the kernels, resulting in loss of viability.

M. VAN SLAGEREN: Do *Aegilops speltoides* variants grow in different habitats? In western Syria I have observed them growing together but they are easily distinguished, and there are very few intermediate forms present.

J. G. WAINES: Yes, they do grow together, but the *ligustica* type variants are more frequent at high elevation and less so at sea level, at least in Turkey. We really know very little about wild wheats and their ecological distribution.

M. VAN SLAGEREN: Why is there no distinction between *boeoticum* and *urartu* in the Flora of West Asia?

J. G. WAINES: In the Flora of Turkey, *Triticum* was not treated by a wheat specialist and omissions were caused by lack of herbarium specimens and poor literature surveys in Ankara and Edinburgh where the Flora was prepared. Another reason may be that although *urartu* and *boeoticum* tend to grow together in mixed populations, the former tends to flower and mature a little earlier. So if one arrives at a collection site slightly later, then only *boeoticum* gets sampled, the *urartu* having shattered. But once you get to know both materials, the differences are obvious. The seed colour in *boeoticum* is blue or yellow, whereas in *urartu* it is red; this could be the origin of the red grain colour in tetraploid wheats. The shape of the grain is also quite distinct. Quite often, there is a third awn in the spikelet, so there are a series of characters which, when you put them together, can distinguish the two species. Crosses can be made only one way, invariably with *boeoticum* as a female parent. If crossed the other way around, then you do not get a hybrid embryo. If *urartu* is crossed with *monococcum*, with the former as female parent, then again the hybrid is sterile and this sterility continues in backcrosses 1 and 2. So they are reproductively isolated species, although they grow together. The populations of wild diploid wheats in Turkey are composed of 90 per cent *boeoticum* and 10 per cent *urartu*. There is some mechanism that is keeping them apart.

Wheat Genetic Resources: Meeting Diverse Needs
Edited by J. P. Srivastava and A. B. Damania

Published by John Wiley & Sons

# 11

# *Utilization of Unreplicated Observations of Agronomic Characters in a Wheat Germplasm Collection*

R. J. GILES

The Agricultural and Food Research Council (AFRC) Cereals Collection is maintained at the Institute of Plant Science Research (IPSR) in Cambridge, UK. It represents the breeding and research histories of the Plant Breeding Institute (PBI), Cambridge, the Scottish Crop Research Institute (SCRI) and the Welsh Plant Breeding Station (WPBS).

Observation data on agronomic and pathologic traits have been recorded in the collection nurseries planted at the PBI over the past 20 years or more. The introduction of computer database software during this period has enabled this type of data to be stored in databases, permitting electronic screening of the germplasm for these characters. For each accession in the collection, it is desirable to summarize the field scores for a particular trait as a single value. Thus, the characteristic location of the variety can be fixed within the overall range of variation of that trait. This is often done by converting observations into ranks based on a 1 to 9 scale (1 being low expression of the character, and 9 high expression), which provides a set of broad categories for screening.

## PROBLEMS IN SUMMARIZING EVALUATION DATA

The attempt to summarize evaluation data using the method described above has given rise to a number of practical problems.

## Non-orthogonality

Each accession has been grown in the collection nurseries for only a few of the seasons during which data have been recorded. Thus germplasm acquired early in the history of the collection tends to be represented only by data from the distant past. Similarly, as the collection has increased in size, the data from new entries are confined to the few seasons following their addition to the collection.

In the mid to late 1960s, the practice at the PBI was to grow the entire wheat, barley and oat collections in both autumn and spring sowings. Consequently, there is a significant block of data for each of these crops, with a high level of replication over seasons. These blocks may yield valuable information, but they represent only a small proportion of the present collection. Between the late 1960s and the late 1970s, only newly acquired accessions were grown, and then for only three or four consecutive seasons. Recently, the main purpose of the nurseries has been to grow new material and accessions with depleted or low-quality seed stocks for regeneration. Because of the rapid expansion of the collection, it has not been possible with the resources available to replicate accessions over seasons for comparative purposes. There is therefore a problem with germplasm which has been grown and observed only once. Without replication, it has not been possible hitherto to determine how their data relate to varieties grown in different seasons.

## Seasonal incompatability

In the past, the practice was to calculate varietal averages for traits over all available seasons, irrespective of which seasons' data were used. The results provided the basis of the ranks which were subsequently stored in the computerized database. It is clear that if, for example, the phenotypic expression of a particular trait were to be depressed in the seasons over which the variety average was calculated, the final result would be depressed in comparison with values for similar varieties grown in seasons where growth was more normal or where phenotypic expression was inflated. It would be possible, therefore, for dissimilar varieties to appear to be similar, and vice versa, if their means were calculated from data which had been derived from different groups of seasons.

For example, the five accessions in Table 11.1 with rank 5 based on unadjusted straw length, in column 'Rank (e)', have means with a range of 7.92 (81.33 to 89.25). However, after adjustment, they appear more diverse, with means over a range of 13.95 (89.49 to 103.44) and with ranks ranging from 4 to 6. Conversely, in Table 11.2 (*overleaf*) the six examples with rank 6 based on adjusted Julian dates (days after 1 January), in column 'Rank (g)', have means with a range of 3.29 (155.01 to 158.30). However, before adjustment they appear more diverse, with ranks ranging from 5 to 7 and means having a range of 12.0 (152.33 to 164.33).

**Table 11.1** Means of observed and adjusted straw length data from Table 11.3. Key to the rank columns is as follows: (e) ranks are based on the original PBI scale (*see* Table 11.9), (f) ranks are derived from the means of column (d) in Table 11.3; (g) ranks are derived by equal increment partition of the data range after adjustment

| Accession | Observed | | | | Adjusted Data | | |
|---|---|---|---|---|---|---|---|
| No. | Means | SD [a] | Rank (e) | Rank (f) | Mean | SD | Rank (g) |
| 264 | 113.67 | 17.13 | 8 | 7 | 125.51 | 4.73 | 7 |
| 271 | 89.25 | 12.91 | 5 | 4 | 103.44 | 5.29 | 5 |
| 277 | 110.67 | 15.76 | 8 | 7 | 22.85 | 3.40 | 7 |
| 303 | 73.67 | 12.23 | 4 | 3 | 89.80 | 9.07 | 4 |
| 590 | 35.67 | 6.34 | (1) | 1 | 45.96 | 15.20 | 1 |
| 788 | 84.33 | 4.78 | 5 | 3 | 89.49 | 4.09 | 4 |
| 789 | 107.67 | 8.34 | 7 | 5 | 113.96 | 6.82 | 6 |
| 929 | 81.33 | 1.25 | 5 | 3 | 89.66 | 5.00 | 4 |
| 988 | 132.67 | 8.38 | 9 | 7 | 134.47 | 2.69 | 8 |
| 1189 | 131.00 | 15.51 | 9 | 8 | 135.22 | 4.93 | 8 |
| 1567 | 68.33 | 8.50 | 3 | 2 | 82.93 | 6.86 | 4 |
| 1571 | 79.00 | 2.95 | 4 | 3 | 92.81 | 2.54 | 4 |
| 2079 | 96.33 | 8.18 | 6 | 5 | 106.91 | 7.68 | 6 |
| 2098 | 83.00 | 8.04 | 5 | 4 | 95.56 | 1.14 | 5 |
| 2194 | 78.67 | 4.50 | 4 | 3 | 92.35 | 4.15 | 4 |
| 2208 | 54.00 | 2.95 | 2 | 1 | 76.58 | 0.33 | 3 |
| 2217 | 103.33 | 6.18 | 7 | 6 | 107.54 | 2.58 | 6 |
| 2495 | 82.33 | 6.13 | 5 | 3 | 92.62 | 3.82 | 4 |
| 2499 | 111.33 | 6.60 | 8 | 6 | 110.51 | 2.16 | 6 |

a SD = Standard Deviation

## Fixed descriptor states

The scales by which ranks are assigned for traits with continuous variability are usually fixed without reference to data derived from any particular season. Consequently, these scales may not accommodate the recorded data precisely (*see* Table 11.3 *overleaf*, columns (b), where the ranks in parenthesis lie outside the range of the ranking scale).

A typical approach adopted to address such problems is to disregard all previous data and plant a nursery containing all the current accessions, using the data collected as the basis for database information. Apart from the sheer scale of an evaluation nursery for a large collection, this solution itself raises problems. Firstly, more recent acquisitions would be excluded from the database until the exercise could be repeated. Secondly, any traits recorded would be susceptible to the peculiarities of the season in which the nursery was grown.

**Table 11.2** Variety means of observed and adjusted ear emergence data from Table 11.4a and 11.4b. Key to the rank columns is as follows: (e) ranks are based on the original PBI scale (*see* Table 11. 9); (f) ranks are derived from the means of column (d) in Table 11.4; (g) ranks are derived by equal increment partition of the data range after adjustment

| Accession | Days after sowing [a] | | | Observed Julian date [b] | | | Adjusted Julian date [c] | | |
|---|---|---|---|---|---|---|---|---|---|
| No. | Mean | SD [d] | Rank (e) | Mean | SD | Rank (f) | Mean | SD | Rank (g) |
| 264 | 224.60 | 5.54 | 5 | 150.80 | 3.25 | 5 | 151.81 | 1.50 | 5 |
| 271 | 225.60 | 5.95 | 6 | 152.00 | 3.29 | 5 | 153.12 | 1.99 | 5 |
| 277 | 216.20 | 4.83 | 4 | 142.40 | 3.56 | 3 | 143.16 | 2.14 | 3 |
| 303 | 230.80 | 6.85 | 7 | 157.20 | 1.83 | 6 | 158.30 | 1.58 | 6 |
| 590 | 239.40 | 8.73 | 8 | 161.40 | 2.42 | 7 | 163.50 | 2.06 | 7 |
| 788 | 237.75 | 3.77 | 8 | 157.00 | 2.12 | 6 | 157.56 | 0.80 | 6 |
| 789 | 232.25 | 2.95 | 7 | 151.50 | 2.87 | 5 | 151.99 | 0.47 | 5 |
| 929 | 235.67 | 2.36 | 8 | 152.33 | 3.40 | 5 | 155.01 | 2.11 | 6 |
| 988 | 244.40 | 6.41 | 9 | 166.00 | 2.48 | 8 | 165.90 | 2.56 | 8 |
| 1189 | 240.00 | 4.86 | 8 | 162.00 | 6.07 | 7 | 161.78 | 3.88 | 7 |
| 1567 | 231.00 | 6.38 | 7 | 160.33 | 5.44 | 7 | 159.42 | 1.29 | 7 |
| 1571 | 230.67 | 7.72 | 7 | 160.00 | 7.35 | 7 | 159.20 | 1.34 | 7 |
| 2079 | 230.33 | 0.94 | 6 | 155.67 | 5.56 | 5 | 155.21 | 0.77 | 6 |
| 2098 | 208.33 | 4.50 | 2 | 138.67 | 3.68 | 1 | 138.41 | 2.50 | 2 |
| 2194 | 234.33 | 3.86 | 7 | 164.33 | 3.40 | 7 | 158.98 | 2.01 | 6 |
| 2208 | 229.00 | 2.83 | 6 | 159.00 | 2.16 | 5 | 153.36 | 0.79 | 5 |
| 2217 | 231.00 | 1.63 | 7 | 161.00 | 2.45 | 6 | 155.35 | 0.74 | 6 |
| 2495 | 222.33 | 6.65 | 5 | 155.00 | 2.83 | 5 | 150.05 | 0.86 | 5 |
| 2499 | 225.67 | 6.18 | 6 | 158.33 | 2.62 | 6 | 153.61 | 1.30 | 5 |

a Table 11.4, column (a); b Table 11.4, column (c); c Tables 11.4, column (c) in parenthesis
d SD = Standard Deviation

An alternative approach might be to adopt a series of control varieties. This would be sound practice, but cannot be done retrospectively. In addition, it may be difficult to find an adequate set of controls because the pattern of variation may change over time with breeding pressure. For example, it would have been difficult in the 1960s to find a group of varieties to adequately represent the range of straw length available at the present time, when there is a bias towards short-strawed types.

Tables 11.3 and Tables 11.4a and 11.4b (*overleaf*) illustrate different ways of deriving ranks 1 to 9 for straw length and ear emergence date, respectively, in the wheat collection. The 19 accessions used were chosen to represent the range of variation found in the collection for these traits. Table 11.5 (*page 120*) gives their identities.

Columns (b) in Tables 11.3, 11.4a and 11.4b show ranks for individual observations using the original method of comparison with fixed descriptor states. Clearly, seasons differ in certain respects and thus exert a marked influence on the stability of the ranks

**Table 11.3** Straw length data from 19 accessions in the AFRC Wheat Collection illustrating the effects of different methods of assigning ranks on a 1 to 9 scale. Key to columns b to d as follows: b) rank according to the original PBI scale (*see* Table 11.5) (ranks in parenthesis fall outside the range of the set scale); c) straw length (excluding spike) in cm (values in parenthesis are the adjusted values from which the final variety means are calculated); d) rank according to equal increment partition of the data range

| Season | 1970 | | | 1971 | | | 1972 | | | 1973 | | | 1974 | | | 1975 | | |
|---|---|---|---|---|---|---|---|---|---|---|---|---|---|---|---|---|---|---|
| Sowing date | 23.10.69 | | | 14.10.70 | | | 28.10.71 | | | 13.10.72 | | | 2.10.73 | | | 16.10.74 | | |
| Order of incorporation | 3 | | | 2 | | | 1 | | | 4 | | | 5 | | | 6 | | |
| Accession No. | b | c | d | b | c | d | b | c | d | b | c | d | b | c | d | b | c | d |
| 264 | 9 | 121 (127.55) | 7 | 5 | 90 (118.97) | 6 | 9 | 130 — | 7 | | | | | | | | | |
| 271 | 5 | 90 (103.19) | 4 | 3 | 68 (100.55) | 4 | 6 | 98 — | 5 | 7 | 101 (112.03) | 4 | | | | | | |
| 277 | 8 | 117 (124.41) | 7 | 5 | 89 (118.13) | 6 | 9 | 126 — | 7 | | | | | | | | | |
| 303 | 5 | 86 (100.05) | 4 | 2 | 57 (91.35) | 3 | 4 | 78 — | 3 | | | | | | | | | |
| 590 | | | | (1) | 27 (66.24) | 1 | 1 | 42 — | 1 | | | | (1) | 38 (29.64) | (1) | | | |
| 788 | | | | | | | | | | 4 | 80 (88.40) | 3 | 5 | 82 (85.11) | 4 | 6 | 91 (94.95) | 3 |
| 789 | | | | | | | | | | 6 | 96 (106.40) | 4 | 8 | 112 (122.92) | 6 | 8 | 115 (112.55) | 5 |
| 929 | | | | | | | | | | | | | 5 | 81 (83.85) | 4 | 5 | 83 (89.08) | 2 |
| 988 | | | | | | | | | | | | | 9 | 124 (138.05) | 7 | 9 | 144 (133.81) | 7 |
| 1189 | | | | | | | | | | | | | 8 | 120 (133.01) | 7 | (9) | 149 (137.48) | 8 |
| 1567 | | | | | | | | | | | | | | | | | | |
| 1571 | | | | | | | | | | | | | | | | | | |
| 2079 | | | | | | | | | | | | | | | | | | |
| 2098 | | | | | | | | | | | | | | | | | | |
| 2194 | | | | | | | | | | | | | | | | | | |
| 2208 | | | | | | | | | | | | | | | | | | |
| 2217 | | | | | | | | | | | | | | | | | | |
| 2495 | | | | | | | | | | | | | | | | | | |
| 2499 | | | | | | | | | | | | | | | | | | |

| Season | 1976 | | | 1977 | | | 1978 | | | 1979 | | | 1980 | | |
|---|---|---|---|---|---|---|---|---|---|---|---|---|---|---|---|
| Sowing date | 9.10.75 | | | 24.10.76 | | | 20.10.77 | | | 24.10.78 | | | 1.11.79 | | |
| Order of incorporation | 7 | | | 8 | | | 9 | | | 10 | | | 11 | | |
| Accession No. | b | c | d | b | c | d | b | c | d | b | c | d | b | c | d |
| 264 | | | | | | | | | | | | | | | |
| 271 | | | | | | | | | | | | | | | |
| 277 | | | | | | | | | | | | | | | |
| 303 | | | | | | | | | | | | | | | |
| 590 | | | | | | | | | | | | | | | |
| 788 | | | | | | | | | | | | | | | |
| 789 | | | | | | | | | | | | | | | |
| 929 | 4 | 80 (96.06) | 4 | | | | | | | | | | | | |
| 988 | | | | 9 | 130 (131.55) | 7 | | | | | | | | | |
| 1189 | 7 | 110 (128.59) | 7 | 9 | 145 (141.80) | 9 | | | | | | | | | |
| 1567 | 2 | 60 (74.37) | 2 | | | | | | | 3 | 65 (83.28) | 2 | 4 | 80 (91.16) | 3 |
| 1571 | 4 | 80 (96.06) | 4 | | | | | | | 4 | 75 (89.85) | 3 | 5 | 82 (92.53) | 3 |
| 2079 | 6 | 100 (117.74) | 6 | 5 | 85 (100.81) | 3 | 7 | 104 (102.17) | 6 | | | | | | |
| 2098 | 4 | 80 (96.06) | 4 | 4 | 75 (93.98) | 3 | 6 | 94 (96.63) | 5 | | | | | | |
| 2194 | | | | 4 | 75 (93.98) | 3 | 4 | 76 (86.65) | 3 | 5 | 85 (96.42) | 4 | | | |
| 2208 | | | | 1 | 50 (76.91) | 1 | 2 | 57 (76.12) | 1 | 2 | 55 (76.71) | 1 | | | |
| 2217 | | | | 6 | 100 (111.06) | 5 | 8 | 112 (106.60) | 7 | 6 | 98 (104.96) | 5 | | | |
| 2495 | | | | | | | 5 | 82 (89.98) | 3 | 4 | 75 (89.85) | 3 | 5 | 90 (98.02) | 4 |
| 2499 | | | | | | | 8 | 120 (111.04) | 7 | 7 | 110 (112.85) | 6 | 7 | 104 (107.63) | 6 |

**Table 11.4a** Ear emergence data from 19 accessions in the AFRC Wheat Collection illustrating the effects of different methods of assigning ranks on a 1 to 9 scale. Key to columns a to d is as follows: a) number of days from sowing to ear emergence. b) rank of (a) according to the original PBI scale (*see* Table 11.5) (ranks in parenthesis fall outside the range of the set scale); c) date of ear emergence (numbers in parenthesis are the adjusted values from which the final variety means are calculated); d) rank of (c) according to equal increment partition of the range of dates. *Occasional discrepancies between columns (a) and (c) are due to a spread of sowing over a number of days (e.g. in autumn 1969, sowing was spread over 3 days from 22 to 24 October). The sowing dates indicated are those on which the majority of plots were sown

| **Season** | **1970** | | | | **1971** | | | | **1972** | | | | **1973A** | | | | **1973B** | | | | **1974** | | | |
|---|---|---|---|---|---|---|---|---|---|---|---|---|---|---|---|---|---|---|---|---|---|---|---|---|
| **Sowing Date** | **23.10.69** | | | | **14.10.70** | | | | **28.10.71** | | | | **13.10.72** | | | | **15.10.72** | | | | **2.10.73** | | | |
| **Order of incorporation** | **5** | | | | **4** | | | | **3** | | | | **2** | | | | **1** | | | | **6** | | | |
| **Accession No.** | **a** | **b** | **c** | **d** | **a** | **b** | **c** | **d** | **a** | **b** | **c** | **d** | **a** | **b** | **c** | **d** | **a** | **b** | **c** | **d** | **a** | **b** | **c** | **d** |
| 264 | *226 | 6 | 156 | 5 | 225 | 5 | 146 | 4 | 214 | 3 | 150 | 5 | 229 | 6 | 150 | 5 | 229 | 6 | 152 | 6 | | | | |
| | | | (154.40) | | | | (150.23) | | | | (152.01) | | | | (150.42) | | | | — | | | | | |
| 271 | *224 | 5 | 155 | 5 | 225 | 5 | 146 | 4 | 216 | 4 | 152 | 5 | 234 | 7 | 155 | 6 | 229 | 6 | 152 | 6 | | | | |
| | | | (153.28) | | | | (150.23) | | | | (153.84) | | | | (156.24) | | | | — | | | | | |
| 277 | 218 | 4 | 148 | 3 | 216 | 4 | 137 | 2 | 207 | 2 | 143 | 3 | 220 | 4 | 141 | 3 | 220 | 4 | 143 | 3 | | | | |
| | | | (145.41) | | | | (141.83) | | | | (145.58) | | | | (139.96) | | | | — | | | | | |
| 303 | 227 | 6 | 159 | 6 | 236 | 8 | 157 | 6 | 219 | 4 | 154 | 6 | 236 | 8 | 157 | 6 | 236 | 8 | 159 | 8 | | | | |
| | | | (157.78) | | | | (160.50) | | | | (155.68) | | | | (158.56) | | | | — | | | | | |
| 590 | | | | | 236 | 8 | 157 | 6 | 226 | 6 | 161 | 7 | 241 | 9 | 162 | 7 | 241 | 9 | 164 | 9 | 253 | (9) | 163 | 8 |
| | | | | | | | (160.50) | | | | (162.10) | | | | (164.38) | | | | — | | | | (166.53) | |
| 788 | | | | | | | | | | | | | 236 | 8 | 157 | 6 | 234 | 7 | 157 | 7 | 244 | 9 | 154 | 6 |
| | | | | | | | | | | | | | | | (158.56) | | | | — | | | | (158.10) | |
| 789 | | | | | | | | | | | | | 231 | 7 | 152 | 5 | 229 | 6 | 152 | 6 | 237 | 8 | 147 | 4 |
| | | | | | | | | | | | | | | | (152.75) | | | | — | | | | (151.54) | |
| 929 | | | | | | | | | | | | | | | | | | | | | 239 | 8 | 149 | 5 |
| | | | | | | | | | | | | | | | | | | | | | | | (153.42) | |
| 988 | | | | | | | | | | | | | | | | | | | | | 255 | (9) | 165 | 9 |
| | | | | | | | | | | | | | | | | | | | | | | | (168.40) | |
| 1189 | | | | | | | | | | | | | | | | | | | | | 244 | 9 | 154 | 6 |
| | | | | | | | | | | | | | | | | | | | | | | | (158.10) | |

**Table 11.4b** For explanation *see* Table 11.4a

| Season | 1975 | | | | 1976 | | | | 1977 | | | | 1978 | | | | 1979 | | | | 1980 | | | |
|---|---|---|---|---|---|---|---|---|---|---|---|---|---|---|---|---|---|---|---|---|---|---|---|---|
| Sowing Date | 16.10.74 | | | | 9.10.75 | | | | 24.10.76 | | | | 20.10.77 | | | | 24.10.78 | | | | 1.11.79 | | | |
| Order of incorporation | 7 | | | | 8 | | | | 17 | | | | 16 | | | | 15 | | | | 14 | | | |
| Accession No. | a | b | c | d | a | b | c | d | a | b | c | d | a | b | c | d | a | b | c | d | a | b | c | d |
| 788 | 237 | 8 | 160 (156.59) | 6 | | | | | | | | | | | | | | | | | | | | |
| 789 | 232 | 7 | 155 (151.66) | 5 | | | | | | | | | | | | | | | | | | | | |
| 929 | 234 | 7 | 157 (153.63) | 5 | 234 | 7 | 151 (157.99) | 5 | | | | | | | | | | | | | | | | |
| 988 | 246 | (9) | 169 (165.46) | 8 | 245 | 9 | 162 (169.25) | 8 | 236 | 8 | 168 (162.70) | 8 | 240 | 8 | 167 (163.71) | 8 | | | | | | | | |
| 1189 | 246 | (9) | 170 (166.45) | 8 | 241 | 9 | 158 (165.15) | 7 | 236 | 8 | 168 (162.70) | 8 | 233 | 7 | 160 (156.51) | 6 | | | | | | | | |
| 1567 | | | | | 236 | 8 | 153 (160.03) | 6 | | | | | | | | | 235 | 7 | 166 (157.63) | 7 | 222 | 5 | 162 (160.61) | 9 |
| 1571 | | | | | 234 | 7 | 151 (157.99) | 5 | | | | | | | | | 238 | 8 | 169 (161.07) | 8 | 220 | 4 | 160 (158.53) | 8 |
| 2079 | | | | | 231 | 7 | 148 (154.92) | 4 | 229 | 6 | 161 (156.27) | 6 | 231 | 7 | 158 (154.45) | 6 | | | | | | | | |
| 2098 | | | | | 202 | 1 | 134 (140.58) | 1 | 211 | 3 | 143 (139.74) | 2 | 212 | 3 | 139 (134.91) | 1 | | | | | | | | |
| 2194 | | | | | | | | | 229 | 6 | 161 (156.27) | 6 | 236 | 8 | 163 (159.59) | 7 | 238 | 8 | 169 (161.07) | 8 | | | | |
| 2208 | | | | | | | | | 225 | 5 | 157 (152.60) | 5 | 231 | 7 | 158 (154.45) | 6 | 231 | 7 | 162 (153.04) | 5 | | | | |
| 2217 | | | | | | | | | 229 | 6 | 161 (156.27) | 6 | 231 | 7 | 158 (154.45) | 6 | 233 | 7 | 164 (155.33) | 6 | | | | |
| 2495 | | | | | | | | | | | | | 226 | 6 | 153 (149.31) | 4 | 228 | 6 | 159 (149.60) | 4 | 213 | 3 | 153 (151.25) | 6 |
| 2499 | | | | | | | | | | | | | 229 | 6 | 156 (152.40) | 5 | 231 | 7 | 162 (153.04) | 5 | 217 | 4 | 157 (155.41) | 8 |

**Table 11.5** Wheat varieties from the AFRC Cereals Collection used in the examples cited in Tables 11.1 to 11.4. All except those marked [a] are available on request from the author

| Accession No. | Cultivar name | Seasonality | Country of origin |
|---|---|---|---|
| 264 | Lathrop | Spring | United States |
| 271 | Sirius (Von Rumker 4/55) | Spring | West Germany |
| 277 | Pinyte | Spring | Tunisia |
| 303 | TB 435/119/10/9 | Spring | England |
| 590 | Dwarf Ministre | Winter | Belgium |
| 788 | Bezostaja 2 | Winter | USSR |
| 789 | Iljicovka | Winter | USSR |
| 929 | Benoist 5780 | Winter | France |
| 988 | Rouge d'Ecosse | Winter | England |
| 1189 | Zidlochovice Jubliee | Winter | Czechoslovakia |
| 1567 [a] | TJB 368/259/997/6559 | Winter | England |
| 1571 [a] | TJB 238/926/3/6457 | Winter | England |
| 2079 | Clement (Cebeco 145) | Winter | Netherlands |
| 2098 | Bei Jing 11 | Winter | China |
| 2194 [a] | SR 4668-8 | Winter | England |
| 2208 | C 645 | Winter | Poland |
| 2217 [a] | Stupice 729 | Winter | Czechoslovakia |
| 2495 | Probstdorf 4258 | Winter | Austria |
| 2499 | Probstdorf 6151 | Winter | Austria |

assigned in this way (*see* Tables 11.6 and 11.7, for data parameters for each season). For example, a comparison of columns (c) in Table 11.3 for seasons 1971 and 1972, which differ by 14 days in terms of sowing date, shows a tendency towards higher straw length values observed in the later-sown 1972 season. Similarly, a comparison of columns (c) in Tables 11.4a and 11.4b for seasons 1979 and 1980 shows a tendency towards later flowering in the earlier-sown 1979 nursery. Originally, ear emergence time was recorded as the number of days after sowing. This is acceptable for spring-sown material, but is meaningless for autumn-sown nurseries, as illustrated by the reversal in flowering behaviour cited above. The reason for this reversal is related to the post-winter environment factors operative in each season, which bear no relationship to sowing date. A more satisfactory basis for recording this trait is to record the date on which ear emergence occurred. Table 11.2 shows smaller standard deviations for the means of the observed Julian date scores, as compared with scores for days after sowing, in 15 out of the 19 cases. This suggests that the date basis of scoring is more reliable. Sowing date is unreliable as an index of flowering time here because these data derive from autumn-sown nurseries, which have been subjected to unequal periods of growth suppression during winter months. The Julian date, as distinct from other methods of date recording, has been adopted for convenience in calculation.

**Table 11.6** Parameters of straw length data showing the effects of adjustment and the increments and factors applied. The increments are the negative of the y-axis intercepts, and the factors are the reciprocals of the slopes of the first principal component axes. Number in common indicates the number of varieties in each season's data set which also occur in any of the data sets previously included

| Season | 1972 | 1971 | 1970 | 1973 | 1974 | 1975 | 1976 | 1977 | 1978 | 1979 | 1980 |
|---|---|---|---|---|---|---|---|---|---|---|---|
| **Order of incorporation** | **1** | **2** | **3** | **4** | **5** | **6** | **7** | **8** | **9** | **10** | **11** |
| **a) Before adjustment:** | | | | | | | | | | | |
| Mean | 108.30 | 76.35 | 97.26 | 96.32 | 107.90 | 120.03 | 98.60 | 94.70 | 96.33 | 84.69 | 85.59 |
| SD [a] | 17.16 | 19.69 | 19.65 | 15.93 | 14.58 | 18.60 | 12.65 | 23.31 | 17.40 | 15.62 | 14.95 |
| Minimum | 42 | 25 | 42 | 44 | 38 | 65 | 50 | 50 | 53 | 50 | 62 |
| Maximum | 160 | 130 | 151 | 176 | 154 | 169 | 135 | 155 | 141 | 140 | 133 |
| No. in common | 814 | 728 | 617 | 320 | 419 | 822 | 739 | 166 | 200 | 300 | 280 |
| **b) After adjustments:** | | | | | | | | | | | |
| Mean | — | 107.54 | 108.89 | 106.76 | 117.75 | 116.24 | 116.22 | 107.44 | 97.92 | 96.22 | 94.99 |
| SD | — | 16.48 | 15.44 | 17.92 | 18.38 | 13.64 | 13.72 | 15.92 | 9.64 | 10.27 | 10.26 |
| Minimum | — | 64.56 | 65.48 | 47.90 | 29.64 | 75.88 | 63.52 | 76.91 | 73.91 | 73.42 | 78.80 |
| Maximum | — | 152.45 | 151.12 | 196.41 | 175.87 | 152.15 | 155.70 | 148.63 | 122.67 | 132.56 | 127.54 |
| No. in common | — | 658 | 560 | 181 | 110 | 358 | 617 | 63 | 103 | 163 | 198 |
| **c) Adjustment increments and factors (see text for explanation):** | | | | | | | | | | | |
| Increment 1 | — | 51.9708 | 44.6364 | 4.9977 | -18.3724 | 40.5454 | 11.5347 | 62.5838 | 103.3651 | 67.5520 | 56.2584 |
| Factor 1 | — | 0.8414 | 0.7703 | 1.0514 | 1.3229 | 0.7223 | 1.0547 | 0.6849 | 0.4898 | 0.6301 | 0.6704 |
| Increment 2 | — | 0.1395 | -2.5445 | -6.7565 | 5.1417 | -1.4933 | -3.1204 | 0.0117 | -11.2636 | -3.6690 | -2.3316 |
| Factor 2 | — | 0.9948 | 1.0200 | 1.0701 | 0.9529 | 1.0153 | 1.0282 | 0.9972 | 1.1314 | 1.0429 | 1.0241 |

a SD = Standard Deviation

**Table 11.7** Parameters of ear emergence date data showing the effects of adjustments and the increments and factors applied. The factors are applied to data sets with the mean adjusted to zero. Number in common indicates the number of varieties in each season's data set which occur in any of the data sets previously included. Note that the full sequence of data sets is not shown: a discontinuity occurs after the eighth set incorporated

| Season | 1973B | 1973A | 1972 | 1971 | 1970 | 1974 | 1975 | 1976 | 1980 | 1979 | 1978 | 1977 |
|---|---|---|---|---|---|---|---|---|---|---|---|---|
| **Order of incorporation** | **1** | **2** | **3** | **4** | **5** | **6** | **7** | **8** | **14** | **15** | **16** | **17** |
| **a) Before adjustment:** | | | | | | | | | | | | |
| Mean | 153.62 | 153.36 | 151.83 | 149.50 | 155.81 | 155.89 | 159.59 | 149.37 | 155.53 | 162.23 | 156.45 | 159.50 |
| SD[a] | 6.99 | 6.75 | 9.55 | 10.23 | 6.28 | 9.47 | 8.46 | 7.29 | 5.95 | 4.15 | 5.92 | 6.72 |
| Minimum | 134 | 132 | 129 | 130 | 140 | 130 | 134 | 132 | 132 | 150 | 139 | 138 |
| Maximum | 166 | 172 | 173 | 173 | 170 | 169 | 177 | 169 | 164 | 172 | 172 | 173 |
| No. in common | 939 | 833 | 819 | 740 | 620 | 614 | 877 | 1057 | 280 | 300 | 296 | 172 |
| **b) After adjustment:** | | | | | | | | | | | | |
| Mean | — | 154.34 | 153.68 | 153.50 | 154.20 | 158.00 | 156.18 | 156.32 | 153.88 | 153.30 | 152.86 | 154.71 |
| SD | — | 7.37 | 8.77 | 9.54 | 7.06 | 8.87 | 8.34 | 7.46 | 6.19 | 4.76 | 6.09 | 6.17 |
| Minimum | — | 129.50 | 132.73 | 135.30 | 136.41 | 135.62 | 130.96 | 138.54 | 129.41 | 139.27 | 134.91 | 135.15 |
| Maximum | — | 176.00 | 173.12 | 175.44 | 170.15 | 172.14 | 173.35 | 176.41 | 162.69 | 164.51 | 168.85 | 167.29 |
| No. in common | — | 673 | 436 | 375 | 287 | 97 | 275 | 373 | 102 | 109 | 176 | 76 |
| **c) Adjustment increments and factors (*see* text for explanation):** | | | | | | | | | | | | |
| Initial increments | — | 1.59584 | 1.74182 | 4.00356 | -1.46580 | 4.31082 | -3.43113 | 7.10180 | -1.54024 | -8.63572 | -3.52644 | -4.63123 |
| Factor 1 | — | 1.02091 | 0.93666 | 0.93295 | 1.09175 | 0.96084 | 0.98929 | 1.00478 | 1.01194 | 1.07020 | 1.01344 | 0.91834 |
| Increment 2 | — | -0.62144 | 0.11404 | -0.00301 | -0.15339 | -0.08529 | 0.01730 | -0.15019 | -0.08464 | -0.29118 | -0.06468 | 0.04090 |
| Factor 2 | — | 1.13843 | 0.98008 | 1.00053 | 1.03026 | 1.02340 | 0.99783 | 1.01851 | 1.02166 | 1.07179 | 1.01467 | 0.99995 |
| Increment 3 | — | -0.00139 | 0.00008 | -0.00000 | 0.00001 | 0.11079 | 0.00672 | -0.00289 | -0.01886 | -0.00170 | -0.00096 | 0.00001 |
| Factor 3 | — | 1.00027 | 0.99998 | | | 0.97030 | 0.99916 | 1.00035 | 1.00472 | 1.00043 | 1.00021 | |
| Increment 4 | — | -0.00001 | -0.00001 | | | -0.14392 | 0.00261 | -0.00005 | -0.00421 | -0.00002 | -0.00001 | |
| Factor 4 | — | | | | | 1.03977 | 0.99967 | | 1.00104 | | | |
| Increment 5 | — | | | | | 0.18693 | 0.00101 | | -0.00093 | | | |
| Factor 5 | — | | | | | 0.95033 | 0.99987 | | 1.00023 | | | |
| Increment 6 | — | | | | | -0.24279 | 0.00039 | | -0.00021 | | | |
| Factor 6 | — | | | | | 1.06789 | 0.99995 | | 1.00005 | | | |
| Increment 7 | — | | | | | 0.31538 | 0.00015 | | -0.00004 | | | |
| Factor 7 | — | | | | | 0.93404 | 0.99998 | | 1.00001 | | | |
| Increment 8 | — | | | | | -0.32720 | 0.00006 | | -0.00001 | | | |
| Factor 8 | — | | | | | 1.07393 | 0.99999 | | | | | |
| Increment 9 | — | | | | | 0.34252 | 0.00002 | | | | | |
| Factor 9 | — | | | | | 0.92722 | 1.00000 | | | | | |
| Increment 10 | — | | | | | -0.36215 | 0.00001 | | | | | |

a SD = Standard Deviation

## TOWARDS A SOLUTION

It was argued above that planting a single nursery from which all the required evaluation data could be collected was not a satisfactory solution. It would also be unsatisfactory because a large amount of data accumulated over the years would become redundant, and to discard these data would be a waste of resources. Moreover, it would not provide the level of replication over seasons which is necessary to establish confidence in the data, although replication could be imposed within the nursery. It would also fail to reveal interactions with seasonal variables — genotype x environment (G x E) interactions — which might appear from observations over a number of seasons. Given that the ideal would be a completely orthogonal design over seasons, and that the data in hand have some level of replication, it becomes desirable to harness this contribution to orthogonality, and to adjust the data as necessary to enable calculation of reliable varietal means.

The basis for using the accumulated data lies in considering how its statistical distribution is modified by seasonal effects, and how this modification can be recognized and used to transform the data into a usable form. The objective of such a transformation is to minimize the variance of the varietal means across seasons in order to obtain a maximum likelihood estimation of the varietal means. It is outside the scope of this paper to attempt to describe the mechanisms of the effects of environmental factors during the seasons.

### Assumptions

The statistical distributions of the scores are assumed to be approximately normal. This generally appears to be the case. The collection nurseries are large enough to be treated as populations with infinite degrees of freedom. Individual populations may have skewed distributions, evidently caused by bias towards particular types of material. For example, in 1974, when the data were biased towards later and taller types, the dominant component was old European varieties. In 1975 and 1976, the majority of entries were from Nepal, and thus the bias was towards medium-length straw. In the four seasons from 1977 to 1980, the emphasis was on advanced breeding lines from the PBI, and the bias was therefore towards short-straw and later-flowering types (*see* Tables 11.6 and 11.7). In addition, environmental factors such as temperature and moisture fluctuations, which accelerate or retard the expression of time-related characteristics within the overall population, modify the statistical parameters of the data.

It is assumed that each genotype occupies its own characteristic place relative to others within the range of variation of a particular trait. Thus, if two similarly constituted populations were to be grown in early and late seasons respectively, it is expected that any particular variety would have a similar relationship to the mean of its population without respect to season. Dissimilarity between seasons in terms of environmental factors may reveal other G x E interactions which modify the relationships within populations. It must also be assumed, therefore, that the presence of such interactions will be disclosed by the incidence of varietal means with relatively large variances. Thus it should be possible to identify genotypes showing G x E interactions by screening the standard deviations of the means. For example, flowering time is strongly dependent on the overall earliness or lateness of a population. In an extreme case, therefore, it is expected that the variance for a winter variety included in a spring-sown nursery will be significantly larger than that for a variety adapted to a spring sowing.

## Manipulation of population statistics

If it is true that varieties hold characteristic positions within population distributions, then, by cross-referencing with corresponding accessions between a pair of nurseries, it should be possible to find factors and increments whereby the two populations may be aligned or 'matched'.

In this study, the strategy has been first to select a primer data set which could be used as a template to which all the other available data sets are matched progressively. A second set is matched to the primer, and the varietal means are calculated over the two; then a third is matched to the combined first and second, and so on. The varietal means thus become the values of the composite data set. By adopting this type of approach, with the exception of the first pair matched, any correspondence is not between individual pairs of data sets, but between each new data set and the composite of data sets which have already been combined.

It is important to note that adjustment to the parameters of data sets is made on the basis of corresponding accessions. Varieties which are not common to both data sets are eliminated from the calculation of adjustment parameters. If there are no varieties common to two populations, it might be possible to make an adjustment on the basis of the ratio of the means of the two populations; however, this would yield a satisfactory result only if the spectrum of types were similar and the number of entries large. Conversely, if the number of corresponding varieties were small, or if their positions within the overall data distribution were uneven, there might be insufficient information on which to calculate an adequate adjustment, and genotypes with strong G x E interactions could exert undue influence.

The strategy adopted to maximize the correspondence of data points is first to find the pair of populations with the largest number of entries in common, taking the larger of the two as the primer set. The computerized algorithm employed then searches for successive populations with the highest number of accessions corresponding with the joint array of those already included. A disadvantage of using this approach is that the primer data set may be atypical in some way. It may therefore be preferable to deliberately select the primer set and control the order in which subsequent data sets are merged.

Although the adjustment factors or increments are based on corresponding accessions only, they are applied to the whole of the data set to be added. Consequently, the data for varieties not previously included in the collation will also be adjusted, bringing them into line on a whole population basis. Thus there is a reasonable expectation that varieties will find their correct location within the composite population despite having no prior correspondence with the composite population which has been built up by the collation process.

Two evaluation characters are used to illustrate this approach. Their data structures differ in important respects, demanding different treatments for each. Most evaluation characters probably conform to the first type — illustrated by straw length; the second type — illustrated by ear emergence date — is fairly rare.

### *Straw length*

The model for the first data type has variation with its origin close to zero and its variance proportional to the distance from the origin. Clearly, there is a minimum length of straw needed to support a flowering spike which prevents the origin being at zero. It is assumed that the phenotype will be modified proportionately by environmental factors. Thus, values in the low range should be less variable than values from the upper range of the variation spectrum. This can be illustrated by a joint scatter diagram of the values from two data sets, where the distribution of points is seen to be a wedge-shape which becomes broader as values increase (*see* Figure 11.1 *overleaf*). The data in hand often have a non-zero origin. Thus there seems to be an additional component in the relationship between pairs of seasons which should be taken into account.

To match the data from one seasonal distribution of straw length data to the selected primer data set, each value in the second data set is divided by the slope of the line fitted to the joint distribution. This brings the slope of the joint distribution to unity. The method used for fitting the line is principal components (PCs) analysis. The first PCs axis gives an unbiased, symmetrical fit to the bi-variate distribution and maximizes the available variance. Unlike the regression method of line fitting, it makes no assumptions about the dependence of one variable on the other. The method is used purely as a means of fitting a line, and no further inferences about the data are drawn. Therefore, the slope of the line is the ratio of the eigenvectors of the joint sums of the squares and products matrix formed from the primer and second data sets.

However, it is necessary first to impose a zero origin on the joint distribution, otherwise the fitted line will rotate away from the PC axis. The data of the second set is therefore decremented by the intercept of the y axis prior to the adjustment of slope (intercept = means of set 2 - [slope x mean of set 1]). The primer set remains unaltered, but the means and variances of the two data sets are brought closer together. Corresponding data points therefore converge, permitting the calculation of varietal means with smaller variances (compare the standard deviations between the means of

**Figure 11.1** Scatter diagrams of unadjusted data points for (a) straw length and (b) ear emergence date illustrating the characteristics of each data type

(a) Straw length

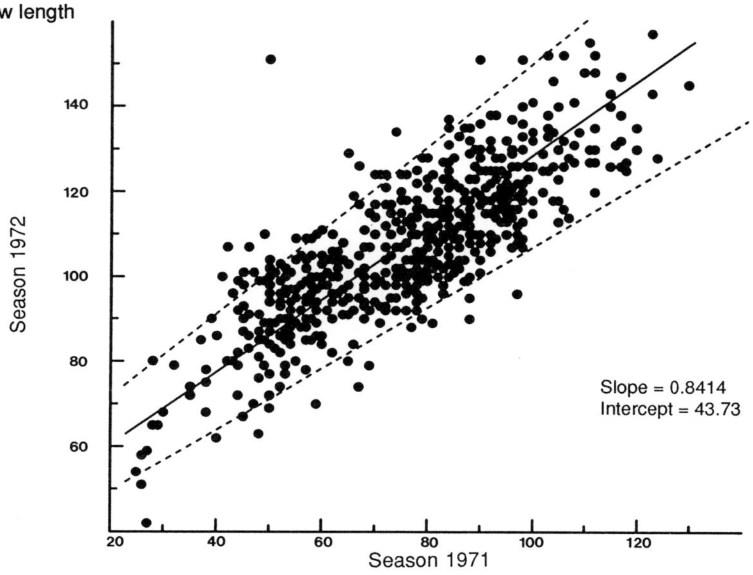


(b) Ear emergence date

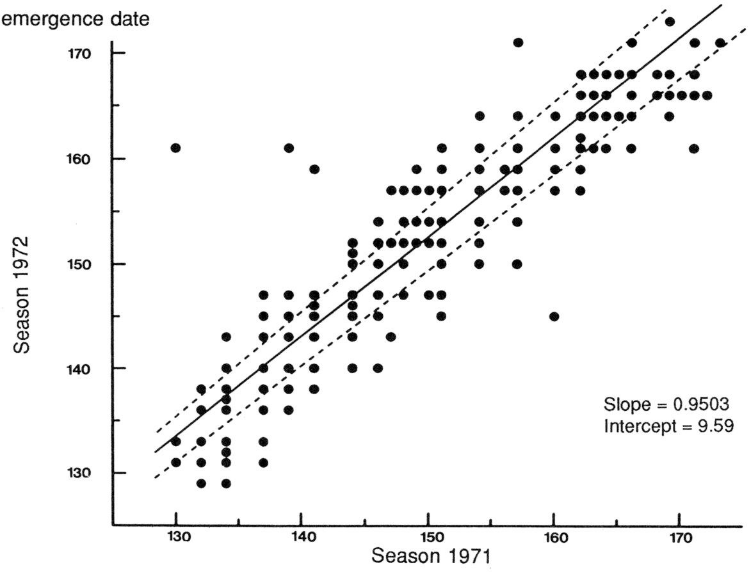


The solid line shows the fitted first principal components axis; the dotted lines show the 90% and 50% confidence intervals for straw length and ear emergence date respectively the observed and adjusted data in Table 11.1). The process may be repeated, after outlying values have been discarded, to provide a fine adjustment (*see below*). The adjustment increments and factors used for straw length, together with their effects on the data represented in Table 11.3, are shown in Table 11.6. Since the order in which the data sets are collated is not predetermined, these values are not absolute.

Clearly, there are problems associated with this method, particularly in relation to dwarf varieties. For accession 590 (Dwarf Ministre), the adjusted value for 1971 in Table 11.3 (66.24) is clearly unrealistic, because this variety could not attain such a height. The variance of the adjusted mean is high, although for the observed mean it is not unusual (*see* Table 11.1). It seems likely that this result may reflect an inherent instability in dwarf types, as well as an inadequacy in the method, which seems to make shorter-strawed types more sensitive to the adjustment process. Table 11.8 shows, for both traits under consideration, how the variance of the variety mean tends to increase with distance from the median of the overall distribution of data. This may be explained in terms of the numbers of accessions common to the two data sets being predominantly from the median region. The numbers of corresponding points in the more distal regions are fewer, and thus have less influence on the adjustment calculations. Inevitably, therefore, the more distant values will receive less accurate adjustment.

Similarly, the three adjusted values for accession 2208 (C 645), which contains two dwarfing genes, are unexpectedly high. The observed values are realistic, and in each of the three seasons are either the lowest values recorded or are no more than 5 cm greater than the lowest. Their mean, however, has a very low standard deviation. The problem here is likely to be related to the fact that there are only 63 varieties common to the four seasons 1977 to 1980 and to the six seasons which precede them in the

**Table 11.8** Variation in the standard deviation of varietal means related to ranking of straw length and ear emergence date data in the wheat collection for accessions with means calculated over two or more observations

| Rank | Straw length | | Ear emergence date | |
|---|---|---|---|---|
| | No. of accessions | Average SD [a] | No. of accessions | Average SD |
| 1 | 2 | 11.19 | 0 | — |
| 2 | 9 | 6.10 | 31 | 2.42 |
| 3 | 42 | 4.32 | 234 | 2.52 |
| 4 | 532 | 2.98 | 464 | 1.98 |
| 5 | 815 | 3.84 | 630 | 1.76 |
| 6 | 888 | 4.39 | 833 | 1.78 |
| 7 | 424 | 5.16 | 548 | 1.70 |
| 8 | 164 | 6.01 | 204 | 2.08 |
| 9 | 26 | 7.21 | 74 | 1.60 |

a S.D. = Standard Deviation

collation process. If the data for these common varieties were restricted to only a portion of the range of variation of straw length, or if parts of the range were poorly represented, the orientation of the fitted PC axis is unlikely to be reliable. Table 11.6 shows the first increment (that is, the negative of the intercept of the fitted line) and the slope of the fitted line (that is, the inverse of the first adjustment factor shown in Table 11.6) for the two seasons 1977 and 1978 to be the largest encountered overall. The conclusion to be drawn is that the spread of the data within the population distributions for the varieties in common in these seasons is not adequate for the success of the method. The remedy is to design a new primer nursery with a more adequate sample of varieties from the whole range of seasons.

### *Ear emergence date*

The second data type has no logical zero or near-zero origin. The means of contrasting data sets may differ, but their variances tend to be similar, and the scatter of points when plotted in opposition occupies a parallel-sided band with a slope approaching unity (*see* Figure 11.1).

To manipulate ear emergence date, first the mean of the second data set is relocated to match that of the primer set. This is achieved by incrementing all its values by the mean difference of corresponding data points. This nullifies the mean difference and, in effect, slides the data sets together without changing their respective variances. This first step is the most significant adjustment (*see* 'Initial Increments', Table 11.7), but it is possible to bring the variances of the two data sets into greater conformity by applying a factor to the second data set, either to expand or to contract its variance as appropriate. The means of both the primer and the second data set are equated with zero, and the adjustment factor, which is applied to the zeroed data set, is the ratio of the means of either the positive or negative halves of the resulting distributions (either half producing the same result). The original means, which remain the same for both data sets, are then restored. Further precision can be achieved by cycling iteratively through this 'matching' and 'stretching' process until a satisfactory level of conformity of the means and variances has been reached. Here the cycle was stopped when the adjustment factor lay within the range $1.0 \pm 10^{-5}$ and the adjustment increment in the range $0.0 \pm 10^{-5}$ or when nine cycles had been completed. Table 11.7 shows the factors and increments used and their effects on the data shown in Tables 11.4a and 11.4b. In most cases, iterations cease after two or three cycles, and in only two comparisons did the number of cycles reach the computer programme's set maximum.

A problem similar to that found in the manipulation of the straw length data appears in Table 11.7. Only 97 accessions are common to the first five seasons. It is at the point of matching season 1974 that the greatest difficulty in calculating the adjustments occurs. The adjustment increments and factors for both 1974 and 1975 fail to approximate adequately to zero and unity, respectively, within the constraints of the computer programme.

### Elimination of extreme values

Although the adjustments to the data sets are calculated on the basis of corresponding accessions only, a problem remains if extreme values exert undue weight upon the calculation of population means and mean differences. Extreme values may arise through G x E interactions or inaccurately recorded or transcribed observations, and are outliers in the scatter of data from two sets (*see* points lying outside the confidence intervals, Figure 11.1). Given sufficiently large numbers, such effects are likely to be random and would mutually cancel out. However, the situation can be envisaged where the extreme values all tend to lie in the same direction. For example, in relation to straw length, the presence of serious lodging would seriously distort the data.

Outliers are eliminated from the calculation by excluding accessions whose differences fall outside a pre-determined confidence interval based on the mean of the differences. The confidence intervals used to derive the results presented were 90 per cent (=1.645 sodium dodecyl sulphate) for ear emergence, and 50 per cent (= 0.674 SDS) for straw length. These intervals were chosen through trial-and-error; they were found to give optimum reduction in variance for the variety means for each character respectively. A weighting factor is applied to eliminate proportional effects in the data (that is, increase in variance of the data related to increase in magnitude of value), and is the reciprocal of the mean of a pair of values. Thus, for the corresponding data points a and b, the weighted difference is 2 (a-b)(a+b).

### Assignment of ranks

Ranks are assigned by equal increment partition of the distribution of the final means as delimited by the extreme upper and lower values. Thus the data stored in the computerized database accurately reflect the range of variation for the trait without wastage of categories either in part or whole. Table 11.9 (*overleaf*) gives the rank definitions derived from the present study.

## CONCLUSION

No claim is made about the validity of the method described, and refinements or alternatives may be envisaged. However, as a pragmatic approach to a problem which faces many evaluators of germplasm, it is offered for consideration.

No other traits have been treated in this way to date, but it is assumed that most which show continuous variability would be susceptible to this type of manipulation. Obvious exceptions are traits which have strong G x E interactions involving other variables (notably, disease susceptibility traits which are subject to changing virulence patterns from season to season).

**Table 11. 9** Scales used for assigning ranks to evaluation data, comparing past with present values

| Rank | Straw length | | Ear emergence date | |
|---|---|---|---|---|
| | Original PBI (cm) | Derived (cm) | Original PBI (days after sowing) | Derived (Julian date) |
| 1 | ≤ 50 | < 57.88 | ≤ 205 | < 134.34 |
| 2 | ≤ 60z | < 69.81 | ≤ 210 | < 139.26 |
| 3 | ≤ 70 | < 81.73 | ≤ 215 | < 144.19 |
| 4 | ≤ 80 | < 93.66 | ≤ 220 | < 149.12 |
| 5 | ≤ 90 | < 105.57 | ≤ 225 | < 154.05 |
| 6 | ≤ 100 | < 117.50 | ≤ 230 | < 158.98 |
| 7 | ≤ 110 | < 129.43 | ≤ 235 | < 163.91 |
| 8 | ≤ 120 | < 141.35 | ≤ 240 | < 168.84 |
| 9 | > 120 | ≥ 141.35 | > 240 | ≥ 168.84 |
| Increment | 10 | 11.92 | 5 | 4.93 |

There may be concern that the final ranks allocated to varieties have no relation to predetermined descriptor states. However, it is suggested that the ranks resulting from the method described fit the range of variation of a trait more accurately and therefore offer a more meaningful discriminatory instrument for screening purposes. Indeed, not only may data from different seasons be combined, but also data from different locations, where the differences encountered may mimic seasonal differences.

## *Discussion*

J. P. Srivastava: Your results are very interesting. Could you briefly give your suggestions for the handling of data from different seasons at the same site and multi-site evaluations?

R. Giles: I think the approach I have taken could also be applied to between-site observations, as well as to between-season observations at a single site, but I do not have sufficient data to test this. I presume that so long as G x E interactions, especially photoperiod responses, do not distort the picture, the method should be more generally applicable.

L. Tanasch: I have great misgivings about the unreplicated trials you mentioned in your germplasm evaluation. Do you think it is advisable to have standard checks over the different years and adjust your values on the basis of performance of these checks?

R. Giles: I agree it would be necessary to use standard checks. But I need to repeat that I was looking at data recorded in the past by other scientists where it was not possible to do that. I presume it would be possible to identify, to a limited extent, varieties which were present in the experiment in more than one year. In this case, they were not there.

Wheat Genetic Resources: Meeting Diverse Needs
Edited by J. P. Srivastava and A. B. Damania

Published by John Wiley & Sons

# 12

# *The Significance of Taxonomic Methods in Handling Genetic Diversity*

M. VAN SLAGEREN

Taxonomy can be seen as the organization of a coherent body of information, ideas and principles about biological organisms, which provides basic data for many biological sciences. The scientific names of organisms are the key to locating previously recorded information. They are arranged in a hierarchy of categories, which provide a mechanism for the communication of information (Rossman et al., 1988).

Taxonomists draw up a classification system that reflects their various objectives. When, through lack of additional research, this system is primarily, or even solely, based on morphological and anatomical similarities, it is called 'phenetical'. The result may be practical, but it carries the risk of being artificial to some extent, thereby possibly overlooking patterns that might be revealed by, for example, genome analysis or evolutionary studies. However, in a group as thoroughly studied as the wheats and their wild relatives, the current classifications reflect relationships through inheritance and can be used subsequently to predict, for example, which organisms have potential usefulness as germplasm. As taxonomy is related particularly to the description and study of biological diversity, it can provide basic data when discovering, storing and using germplasm.

Within the classical concept, a distinction is often made between the terms 'taxonomy' and 'systematics'. Taxonomy refers to the delimitation of units, and is based primarily on morphological and anatomical data — that is, readily observable data. The resulting classification is of a phenetical nature. Systematics, on the other hand, starts with the analysis of the phenetical data of a group, but classifications aim at reflecting a reconstruction of the evolutionary history of the group involved. Characters which are present in more than one form are scored for the state in which they appear; data on geography, phytochemistry and so on are also taken into account. Such classifications are called 'phylogenetical' and aim to define so-called monophyletic groups.

The genus *Triticum*, as it is generally delineated, comprises various genome types of quite different origin, up to the point that the tetraploid and hexaploid wheats can be viewed as 'bigeneric hybrids' (Gupta and Baum, 1986). Thus the genus might be considered polyphyletic in the above-mentioned, evolutionary sense.

In this paper, as I shall not be referring to a strict phylogenetical system for the wheat group, the term 'taxonomy' is used rather than 'systematics'. The paper discusses some aspects of classical taxonomy, roughly similar to what Hawkes (1986) calls the 'morpho-geographical' approach. This is followed by a brief analysis of the numeric-taxonomical treatment of the Triticeae as outlined by Baum (1977, 1978a, 1978b), and by the genomic classification drawn up by Löve (1982, 1984).

## CLASSICAL TAXONOMY

### Splitting and lumping

In genebank activities, a clear distinction of the taxa involved is considered essential, particularly with reference to the problem of how and if intraspecific diversity should receive formal taxonomical recognition. Until early this century, this problem was usually solved by formal recognition of intraspecific entities such as varieties and formae. For most of these small taxa, it is now recognized that the particular feature involved is linked with, for example, an environmental effect, which does not presuppose a genetic basis for it (Hawkes, 1978). Thus, the classifications of *Aegilops* and *Triticum* now usually show the effect of 'lumping' rather than of 'splitting'. Table 12.1 illustrates this with the example of *Ae. triuncialis* in a comparison with a treatment in the Flora of Turkey (Davis, 1985) and the monograph by Hammer (1980). The latter shows a narrow definition of taxa, resulting in many intraspecific units.

In *Triticum*, the number of species varies between four (Morris and Sears, 1967) and 27 (Dorofeev and Korovina, 1979). Within *T. aestivum* (bread wheat), subdivisions have gone much further. Morris and Sears (1967) do not differentiate in a formal way at all, distinguishing only five 'groups', whereas MacKey (1966) distinguishes six formal subspecies. Dorofeev and Korovina (1979), however, subdivide *T. aestivum* (*sensu stricto* — that is, excluding such groups as *spelta* and *sphaerococcum*) into two subspecies, three convarieties, four subconvarieties and 194 varieties! It is doubtful whether this system is still practical.

It should also be noted that most Floras of West Asia and North Africa — for example, Bor (1968) for Iraq, Mouterde (1966) for Lebanon and Syria, Davis (1985) for Turkey, and Täckholm (1974) for Egypt — subdivide *Triticum* only at the species level and keep *Aegilops* as a separate genus. While most authors show their awareness of the recent monographs on the wheat group, the undoubtedly practical nature of floristic work accounts for this treatment.

**Table 12.1** Comparative classifications of *Aegilops triuncialis*

| ***Aegilops triuncialis*** | |
|---|---|
| species *triuncialis* | species *triuncialis* |
| | var. *triuncialis* |
| | forma *triuncialis* |
| | forma *nigro-albescens* |
| | forma *nigro-feruginea* |
| | forma *brunnea* |
| | var. *flavescens* |
| | forma *flavescens* |
| | forma *nigro-flavescens* |
| | forma *rubiginosa* |
| | forma *nigro-rubiginosa* |
| | var. *constantinopolitana* |
| species *persica* | species *persica* |
| | var. *persica* |
| | var. *assyriaca* |
| | var. *anathera* |

Source: P.H. Davies in Flora of Turkey, Vol 9, 1955 and K. Hammer in Fedes Report, 91, 1980

## Nomenclature

Formal recognition of taxa, by whatever method, invokes the application of the rules of botanical nomenclature. These are laid down in the International Code of Botanical Nomenclature (ICBN) (Greuter et al., 1988). The Code states that giving names to taxa only serves the purpose of supplying a means of reference, and that the aim should be to provide a stable method of doing so, avoiding or rejecting the use of names which may cause error or ambiguity.

An additional aim is to avoid the superfluous creation of names. Names of taxa are the most tangible result of the application of the Code, and any revisions must be unambiguous. Many local revisions, as well as monographs, have been published on the *Aegilops-Triticum* group, and the results have been confusing. This is partly because non-taxonomists, such as (cyto)geneticists, have proposed various rearrangements and classifications of the group.

Some of the ICBN rules are illustrated in the following examples:

1. *Priority principle:* When two or more names are given to a taxon, the oldest legitimate one should be followed (principle IV, Art. 11.3). Thus *Ae. biuncialis* (1842) is correct, and *Ae. lorentii* (1845) must be rejected. Here Art. 42.2 (note 2) and Art. 44.1 apply: an illustration with analysis serves as valid publication. *Triticum peregrinum* (Hackel, in Fraser, 1907) is correct, and thus *Ae. variabilis*

(Eig, 1929) must be rejected (the latter name is still frequently encountered). Note that the association of *peregrina* with *Aegilops* (Maire and Weiller, 1955) is more recent than that of *variabilis*. This indicates that the species is considered the basic unit in the plant kingdom, and not the genus to which it is assigned, since a genus is thought to be artificial.

2. *Simultaneous publication:* The simultaneous publication of so-called alternative names is not valid (Art. 34.4). Thus the publication of the important *Ae. searsii*, which was combined with *Aegilops* and *Triticum* at the same time, is invalid. This was corrected by Hammer in 1980.

Applying the ICBN rules can be problematic, however, when their interpretation is not clear. Problems also arise from the ambiguous use of the species concept by previous authors. Examples within *Aegilops* are the simultaneous use of *Ae. ovata* and *Ae. geniculata* and of *Ae. caudata* and *Ae. markgrafii*. A choice has to be made here which will best comply with the rules. These matters will be discussed in a future revision of *Aegilops*.

## NUMERICAL TAXONOMY

The Triticeae have been analysed using numerical techniques in a series of papers by Baum (1977, 1978a, 1978b). His purpose was to assess the phenetic relationships of the tribe at a generic level on a worldwide basis. His analysis was based on, firstly, as many clearly defined morphological characters as possible, and secondly, as much information as possible on intergeneric crosses and hybrids. The resulting classification (Baum, 1978a) is therefore phenetical, and does not reflect any evolutionary (phylogenetical) history of the tribe.

The starting point was the identification of 28 operating taxonomic units (OTUs). Here *Aegilops*, *Triticum* and *Amblyopyrum* were distinguished. For all OTUs, characters were taken from previous descriptions; herbarium material was used to make the data set complete, using a selected number of species per genus.

The next step was to carry out a cluster analysis of the characters to establish the relative importance of (clusters of) characters. This mathematical approach is not only widely considered to be a more objective, repeatable way of character assessment, but can also overcome the difficulty of reviewing the amount of data produced by having so many taxa (Baum, 1978a).

After completion of the data matrix, clustering the OTUs was carried out (Baum, 1978a). Because of the intricate patterns of relationships in the Triticeae, a total of 43 spanning trees became visible. All clusters, however, showed *Aegilops* and *Triticum* in the same group, often joined by *Amblyopyrum* and, to a lesser extent, by *Henrardia* and *Dasypyrum*. For the purposes of classification, Baum proposed six groups within the Triticeae, and united *Aegilops*, *Triticum*, *Amblyopyrum* and *Henrardia* in one of them. This classification represented one of his spanning trees, namely the one he considered best in terms of its information content. The close relationship of wheat and its wild relatives, based solely on a comparison of morphological characters, was underlined.

Baum's study thus confirmed, by using more 'objective' methods, what has been concluded in classical taxonomy for a long time. His suggestion that *Amblyopyrum* could be united with *Aegilops* (Baum, 1978a) reflects a problem that has been solved in one of two ways by most authors: monographic works on *Aegilops* usually include *Amblyopyrum* in *Aegilops*, as *Ae. mutica* (for example, Eig, 1929 and Hammer, 1980); flora treatments, however, keep *Amblyopyrum* apart from *Aegilops* (for example, Bor, 1968 and Davis, 1985).

## GENOMIC CLASSIFICATIONS

The undue emphasis on genome constitution in the Triticeae has recently led to proposals for a rather different classification of the group. In the early 1980s, classification proposals for annuals were put forward by Löve (1982,1984) and for perennials by Dewey (1984). As the wheat group consists of annuals, only Löve's papers are discussed here.

The basic idea was to define genera in direct relationship to their genomic constitution. This led to the distinction of 38 genera in the Triticeae (*see* Table 12.2).

**12.2** Genomic classification of the Triticeae, illustrated by *Aegilops (pro parte)*

| Classical concept | Genome | Löve 1982, 1984 | Genome |
|---|---|---|---|
| *Aegilops bicornis* | Sb | *Sitopsis bicornis* | B |
| *Ae. speltoides* | S | *S. speltoides* | B |
| *Ae. longissima* | Sl | *S. longissima* | B |
| *Ae. searsii* | Ss | *S. searsii* | B |
| *Ae. sharonensis* | Sl | *S. sharonensis* | B |
| *Ae. caudata* | C | *Orrhopygium caudatum* | C |
| *Ae. squarrosa* | D | *Patropyrum tauschii* | D |
| *Ae. comosa* | M | *Comopyrum comosum* | M |
| *Ae. mutica* | Mt | *Amblyopyrum muticum* | Z |
| *Ae. uniaristata* | Un | *Chennapyrum uniaristatum* | L |
| *Ae. umbellulata* | U | *Kiharapyrum umbellulatum* | U |
| *Ae. ovata* | UM | *Ae. ovata* | UM |
| *Ae. ovata* | UM | *Ae. geniculata* | UM |
| *Ae. biuncialis* | UM | *Ae. lorentii (biunc.)* | UM |
| *Ae. columnaris* | UM | *Ae. columnaris* | UM |
| *Ae. triaristata* | 4nUM | *Ae. triaristata* | 4n UMM |
| | 6nUMUn | | 6n UMM |

The genomic concept is considered useful as a starting point for phylogenetic studies (Gupta and Baum, 1986) and, with the help of advanced techniques, for the interpretation of molecular data (West et al.,1988). The various concepts of the evolution of the tetraploid and hexaploid wheats (for example, Feldman, 1976) may serve as an example.

However, as West and his co-authors recognize, Löve's classification presents many problems from a taxonomic as well as a practical point of view. In the case of *Aegilops*, the well-delineated 'classical concept' is completely lost, and Löve's definition confines the genus to the genomic type UM which is present in the type species *Ae. ovata* and a few others. More confusion arises from his use of genomic formulae which differ partially with the widely accepted codes proposed by Kimber and Sears (1983); for example, compare the codes for the 'classical' species *Ae. mutica* and *Ae. uniaristata* with Love's designations in Table 12.2.

Moreover, I tend to agree with Gupta and Baum (1986) that in this genomic system the very purpose of using taxonomy and giving names to biological units has been ignored. As the ICBN states, the purpose of giving names is to provide a means of reference and not to indicate the characters or history of the taxa involved.

Love (1984) himself admits that his system is not perfect and that he partially included taxa using the classical morpho-geographical definitions. In *Aegilops*, this is illustrated by *Ae. triaristata,* of which tetraploid (UM) and hexaploid (UMM) forms exist. Some authors have suggested separating the two forms into *Ae. neglecta* (4n) and *Ae. recta* (6n), respectively. Kimber and Feldman (1987) suggest the presence of a sterile uppermost spikelet in *Ae. neglecta* and a fertile one in *Ae. recta*. In the classical sense, this is already hazardous, and my survey of the material at the International Center for Agricultural Research in the Dry Areas (ICARDA), characterized in the field in the 1987-88 season, showed both forms readily present in one group of tillers, grown out of one seed. Although Love's classification may be useful for certain types of research, I would not advocate it for genetic resources activities, however closely linked they may be to the study of genetic variation.

The last issue to be dealt with here is the question of whether the inclusion of *Aegilops* in *Triticum* contributes to a better taxonomy of the group. Ever since Bowden (1959) proposed it, it has been a much debated issue.

The idea of merging *Aegilops* and *Triticum* is not new. For example, the group was treated as *Triticum sensu lato* by Grenier and Godron (1856) in the Flora of France, by Hackel (1887) in a worldwide treatment of grasses, and by Ascherson and Graebner (1898-1902) in the Flora of Central Europe. On the other hand, there have been just as many taxonomic and floristic treatments that keep the two genera apart, particularly in recent years. Bowden's 1959 paper was in fact the start of a new revision of the group, based on genetic linkage and crossability.

Cytogeneticists now show a preference for unity under *Triticum*, thereby emphasizing the genetic links between the groups. However, it is suggested that genomic data, as well as cytogenetical, biochemical and physiological data, should be used only insofar as they can be correlated with morphological characters. In this way, they can be of great use in helping to distinguish and define taxa (Hawkes, 1978). Thus it is not necessary to unite *Aegilops* with *Triticum*, despite the close genetic relationship.

Bowden's argument for merging *Aegilops* and *Triticum* derived from the conclusion that two of the three genomes of hexaploid wheat came from *Aegilops*. He felt that it was nomenclaturally incorrect to include the parental diploid wheat species in *Triticum*, together with the allotetraploid and allohexaploid species, and to place the other two parents in *Aegilops*. He proposed a merger of the two genera in an enlarged genus *Triticum*, underlining its hybrid origin but without treating it as a so-called nothogenus. It must be noted that Bowden's argument applies only to nothogenera and, since he treated *Triticum* as a botanical taxon, his preference for a merger has been considered neither necessary nor appropriate (Gupta and Baum, 1986). A note in the ICBN (under Art. H. 3.4) also states that it is not necessary to designate nothotaxa for taxa for which it is shown at one stage that they are of hybrid origin.

Apart from the argument over the nomenclature, Bowden has also been criticized in relation to crossability. Many genera of the tribe Triticeae and all other presently recognized genera of the subtribe Triticineae (*Aegilops*, *Secale*, *Agropyron* and *Haynaldia*) can be hybridized with *Triticum*. Thus, incorporating only *Aegilops* into *Triticum* is not justified (Mackey, 1966). Baum (1977) lists many of the formally recognized intergeneric hybrids.

## CONCLUSION

Analysis of the evolution in the Triticeae indicates that it is a fairly young group in which speciation is still going on and intermediate populations have not yet become extinct (MacKey, 1966). A network of processes, rather than a clear-cut trend, is all that can be detected at present, and thus classifications vary from the proposal to unite all taxa in one genus to the practical distinction of many, albeit sometimes artificial, genera. In view of the nature of genetic resources work, I would advocate separating *Aegilops* from *Triticum* and thus maintaining the classical concepts of both genera; the status of *Amblyopyrum* is considered a minor problem. It is at this practical level that taxonomy can make the greatest contribution to genetic resources activities.

## *Discussion*

S. JANA: In view of the genetic diversity in the Triticeae, what is the significance of taxonomic methods?

M. VAN SLAGEREN: Firstly, it provides the names of the units with which one is working. Secondly, it will reflect, although individually, the ideas of the author on the genetic diversity of the taxa. This can be used as a starting point for genetic studies.

M. N. SANKARY: Are you using autecological information, such as salinity response and root systems, in your proposed re-classification of the *Aegilops* species?

M. VAN SLAGEREN: This is very difficult to answer. I would prefer a system which would be readily usable and easily observable by the users. If there are striking differences in the root systems, for example, it must be mentioned but whether it is practical for field collection is another matter. Of course, species need to be accompanied by as much ecological information as possible, even if it is not possible to discriminate between them on the basis of this information.

M. N. SANKARY: What I meant was a classification which is biologically sound and which will not be controversial in the future.

M. VAN SLAGEREN: I agree, that is possible.

J. G. WAINES: What kind of classification system, then, will you construct — taxonomic or biosystematic?

M. VAN SLAGEREN: A taxonomic one, primarily.

J. G. WAINES: That means that we would still need another classification system, such as a biosystematic one or a combination of both. But that is not necessarily a step forward. I agree with Dr Sankary that we need one which is not controversial. For instance, every time we have a wheat genetic symposium a new system is proposed!

M. VAN SLAGEREN: I agree with you. At the moment I really do not know how it is going to work. No distinction in the absolute sense is possible. A system has to be practical in the first place. However, given the wealth of biosystematic data available on this species it should, of course, be taken into account.

T. T. CHANG: I would like to make a comment. I cannot help comparing wheat with rice. Our studies on rice are primitive, apart from some economic aspects, where we are quite good. At this symposium you really have a marvellous array of talents — cytogeneticists, biochemists, taxonomists and so on — and the greater the number of people working on the same problem the more they tend to go their own ways. That is human nature. In the International Wheat Genetics Symposium you have a wonderful forum for a high-powered discussion on wheat research. Reading the papers prepared for presentation here, I am amazed at the amount of work which has been done in comparison to other crops.

Wheat Genetic Resources: Meeting Diverse Needs
Edited by J. P. Srivastava and A. B. Damania

Published by John Wiley & Sons

# 13

# *The Case for a Wheat Genetic Resources Network*

Y. J. ADHAM and D. H. VAN SLOTEN

At present there are a limited number of long-term storage centres for wheat (base collections). It is necessary to expand this capacity by involving many more parties, particularly scientists and institutes from national programmes. Such action would lead to a truly global wheat genetic resources network. This paper looks first at the important role played by the international agricultural research centres which belong to the Consultative Group on International Agricultural Research (CGIAR) in the conservation and use of plant genetic resources, and then describes the particular role played by the International Board for Plant Genetic Resources (IBPGR), one of the CGIAR centres.

## INTERNATIONAL AGRICULTURAL RESEARCH CENTRES

It is always dangerous to play the numbers game, but I feel it is useful to show how much germplasm is conserved in the international agricultural research centres as a percentage of total world holdings. Table 13.1 (*overleaf*) shows that the following international centres together maintain a total of 419 000 germplasm samples: the Centro Internacional de Agricultura Tropical (CIAT), the Centro Internacional de Mejoramiento de Maíz y Trigo (CIMMYT), the Centro Internacional de la Papa (CIP), the International Center for Agricultural Research in the Dry Areas (ICARDA), the International Crops Research Institute for the Semi-Arid Tropics (ICRISAT), the International Institute for Tropical Agriculture (IITA), the International Livestock Centre for Africa (ILCA), the International Rice Research Institute (IRRI) and the West African Rice Development Association (WARDA).

This represents 35 per cent of the total unduplicated world holdings of 1 200 000 accessions (Van Sloten, 1989). Furthermore, the material in the collections of the international centres is freely available, as illustrated by the large number of samples distributed annually and the repatriation of national collections. In 1988 the CGIAR approved a policy statement on plant genetic resources, and this was published by the IBPGR in English, French and Spanish.

**Table 13.1** Germplasm holdings of international agricultural research centres

| Centre | Rounded No. of accessions |
|---|---|
| CIAT | 40 000 |
| CIMMYT | 48 000 |
| CIP | 10 000 |
| ICARDA | 87 000 |
| ICRISAT | 96 000 |
| IITA | 40 000 |
| ILCA | 9 000 |
| IRRI | 83 000 |
| WARDA | 6 000 |
| Total | 419 000 |
| Percentage of unduplicated world holdings | 35 |

On the technical side, the international agricultural research centres and the IBPGR have introduced an improved planning mechanism for plant genetic resources activities within the CGIAR through meetings of genetic resources staff of all relevant CGIAR centres. The first of these meetings took place in October 1987 at CIMMYT, in Mexico; the second was held in February 1989 at IRRI, in the Philippines.

## INTERNATIONAL BOARD FOR PLANT GENETIC RESOURCES

The IBPGR's activities can be divided into four major categories: Administration, Communication, Research Programmes and Field Programmes. Under the leadership of the Director, the Deputy Director and the three Heads of Programme, the IBPGR currently employs over 20 internationally recruited professional staff; this number is likely to increase to 30 in the near future. In addition, the Board employs nationally recruited professional staff, as well as support staff, at its headquarters in Rome and at its offices around the world.

Most IBPGR staff are based at the regional offices, reflecting the decentralized nature of the Board and the importance it attaches to close links with national programmes, particularly those in the developing world. The Board has regional offices in:

| | |
|---|---|
| Mexico | for Meso-America and the Caribbean, based at CIMMYT; |
| Cali, Colombia | for South America, based at CIAT; |
| Rome, Italy | for Europe, North Africa and South-West Asia, based at the headquarters of the Food and Agriculture Organization (FAO); |
| Niamey, Niger | for West Africa, based at the ICRISAT Sahelian Center; |
| Nairobi, Kenya | for East and Southern Africa, based at the International Laboratory for Research on Animal Diseases (ILRAD); |
| New Delhi, India | for South and South-East Asia, based at National Bureau for Plant Genetic Resources (NBPGR) with a sub-office in Los Baños, the Philippines; |
| Beijing, China | for East Asia, based at the Chinese Academy of Agricultural Sciences (CAAS). |

In addition, the IBPGR has full-time plant collectors based in Cyprus, Niger and Zimbabwe. The distribution of the germplasm collected through IBPGR missions is decentralized, with Seed Handling Units at Kew (UK), and in Singapore; a third unit, serving Latin America, is due to open this year.

The functions of the IBPGR can be summarized as follows:

- service function, for the entire plant genetic resources community, including publications, public affairs and library service;
- research function, combining both strategic research in plant genetic resources and applied research to address immediate problems identified by IBPGR collaborators;
- scientific and technical support, for the development of national genetic resources programmes in the developing world, as well as information services to promote better coordination of genetic resources activities.

It is important to emphasize that the IBPGR is not a technical assistance agency; funding for institution building is not within the remit of the Board. The IBPGR places high priority on the role of the regional offices and, against this background, considers the development of a crop genetic resources network as one of the best tools for stimulating the exchange of information and improving coordination.

## IBPGR AND WHEAT GENETIC RESOURCES

Along with rice and maize, wheat is one of the three major crops in the world, both in terms of area planted and in terms of total production. Since its inception in 1974, the IBPGR has therefore given high priority to wheat.

Meetings have been held regularly to monitor the progress of wheat germplasm collecting. The first of these meetings was the IBPGR Symposium on Wheat Genetic Resources, which was held in Leningrad in 1975. Subsequently, an IBPGR Advisory

Committee on Wheat Genetic Resources, co-sponsored by CIMMYT, was established. This committee has met many times to review what was known about the material in collections and the progress of collecting missions, and to make recommendations for further collecting. A summary of wheat germplasm collected through IBPGR missions is shown in Table 13.2. Since 1976, about 6000 samples of wheat and over 2000 samples of *Aegilops* species have been collected. In 1981 a wheat description list was published (Croston and Williams, 1981); this list was revised in 1985.

For some time, the IBPGR has had a full-time staff member coordinating the work on wheat genetic resources. This has led not only to the collecting of additional wheat germplasm, but also to publications and status reports, namely: *Wheat Collecting: An Inventory 1981-1983* (Chapman, 1984), *A Survey and Strategy for Collection* (Chapman, 1985) and *Genetic Resources of Wheat* (Chapman, 1986).

**Table 13.2** Germplasm collected through IBPGR-supported missions

| Countries | *Triticum aestivum* | *T. durum* | *Triticum* spp. | Total | *Aegilops* spp. |
|---|---|---|---|---|---|
| Afghanistan | — | — | 947 | 947 | 46 |
| Algeria | 107 | 236 | 18 | 361 | 17 |
| Bolivia | 10 | 5 | 13 | 28 | — |
| Bhutan | 24 | — | — | 24 | — |
| Cyprus | — | 80 | — | 80 | 166 |
| Egypt | 11 | 37 | 56 | 104 | — |
| Ethiopia | — | 5 | 155 | 160 | — |
| Greece | 22 | 23 | 243 | 288 | 624 |
| India | 157 | — | — | 157 | — |
| Iran | — | — | 271 | 271 | 118 |
| Iraq | — | — | 17 | 17 | 25 |
| Jordan | — | 316 | — | 316 | 111 |
| Libya | 99 | 33 | — | 132 | 15 |
| Morocco | 160 | 141 | — | 301 | 21 |
| Malta | 11 | — | — | 11 | — |
| Nepal | 119 | — | 287 | 406 | — |
| Oman | 69 | — | 9 | 78 | — |
| Pakistan | 663 | 20 | 113 | 796 | 12 |
| Portugal | 125 | 3 | 68 | 196 | 227 |
| Spain | 246 | 86 | 3 | 335 | 169 |
| Sudan | 20 | — | 30 | 50 | — |
| Syria | 6 | 22 | 225 | 253 | 336 |
| Tunisia | — | 44 | 75 | 119 | 2 |
| Turkey | 52 | 126 | 233 | 411 | 387 |
| Yemen, PDR | — | — | 68 | 68 | — |

## CROP GENETIC RESOURCES NETWORKS

The crop genepool (that is, the cultivated crop variation and the variation in related wild species) is the focal point of the IBPGR's activities. To ensure the conservation and utilization of this genepool, networks of collaborating institutes and scientists are being established for each crop, both major and minor (Marshall, 1989). By encouraging the development of such networks, the IBPGR is aiming at:

- the long-term secure storage of base collections;
- the ready provision of information through databases;
- the availability of samples from active collections;
- the improved use of the crop genepool.

## PROSPECTS FOR A WHEAT GENETIC RESOURCES NETWORK

In order to identify gaps in collections, several factors need to be considered. Firstly, it is impractical to attempt to capture all diversity, and a compromise has to be reached whereby one can state, with reasonable confidence, that sufficient material is conserved. Secondly, it would be wrong to say that once a certain number of accessions has been reached, collecting should cease (unless a taxon has become extinct), because species will continue to evolve. Hence it is difficult to estimate the number of accessions which will provide representative variability, and the current emphasis should clearly be placed on collecting material which is both endangered and obviously under-represented in collections.

To define this material, we need to answer the following questions: How important is a taxon? Where is its greatest diversity? How much has been collected and from where? If a serious gap is found, action needs to taken if the taxon is still available for collection. Thus, areas where little collection activity has taken place or where erosion is considerable need to be identified as top priority collecting regions. This survey provides information for each species and will enable priorities to be more clearly defined than hitherto.

As plant breeders usually require material that is easily used, collection in the past has concentrated on collecting durum and bread wheats, somewhat to the neglect of the other groups (*see* Table 13.3 *overleaf*). The other groups contain taxa which are parental to the polyploid cultivars and need greater attention because they form a valuable genetic resource.

Originally, there were to be six base collections for wheat germplasm: N.I. Vavilov collection, USSR; United States Department of Agriculture (USDA) Collection, USA; Centro Nazionale di Ricerche (CNR), Bari, Italy; Kyoto University, Japan; CIMMYT, Mexico; and ICARDA, Syria. It was envisaged that each of these would have long-term storage facilities and that among them would be accommodated duplicate sets of most of the world's wheat germplasm. It is now proposed to widen the scope of these arrangements to a network of interested institutes and scientists dealing with all aspects

**Table 13.3** *Triticum* and *Aegilops* species in collections

| Continent | *T. aestivum* | *T. durum* | *T.* species | Total | *Ae.* species |
|---|---|---|---|---|---|
| Africa (8)[a] | 4063 | 2307 | 6337 | 12 707 | 55 |
| America(12) | 65 284 | 13 093 | 59 846 | 138 223 | 959 |
| Asia (17) | 49 589 | 22 378 | 46 975 | 118 942 | 3843 |
| Australia (1) | — | — | 22 686 | 22 686 | 811 |
| Europe (24) | 57 319 | 7229 | 163 636 | 228 184 | 10 330 |
| Total | 176 255 | 45 007 | 299 480 | 520 742 | 15 998 |

a Figures in parentheses indicate number of countries

of the conservation and use of wheat genetic resources. It should be recognized that a viable crop genetic resources network can be established and maintained only if there is full participation in the decision-making process and the goodwill of all parties concerned. The advantage of such a wheat genetic resources network are:

- improved coordination and collaborative efforts will be possible for collecting, conserving and using germplasm, leading to the avoidance of duplicated efforts and a better sharing of responsibility;
- new avenues will be opened for better identification of research priorities, and for managing and using germplasm;
- all interested parties will be able to participate, and thus will benefit from increased access to germplasm and related information;
- specific needs within the network can be identified, and the IBPGR and other centres can provide inputs or mobilize support.

It is hoped that this workshop will be the starting point of a wheat genetic resources network.

# PART 3

# Research at National Genebanks

Wheat Genetic Resources: Meeting Diverse Needs
Edited by J. P. Srivastava and A. B. Damania

Published by John Wiley & Sons

# 14

# *Ecogeographical Survey of* Aegilops *in Syria*

M. N. SANKARY

Recognizing the importance of wild relatives of wheat, systematic survey missions were carried out to explore the distribution patterns of the recognized plant associations found in the arid and very arid zones of Syria (Sankary, 1982, 1988) and those of the semi-arid and subhumid zones; the findings of the latter mission have not yet been published. The aim of these surveys was to examine the relationship between *Aegilops* species in Syria and their environment. It is hoped that the strategy described in this paper will improve collecting strategies and provide the necessary framework for establishing and maintaining *Aegilops* species bioreserves.

Each *Aegilops* genome reflects specific ecological amplitudes and potential. Evolutionary advantages are clearly reflected in polyploids derived from specific genomes. A genome not only gives substantial basis as the taxonomical unit incorporating the cytoplasm to which it belongs (Kihara, 1982), but also expresses an ecological basis for environmental tolerance. Table 14.1 (*overleaf*) shows the phytosociological differentiation in terms of plant species and cover abundance values, after protection for more than two decades at the Moslemeiah Arboretum near Aleppo.

## EU-MEDITERRANEAN SPECIES (DIPLOID, M GENOME)

### *Aegilops comosa* Sibth et Sm. (syn. *Ae. heldreichii* Holms, *Triticum comosum* (Sibth et Sm.) Richt.)

*Ae. comosa* is a rare species in Syria. It is found occasionally in the degraded forests and woodlands of the subhumid and humid zones (650-1000 mm rainfall per annum) north-west of Jabal Samaan, extending towards the Cassius mountains and the Turkish border. Its presence in Syria is rather peripheral and represents the extreme eastern distribution limit east of the Mediterranean. This may explain why Chapman's 1985

**Table 14.1** Site phytosociological differentiation in terms of plant species and cover-abundance values, 24 years after protection, at al-Moslemeiah Arboretum, 15 km north of Aleppo, Syria

| Species | Open niches | | Sheltered niches | |
|---|---|---|---|---|
| | Deeper soil | Shallower and stony soil | Southern exposure | Northern exposure |
| Nanophanerophytes | | | | |
| *Capparis spinosa* | + | 2.0 | 1.0 | + |
| Hemicryptophytes | | | | |
| *Onobrychis aurantica* | 1.0 | 2.5 | 0.0 | 0.0 |
| *Astragalus cretaceous* | + | 0.0 | 0.0 | 0.0 |
| *Eryngium creticum* | + | + | + | 0.0 |
| *Eryngium glomeratum* | 1.0 | 2.0 | + | 0.0 |
| *Phlomis orientalis* | 1.0 | 2.0 | + | 0.0 |
| Geophytes | | | | |
| *Asphodelus microcarpus* | 2.0 | + | + | 0.0 |
| *Hordeum bulbosum* | 2.0 | 1.0 | 0.5 | 1.0 |
| Therophytes | | | | |
| *Aegilops vavilovii* | 1.0 | 0.5 | 0.0 | 3.0 |
| *Aegilops triuncialis* | 1.0 | 5.0 | 2.0 | 6.0 |
| *Aegilops columnaris* | 5.0 | 1.0 | 1.0 | 4.0 |
| *Aegilops ovata* | 6.0 | 50.0 | 65.0 | 5.0 |
| *Avena barbata* | 50.0 | 0.5 | 10.0 | 30 |
| *Taeniatherum asperum* | 1.0 | + | + | + |
| *Trifolium globosum* | 1.0 | + | 0.5 | 2.0 |
| *Trifolium campestre* | 1.0 | 0.5 | 2.5 | 3.0 |
| *Trifolium stellatum* | 5.0 | 1.0 | 4.0 | 5.0 |
| *Trifolium pillulare* | 0.5 | 0.5 | 0.5 | 0.5 |
| *Senecio* spp. | 2.0 | 1.0 | 0.5 | 1.0 |
| *Helianthemum* spp. | 5.0 | 1.0 | 1.0 | 0.5 |

map of the distribution of *Ae. comosa* does not include Syria (Chapman, 1985).This species is found on both terra rosa and alluvial soils in degraded *Quercus calliprinos* or *Q. infectorea* woodlands, but more rarely on the edges of *Q. pseudoceris* forests. It generally avoids saline soils.

### *Aegilops uniaristata* Vis. (syn. *Triticum uniaristatum* (Vis.) Richt.)

Although there is no environmental barrier restricting the presence of *Ae. uniaristata* in the coastal mountains of Syria, this species has not been observed in these areas by any plant taxonomist or germplasm collector (Chapman, 1985; Kimber and Feldman, 1987).

## IRANO-TURANIAN SPECIES (DIPLOID, C GENOME)

***Aegilops caudata*** **L.**
**(syn.** ***Ae. markgrafii*** **Greuter,** ***Triticum caudatum*** **(L.) Godr. and Gren,** ***T. dichasians*** **(Zhuk.) Bowden)**

*Ae. caudata* is found sporadically in areas between 300 and 1500 m above sea level, with 400-800 mm rainfall per annum and cold or very cold winters. At lower altitudes, it is found in Maleikeiah and in the fertile north-eastern corner of the Syrian Gazirah; at higher altitudes (1000-1500 m) it is found in the Anti-Lebanon and Jabal al-Arab mountains. It occurs within both the *Pyrus syriaceto-Acer microphylletum* association and *Quercus* species woodlands.

This species grows on calcareous soils with a pH between 7.1 and 8.4. This includes both stony and alluvial terra rosa soils and terra cinnamonica (pale brown) soils. It also grows in abandoned fields and on the edges of cultivated areas. However, it has not been observed in the Syrian Badia or Jabal Abdel-Azeiz mountains or in their flood plains or saline *sebkhas*.

***Aegilops umbellulata*** **Zhuk.**
**(syn.** ***Triticum umbellulatum*** **(Zhuk.) Bowden)**

*Ae. umbellulata* generally spreads to drier sites and habitats than *Ae. caudata*. It overlaps with *Ae. caudata* in the north-eastern, north-western and south-western corners of Syria, but it is found more frequently and abundantly with *Ae. lorentii* and *Ae. ovata* on shallow, compacted or very disturbed calcareous and basaltic soils.

This is a pioneer annual species which belongs to the lower stages of succession in the degraded *Pistacia atlanticeto-Prunus microcarpetum* association (Aleppo al-Huss, Balaas Abdul-Aaziz and the Sinjar mountains) and in degraded *Quercus* species woodlands (west and north-west of Aleppo, Zabadani, Hawran and Jabal al-Arab). Between 300 and 1500 m, it also grows in marginal areas on cinnamonic soils, along roadsides, in abandoned fields and on the edges of cereal fields and vineyards.

***Aegilops triuncialis*** **L.**
**(syn.** ***Triticum triunciale*** **(L.) Raspail,** ***Ae. bushirica*** **Rozhev.)**

*Ae. triuncialis* is a weedy species with wide morphological variations. Post and Dinsmore (1932, 1933) reported two varieties, *anathera* Hausskn. and Bornm and *glabrispica* Eig. The variety *assyriaca* was collected by Witcombe (1983) from northern Syria, and is also found at Moslemeiah, north of Aleppo.

This species occurs in rainfed, disturbed and rocky areas with over 320 mm rainfall per annum, particularly in very degraded woodlands and forests, burned fields, grazed pastures and hill slopes, and sometimes on compacted soils under tree shade and along roadsides. It is common (but dominates in only a few areas) in combination with one or more *Aegilops* species at 300-1600 m elevation (north of Aleppo to the Turkish border, north of Hassakah to Jabal Sinjar and Jabal Abdul-Aziz, Hamma, Homs, Zabadani, Bloudan and Sowaida). Although *Ae. triuncialis* has also been reported from the arid zones of Kuwait, I have not observed it in open rangelands in the arid zones of either Syria or Kuwait.

## KURDISTANIAN SPECIES (DIPLOID, D GENOME)

### ***Aegilops squarrosa*** **L.** **(syn. *Triticum squarrosum* (L.) Raspail, *Triticum tauschii* Coss., *Triticum aegilops* P. Beauv.)**

*Ae. squarrosa* is a rather rare species in Syria, found above 1000 m in the Anti-Lebanon mountains. It is an accidental species in *Pyrus syriaceto-Acer microphylletum* and *Pistacia atlanticeto-Agropyron libanoticetum* associations. Post and Dinsmore (1933) reported it from the al-Masnaa area. It is also found, albeit rarely, in the flood plains and grain fields of Abu Horairah, and in the wadies and on the marly slopes in Jabal Abdul-Aziz and Sinjar.

## SYRIAN SPECIES (DIPLOID, S GENOME)

### ***Aegilops speltoides*** **Tausch** **(syn. *Ae. aucheri* Boiss., *Ae. ligustica* (Savign.) Coss., *Triticum speltoides* (Tausch). Gren. ex Richt.)**

*Ae. speltoides* is found in the wetter parts of Mediterranean-type semi-arid and subhumid areas, mainly where the annual rainfall is between 400 and 750 mm. It is not recorded in any of the 13 potential plant associations and their related plant communities in Syria's arid and very arid areas (Sankary, 1977, 1982, 1988a). Phytosociologically, it has a transitional successional status within the following associations and alliance:

- *Quercus calliprineto-Olea oleasteretum*, on terra rosa soils (Jabal Samaan, Jabal al-Zaweiah and Jabal al-Wastani);
- *Quercus aegilopeto-Celtis australietum* on alluvial terra rosa soils in plains and wadies (Jabal Samaan and Qarqania);
- *Pinus brutia* alliance on marly soils (Jisr al-Shoghor).

Geographically, this species' centre of origin is in Syria, from the wetter areas of Jabal Samaan and Jabal Afrein to Bailan and Lattakia. In the coastal areas it is found in only a few places. All sites have calcareous soils with a pH between 7.0 and 7.8. It avoids the eroded and compacted soils of the shrubless grazing lands of the *Asphodelus microcarpeto-Hordeum bulbosetum* plant community and the successionally related communities of the degraded habitats. It rarely occurs in basaltic, sandy, saline and nutrient-poor soils which have less than 600 ppm and 6 ppm of N and $P_2O_5$, respectively. It has not been observed growing in Mediterranean thickets or beneath the following competitive trees: *Quercus calliprinos*, *Q. aegilops*, *Ceratonia siliqua*, *Pinus brutia*, *P. halepensis*, *Pistacia palestina* and *P. lentiscus*. It is becoming increasingly rare on the edges of cereal fields, but is more common in areas where it has been protected from grazing or fire for 6-10 years; here it competes well with annual grasses and herbs. It is found occasionally in hard limestone pockets with uncompacted soils, and in recently cut or disturbed woodlands at 300-1000 m above sea level.

*Ae. speltoides* is considered a Fertile Crescent species with no ability to penetrate the plant associations and disturbed plant communities of Saudi Arabia, Yemen and the Oman mountains, even at altitudes above 2500 m. Similarly, it does not occur in the *maquis* and *garigues* of Jabal al-Akhdar in Libya, or in the *Pistacia atlantica* and *Pinus halepensis* associations of the Atlas mountains in North Africa.

Because it is a very palatable narrow-stemmed annual, stays green longer than short-stemmed *Aegilops* species and wild *Avena* species and has a better Ca : P ratio than these species, it is considered a valuable energy and phosphorus source in Mediterranean tree-and-shrub grazing systems.

By using systematic sampling, it has been found that *Ae. speltoides* populations are very dispersed and isolated. The total area occupied by this species in Syria is now probably less than 50 ha, 95 per cent of which is occupied by *Ae. speltoides* var. *aucheri* and the remainder by *Ae. speltoides* var. *ligustica*.

It is possible that *Ae. speltoides* is of older genetic stock than other *Sitopsis* members, such as *Ae. longissima* and *Ae. bicornis*.

Figure 14.1 (*overleaf*) illustrates the successional status of *Ae. speltoides* in relation to other *Aegilops* species.

### *Aegilops longissima* Schweinf. and Muschl. (syn. *Ae. searsii* Feld. and Kis., *Ae. sharonensis* Eig., *Triticum longissimum* (Schweinf. and Muschl.) Bowden)

Although it is not abundant, *Ae. longissima* is a more common species in Syria than either *Ae. cylindrica* or *Ae. squarrosa*. However, all three species avoid *Asphodelus microcarpeto-Poa bulbosetum*, *Poterietum spinosi* and halophytic plant communities. *Ae. longissima*, together with *Ae. speltoides*, may be dominant in sporadic sites (relatively mesic limestone rock pockets, with the nutrients and organic matter of rich terra

**Figure 14.1** Relative successional trend of *Aegilops* species at Jabal al-Zaweiah, near Mahambael, Syria (semi-humid with 650 mm rainfall per annum)

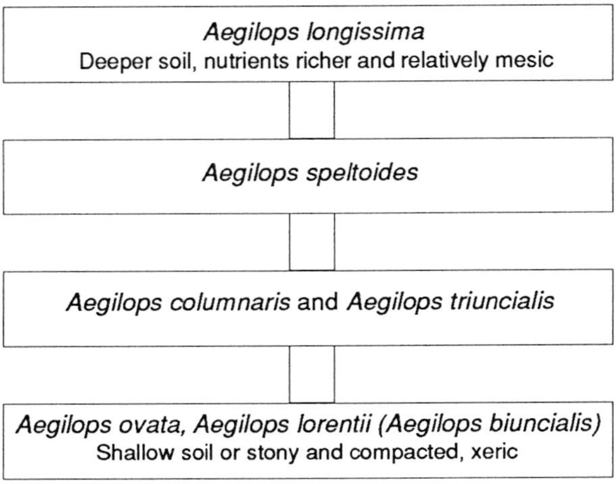


rosa soil) which have not been grazed for 15-20 years (*see* Figure 14.1). At such sites, *Ae. longissima* grows with several perennial grasses (mainly *Stipa bromoides*, *Stipa lagascae*, *Hordeum bulbosum* and *Dactylis glomerata*) and perennial herbs. Generally, *Ae. longissima* retreats with soil compaction, salinization, erosion and grazing, and it avoids heavy and gravelly soils.

In the favourable environment of Jabal al-Zaweiah, between Orem al-Gawz and al-Basqoul at 600-650 m elevation, *Ae. longissima* stays green in late spring longer than *Aegilops* species with C or CM genomes. The latter grow in deteriorated sites dominated by combinations of *Verbascum* species, *Eryngium creticum*, *Teucrium polium*, spiny *Astragalus*, *Paronychia argentia*, *P. kurdica*, *Asphodelus microcarpus*, *Poa bulbosa*, *Cichorium intybus*, *Centaurea iberica*, *Echinops* species and *Carlina racemosa*.

Contrary to the findings of Kimber and Feldman (1987), *Ae. longissima* was not observed growing on grey calcareous steppe soils or other soil types in the arid or very arid areas of Syria, although it does occur on basaltic soils at altitudes above 900 m in Jabal al-Arab. It may occur in degraded woodlands, especially in herbaceous and grassy sites, within the following plant associations:

- *Ceratonia siliqueto-Pistacia lentiscetum* on coastal plains and foothills;
- *Quercus calliprineto-Pistacia palaestinetum* on both terra rosa and basaltic soils;
- *Quercus aegilopseto-Styrax officinalietum;*
- *Pinus brutieto-Myrtus communietum.*

Regardless of their taxonomic status, *Ae. sharonensis* and *Ae. bicornis* have not yet been reported in Syria.

## SEMI-ARID MEDITERRANEAN-IRANO-TURANIAN SPECIES (TETRAPLOID, CM GENOME)

### ***Aegilops ovata*** **L.**
**(syn. *Ae. geniculata* Roth., *Triticum ovatum* (L.) Raspail)**

*Ae. ovata* is a widely distributed species in Syria's semi-arid areas (Aleppo, Moslemeiah, Qaratschock, Homs and Jabal al-Arab) and semi-humid areas (Zabadani, Bloudan, Jabal al-Arab and Tartous), with an ability to penetrate the marginal belt of arid areas (Maaloula, Aaqrabat and Jabal al-Huss) at altitudes between 280 and 1600 m. It is found mainly on compacted, eroded, skeltal or stone-mulched shallow soils, and may occur in many pioneer, disturbed and overgrazed disclimax plant communities, regardless of the soil's parent materials (hard limestone, gypsum, marl and basalt). Its fidelity to specific communities is not clear.

### ***Aegilops lorentii*** **Hochst.**
**(syn. *Ae. biuncialis* Vis., *Triticum macrochaetum*, *Ae. ovata* var. *lorentii* Boiss.)**

The distribution ecology of *Ae. lorentii* is, with a few exceptions, similar to that of *Ae. ovata*. However, *Ae. lorentii* is becoming more abundant and common than *Ae. ovata* and occurs in drier areas (Khanaser, Wadi al-Azieb, Jabal al-Balaas and Hassia). It is related phytosociologically, south and south-west of Aleppo (semi-arid, with cold winters), to the annual grass-legume patches of the *Capparis spinoseto-Verbascetum* and *Asphodelus microcarpeto-Poa bulbosetum* communities on skeltal terra rosa soils.

*Ae. lorentii* and *Ae. ovata* may be found on disturbed, man-made hills, such as Tel Mardekh and similar hills south-west of Aleppo at 450 ± 100 m elevation. Both or either species may become part of very disturbed *Capparis spinoseto-Peganum harmaletum* plant communities. *Ae. lorentii*, *Ae. ovata* and occasionally *Ae. umbellulata* are found in the Balaas intermontane flood plains (180-200 mm per annum). They grow in openings of *Pistacia atlanticeto-Rhamnus palaestinetum* communities, between dwarf shrubs (*Haloxylon articulatum* and *Artemisia herba-alba*).

Like other *Aegilops* species, *Ae. ovata* and *Ae. lorentii* usually avoid thickets, *Pinus brutia* woodlands, semi-closed woodlands or forest openings dominated by perennial grasses. On the steep marly slopes alongside the Jisr al-Shoghour-Lattakia road, *Ae. lorentii* occurs with *Teucrium polium*, *Micromeria juliana* and *Lolium* species only outside native stands or plantations of *Pinus brutia*. Similarly, on the hard limestone hillsides near Messeiaf, *Ae. lorentii* and *Ae. ovata* occur in the degraded *Asphodelus microcarpeto-Poa bulbosetum* community but are absent from the adjacent *Quercus calliprineto-Crataegus monogynetum* community. Severe competition of *Q. calliprinos*, *C. monogyna*, *Phillyrea media*, *Osyris alba* and *Sarcopoterium spinosum* apparently prevent the growth of *Aegilops* species. The same is true east of Tel-Kalakh, where no *Aegilops* species are found in the *Q. aegilopeto-Styrax officinalietum* community.

Generally, *Ae. lorentii* is common in very degraded sites lying between the 200 and 600 mm isohyets in Syria (the Badia mountains, the Anti-Lebanon mountains, Aadra, Qasseioun, Qonaitrah, Jabal al-Zaweiah, Moslemeiah and the Turkish border); it grows in all soil types except saline (terra rosa, rendzina, basalt, conglomerate, gypsum, alluvial). In the arid area (the steppe) it occurs occasionally in *Salsola azaureneto-Stipa barbatetum* communities and, to a lesser extent, in other, related communities (Sankary, 1977).

### *Aegilops columnaris* Zhuk. (syn. *Triticum columnare* (Zuhk.) Morris and Sears)

*Ae. columnaris* is uncommon in Syria. It is found mainly in the northern semi-arid sector (Sinjar, the northern part of Abdel-Aziz, Jabal Samaan and the al-Zaweiah mountains). It may occur in areas between the 250 and 500 mm isohyets, which include the wetter parts of destroyed *Pistacia atlanticeto-Prunus microcarpetum* communities and the drier parts of destroyed *Quercus calliprineto-Pistacia palaestinetum* communities, especially on cinammonic and terra rosa loose soils, but rarely on basalt.

At 470 m elevation, near Babeis village on Jabal Samaan mountain, north-west of Aleppo, *Ae. columnaris* is found with *Ae. triuncialis* on disturbed and uncompacted terra rosa soil in a mixed stand of the disclimax *Notobasis syriaceto-Avena barbatetum* plant community. All members of the community are annuals, the most important of which are: *Notobasis syriaca*, *Avena* species, *Lolium rigidum*, *Bromus* species, *Briza maxima* and *Trifolium stellatum*. In the adjacent, relatively compacted rock pockets, the perennial grasses *Dactylis glomerata*, *Hordeum bulbosum*, *Poa bulbosa* and *Stipa lagascae* are found with very sporadic and stunted relics of *Q. calliprinos*. *Ae. columnaris* has rarely been observed above 1000 m in Syria.

### *Aegilops triaristata* Willd. (syn. *Ae. neglecta* Req. ex Bertool., *Ae. recta* (Zhuk.) Chenn., *Triticum triaristatum* (Willd.) Godr. and Gren, *Ae. ovata* var. *triaristata* Gris.)

Although *Ae. triaristata* is considered a northern Syrian species, it does occur sporadically in southern Syria. Its distribution is linked more or less with the Mediterranean-type semi-arid climate (Jabal Samaan, and near Salkhad in Jabal al-Arab), where it may occur on very disturbed sites as well as moderately disturbed sites. However, it avoids forests and closed or semi-open woodlands. It is sometimes found in the watersheds of the arid northern areas between al-Khatuniya and the Iraqi border, as well as in the intermontane plains of the Anti-Lebanon mountains up to 1100 m.

This species grows well on skeltal soils (15-25 cm deep), derived from hard limestone, calcareous conglomerates, dolomites or basalts, especially in the degraded stages of two potential plant associations in Syria:

- *Pistacia atlanticeto-Prunus microcarpetum*;
- *Quercus calliprineto-Pistacia palaestinetum*.

A comparison of adjacent transect lines shows that *Ae. triaristata* plants growing in ploughed durum wheat fields form better stands, attain a greater height, gain more tillers and produce a higher yield than those growing in open rangelands.

## SAHARAN-ARABIAN SPECIES (TETRAPLOID, CS GENOME)

### ***Aegilops kotschyi* Boiss. (syn. *Ae. peregrina* (Hack.) Maire and Weill., *Ae. variabilis* Eig., *Triticum kotschyi* (Boiss.) Bowden, *T. peregrinum* Hack. and Fraser)**

*Ae. kotschyi* is a semi-arid and arid zone species which has an ability to penetrate the *Haloxylon salicorniceto-Stipagrostis plumosetum* community on sandy gypsiferous soils in the arid zone (Sankary, 1977). It penetrates well into the 110-140 mm rainfall belt. Occasionally, *Ae. kotschyi* will compete well in flood plains and depressions which contain the *Haloxylon articulateto-Hordeum glaucetum* community, where it may dominate *Hordeum glaucum*. In general, *Ae. kotschyi* is the only *Aegilops* species which will grow in the hot desert conditions of Arabia. It has been recorded in some eastern areas of Arabia, including Kuwait (Bor, 1968), and in the Najd and Nafud regions (Migahid, 1978). In wetter habitats, it may occur as an opportunistic species in the degraded stages of the following plant associations at 300-1000 m above sea level:

- *Pistacia atlanticeto-Prunus microcarpetum*;
- *Quercus calliprineto-Pistacia palaestinetum*;
- *Ceratonia siliqeto-Olea oleasteretum*.

*Ae. kotschyi* is found in diverse edaphic conditions, including terra rosa, rendzina, serpentine, calcareous cinnamonic alluvial (Sankary, 1977), calcareous sands, gypsiferous sands, sand stones and conglomerates. In a few places near Palmyra it is found on saline soils.

## NORTH AFRICAN/WEST MEDITERRANEAN SPECIES (TETRAPLOID, DM GENOME)

### ***Aegilops ventricosa* Tausch. (syn. *Triticum ventricosum* Ces., Pass. and Gib.)**

*Ae. ventricosa* has not been reported in Syria, nor in the adjacent countries east of the Mediterranean. Mouterde (1966) reported a specimen from Jaffa in Palestine. Apparently, the D genome embodied in this species enabled it to spread to the arid and semi-arid zones of North Africa.

## SYRIAN-TURCOMANIAN SPECIES (HEXAPLOID, DMS GENOME)

### *Aegilops vavilovii* (Zhuk.) Chenn.
**(syn. *Ae. crassa* Boiss. ssp. *palaestina* Eig., *Ae. crassa* ssp. *vavilovi* Zhuk., *Triticum syriacum* Bowden)**

*Ae. vavilovii* is an endemic and rare species of Roman Syria, with penetration wedges into northern Iraq and into the Sinai. It prefers a semi-arid east Mediterranean-type climate (250 to 400 mm rainfall per annum). In Syria it is found in protected sites (the Moslemeiah Arboretum, north of Aleppo) or in cereal fields (south and west of al-Qameshli).

The assumed parents of *Ae. vavilovii* (the tetraploid *Ae. crassa* and the diploid *Ae. speltoides*) may have crossed naturally into the area north-west and north-east of Aleppo, where this species originated, and spread southwards to Hauran, Syria, and then to Jordan, Palestine and the Sinai, and eastwards to west of Qameshli and northern Iraq.

In the Moslemeiah Arboretum, thi s species grows sympatrically with *Ae. columnaris*, *Ae. triuncialis*, *Ae. ovata* and *Ae. lorentii* (*see* Figure 14.2). Site eco-differentiations, expressed in terms of cover-abundance values (Sankary and Mouchantat, 1986), have been observed some 20 years after protection. Together with *Avena barbata*, *Ae. triuncialis* and *Ae. columnaris*, *Ae. vavilovii* is found in the northern lee of the Cupersus windbreak, which is cool and shady, but it appears to avoid the south-facing side, which is sunny and hotter during the growing season.

This contrasts with the site preference which has been observed for *Ae. ovata* and *Ae. lorentii*. These two species dominate the southern side of the belt, together with the xeric *Pinus brutia*. Generally, *Ae. ovata* and *Ae. lorentii* grow abundantly on sunny skeltal or compacted soils with terra rosa affinities, while *Avena barbata* dominates sunny and deeper soils, leaving the cooler and wetter sites to *Ae. vavilovii* and *Ae. columnaris*.

Phytosociologically, the distribution of *Ae. vavilovii* in Syria is linked with the distribution of the *Pistacia atlanticeto-Prunus microcarpetum* community. Usually, this species avoids *Quercus calliprineto-Pistacia palaestinetum* and *Ceratonia siliqueto-Pistacia lentiscetum* associations as well as the *Pinus brutia* alliance.

### *Aegilops juvenalis* (Thell.) Eig.
**(syn. *Ae. turcomanica* Rosh., *Triticum juvenale* Thell., *T. turcomanicum* (Rosh.) Bowden)**

*Ae. juvenalis* is a rare species in Syria. Mouterde (1966) reported a specimen from Karatchok Dag in 1955. Bor (1968) reported *Ae. juvenalis* growing in areas where there is stony ground, as well as in adobe ruins, on the sides of dry runnels and as a weed in wheat fields.

**Figure 14.2** Relative successional trend of *Aegilops* species at the Moslemeiah Arboretum, north of Aleppo (semi-arid with 350 mm rainfall per annum)

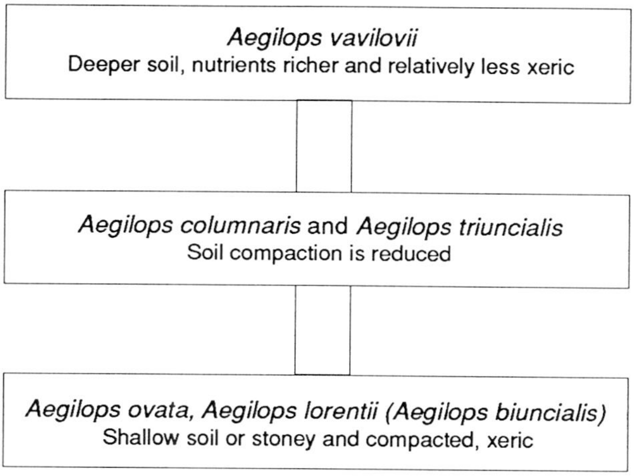


## HALOPHYTIC MESOPOTAMIAN SPECIES (TETRAPLOID AND HEXAPLOID, DM, DDM GENOMES)

### *Aegilops crassa* Boiss. (syn. *Triticum crassum* (Boiss.) Aitch. and Hemsel)

*Ae. crassa* shows a distinctive phytogeographical and ecological distribution in Syria. It does not completely overlap either with *Ae. speltoides* or with many other *Aegilops* species. It is found discontinuously through a transitional climatic belt extending from the Mediterranean-type semi-arid areas into the arid areas (160 to 300 mm precipitation per annum), where it penetrates well into Irano-Turanian territory (300 to 1000 m elevation), but never into the Saharo-Arabian territory. It grows in relatively deep, silty-loam soils and on clay-loam alluvial, calcareous (27-42 per cent $CaCO_3$) gypsiferous soils. Salinity and pH vary between 1 and 18 and 7.1 to 8.5, respectively, while soil nitrogen of the surface layer ranges between 500 and 2000 ppm.

This species is found very sporadically in flood plains (Wadi al-Azeib), in saline areas (Jaboul, Kherbet Howish, al-Haraiq) and in cultivated wheat fields (Breda, 5 km south of al-Rasafah, and in al-Gazirah, north-east Syria). Overgrazing and the cultivation of improved varieties make it a rare grass compared to *Hordeum glaucum*, *Bromus tectorum* and *Koeleria phleoides*.

Protection of 20 ha in the Wadi al-Azeib flood plains of the Syrian arid zone for 14 years yielded only three *Ae. crassa* colonies, covering about 0.2 $m^2$. However, other climax species, namely *Hordeum spontaneum* and *Avena sterilis*, were able to establish many spots covering 60 and 3.5 $m^2$ respectively. In cultivation at the same habitat, *Ae. crassa* performed better in terms of growth and production than other *Aegilops* species, but produced less forage than *Hordeum spontaneum*.

Several plant associations and communities may support the presence of *Aegilops crassa*. These are:

- *Pistacia atlanticeto-Atriplex leucocladetum*, in the mountains and wadies of the arid zone;
- *Atriplex leucocladeto-Hordeum* and its disclimax community, *Haloxylon articulateto-Hordeum glaucetum*, in the flood plains;
- *Artemisia scoparieto-Hordeum glaucetum*, in the flood plains of al-Gazirah and al-Rasafah;
- *Limonium spicateto-Sphenobus divaricatetum* in the saline areas.

## ANATOLIAN-CASPIAN SPECIES (TETRAPLOID, CD GENOME)

### ***Aegilops cylindrica*** **Host.**
### **(syn. *Triticum cylindricum* Ces., Pass. and Gib.)**

*Ae. cylindrica* is an Anatolian-Caspian species. It is found in Syria very sporadically, on the mountains and slopes of Taurus, Sinjar, Anti-Lebanon and al-Arab (Jabal al-Dorouz), at 900 to 1600 m above sea level. It is occasionally found in the north-eastern corner of Syria, where precipitation varies between 450 and 600 mm per annum. Apparently it needs wetter areas or niches than *Ae. crassa*, *Ae. kotschyi*, *Ae. vavilovii*, *A. umbellulata*, *Ae. triuncialis*, *Ae. ovata*, *Ae. lorentii*, *Ae columnaris* and *Ae. triaristata*.

In the Anti-Lebanon mountains it occurs sporadically in *Pyrus syriaceto-Acer microphylletum* and *Pistacia atlanticeto-Agropyron libanoticetum* associations, where it may be found rarely with *Ae. squarrosa* on brown or pale brown soils derived from hard lime stones or dolomites. In other locations it may be found in degraded *Quercus* woodlands, both on basaltic (near Salkhad and Shahba in Jabal al-Arab) and calcareous soils in the north-east of Syria.

## CONCLUSION

Generally, *Aegilops* species are unevenly distributed in the five major zones of Syria (Sankary, 1977). The main centre of distribution is in the north, with microcentres in Hawran and Jabal al-Arab. Al-Hammad and al-Hirar, which are very arid areas, represent the *Aegilops*-free quarter in Syria. The early civilizations in Syria both encouraged the evolution of new polyploid *Aegilops* taxa and helped the spread of *Aegilops* species from less stressed environments to the xeric habitats.

Polyploid *Aegilops* species show wider adaptation and distribution amplitudes than the related diploid *Aegilops* species, and may occupy harsher and less fertile environments. *Hordeum* species, however, show wider distribution and deeper penetration in the arid and saline areas of Syria than do *Aegilops, Triticum, Secale, Heteranthelium, Taeniatherum* and perennial *Agropyron* species.

Although *Aegilops* species show considerable variations in forage yield, they generally produce less total leaf area per plant than the average wheat variety and less forage than the Wadi al-Azeib form of *Hordeum spontaneum* found in arid and marginal areas.

## *Discussion*

L. PECETTI: During your surveys have you ever come across *Aegilops* species growing on saline soils in Syria or elsewhere?

M. N. SANKARY: Yes, there are several sites in the saline areas of Sabkhet Kherbet Howish and Al-Gaboul meadows, as well as in Al-Rasafeh (gypsiferous soils). In particular, *Aegilops crassa* and *Ae. vavilovii* (syn. *Triticum syriacum*) are found to be adapted for tolerance to saline habitats even more than wild barley. In fact, in our study on pre-adaptation, I discovered that *Ae. vavilovii* can tolerate up to 50 per cent sea water in the germination stage. And this is similar to our results with *Hordeum spontaneum*, the wild barley. We could transfer these genes to cultivated wheat rather than use alien species such as *Agropyron elongatum*, for example.

Wheat Genetic Resources: Meeting Diverse Needs
Edited by J. P. Srivastava and A. B. Damania

Published by John Wiley & Sons

# 15

# *Wheat Genetic Resources in Ethiopia and the Mediterranean Region*

P. PERRINO and E. PORCEDDU

Wheat breeding began in earnest at the beginning of this century. The high-yielding, disease-resistant varieties obtained through breeding have gradually replaced the primitive cultivars, which were characterized by greater genetic variability. Thus, while breeders have increased yield potential they have also narrowed the genetic base of cultivated material, so that most of the varieties grown today are closely related.

A high degree of genetic variation is nevertheless necessary for the breeding process itself, since superior genotypes obviously cannot be selected from homogeneous germplasm. Narrow-based local genepools can be greatly enriched by introducing exotic genes from the germplasm held in genebanks for this purpose.

Collecting activities intensified during the 1960s, when it was realized that genetic erosion was taking place at an alarming rate and that several collections, established and maintained by distinguished scholars, had either diminished in size or been lost altogether because of unfavourable weather conditions during continuous rejuvenation and lack of facilities for medium- and long-term conservation. Today, the conditions for preserving plant genetic resources *ex situ* are much improved, though most genotypes and landraces have already disappeared from the world's two most important centres of diversity, Ethiopia and the Mediterranean region. This paper describes the activities of the Germplasm Institute in Bari, Italy in collecting wheat germplasm in these areas.

## GERMPLASM COLLECTION MISSIONS

Since 1971 the Germplasm Institute of the Centro Nazionale di Ricerche (CNR) has organized 60 missions for collecting wheat and other germplasm in the Mediterranean and Ethiopian centres of diversity (Porceddu and Perrino, 1973; Porceddu and Olita, 1974; Perrino et al., 1976a, 1976b, 1981, 1982, 1984a; Hammer et al., 1986). Details of these missions are provided in Table 15.1. Most of the missions conducted outside Italy were organized in collaboration with the Food and Agriculture Organization (FAO), the International Board for Plant Genetic Resources (IBPGR) and national institutions in the countries concerned. Those carried out in Ethiopia were sponsored by the CNR. Since 1980, the International Center for Agricultural Research in the Dry Areas (ICARDA) has collaborated in the missions. Botanists from the genebank at Getersleben in East Germany (ZIGUK) joined the collecting missions in Italy and Libya. Not all the missions listed in Table 15.1 were oriented towards wheat alone; some of them also collected other threatened germplasm, including that of wild species.

**Table 15.1** Germplasm collection missions organized by the Germplasm Institute, Bari and/or in collaboration with the Food and Agricultural Organisation, International Board for Plant Genetic Resources and other national and international organizations

| **Country** | **71** | **72** | **73** | **74** | **75** | **76** | **77** | **78** | **79** | **80** | **81** | **82** | **83** | **84** | **85** | **86** | **87** | **Total** |
|---|---|---|---|---|---|---|---|---|---|---|---|---|---|---|---|---|---|---|
| Algeria | - | - | 1 | - | 2 | 2 | 1 | 1 | - | - | - | - | - | - | - | - | - | 7 |
| Egypt | - | - | - | - | - | - | - | - | 1 | 1 | 2 | 2 | - | - | - | - | - | 6 |
| France | - | - | - | - | - | - | - | - | - | - | - | - | - | - | 1 | 1 | - | 2 |
| Greece | - | - | - | - | - | - | 2 | 1 | 1 | 1 | - | - | - | - | - | - | - | 5 |
| Italy | 1 | 1 | 1 | 2 | 1 | 1 | 2 | 2 | - | 4 | 3 | 1 | 3 | 2 | 2 | 2 | 1 | 27 |
| Libya | - | - | - | - | - | - | - | - | - | - | 1 | - | 1 | - | - | - | - | 2 |
| Morocco | - | - | - | - | - | - | - | - | - | - | - | - | - | 1 | - | - | - | 1 |
| Spain | - | - | - | - | - | 1 | 1 | 1 | - | - | - | - | - | - | 1 | - | - | 4 |
| Tunisia | - | - | - | - | - | 1 | 2 | - | - | - | - | - | - | - | - | - | - | 3 |
| Ethiopia | - | - | 2 | 1 | - | - | - | - | - | - | - | - | - | - | - | - | - | 3 |
| Total | 1 | 1 | 4 | 3 | 3 | 5 | 8 | 5 | 2 | 6 | 6 | 3 | 4 | 3 | 4 | 3 | 1 | 60 |

The extent of the areas covered in each country visited varied according to the importance of the crop and the ecogeographical and climatic conditions (*see* Figure 15.1). There was no relationship between the extent of the explored area and the number of wheat fields or exact area cultivated to wheat. For instance, in the Algerian, Libyan and Tunisian deserts (Perrino et al., 1984b, 1984c) wheat is cultivated only in oases. Some of the areas in these countries were visited more than once, and for the sole purpose of collecting wheat.

The total number of wheat samples collected in all countries was 2171 (*see* Table 15.2). Of these, 1815 were collected in the Mediterranean countries, and 356 in Ethiopia. Nearly 50 per cent of the samples were collected in 2 years, 1973 (466

**Figure 15.1** Areas explored by Germplasm Institute, Bari for genetic resources conservation

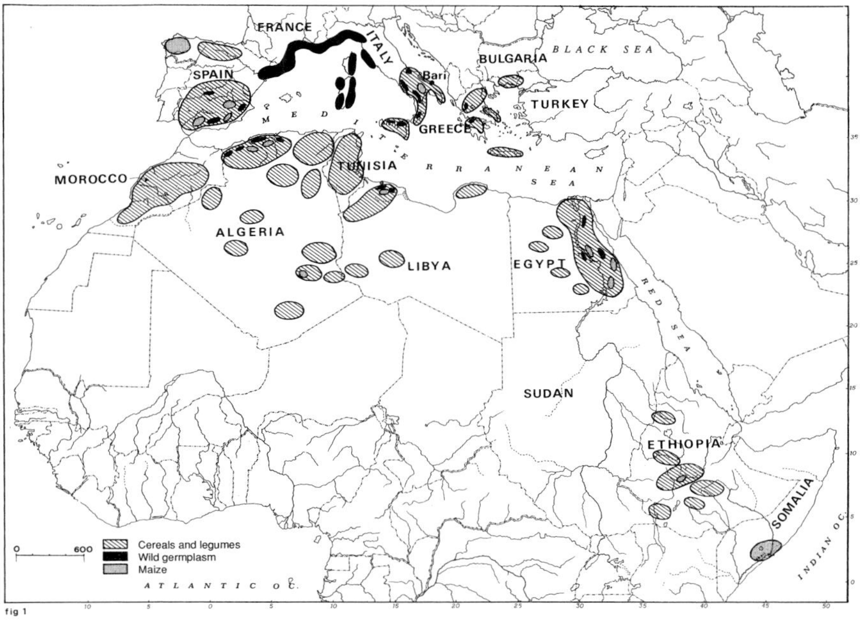


fig 1

**Table 15.2** Number of wheat accessions collected in the Mediterranean region and in Ethiopia from 1971 to 1987

| Origin | 71 | 72 | 73 | 74 | 75 | 76 | 77 | 78 | 79 | 80 | 81 | 82 | 83 | 84 | 85 | 86 | 87 | Total |
|---|---|---|---|---|---|---|---|---|---|---|---|---|---|---|---|---|---|---|
| Algeria | - | - | 70 | - | 125 | 103 | 71 | 10 | - | - | - | - | - | - | - | - | - | 379 |
| Egypt | - | - | - | - | - | - | - | 22 | - | 11 | 55 | 37 | - | - | - | - | - | 125 |
| Greece | - | - | - | - | - | - | 91 | 56 | 49 | 35 | - | - | - | - | - | - | - | 231 |
| Italy | 64 | 1 | 124 | 2 | 6 | - | - | - | - | 42 | 40 | 39 | 21 | 21 | 10 | 8 | 24 | 402 |
| Libya | - | - | - | - | - | - | - | - | - | | 32 | 3 | 61 | - | - | - | - | 96 |
| Morocco | - | - | - | - | - | - | - | - | - | - | - | - | - | 55 | - | - | - | 55 |
| Spain | - | - | - | - | - | | 347 | - | 61 | - | - | - | - | - | - | - | - | 408 |
| Tunisia | - | - | - | - | - | 44 | 75 | - | - | - | - | - | - | - | - | - | - | 119 |
| Ethiopia | - | - | 272 | 76 | - | 8 | - | - | - | - | - | - | - | - | - | - | - | 356 |
| Total | 64 | 1 | 466 | 78 | 131 | 155 | 584 | 88 | 110 | 88 | 127 | 79 | 82 | 76 | 10 | 8 | 24 | 2171 |

samples) and 1977 (584 samples). Most of those collected in 1973 came from Ethiopia (272) and Italy (124). Most of the Mediterranean samples were collected in Spain (408), Italy (402), Algeria (379), Greece (231) and Egypt (125). A high percentage of the samples from Tunisia (119) and Libya (96) came from the oases areas, while those from Morocco (55) were collected mainly in the southern part of the Atlas mountains.

## TOPOGRAPHICAL AND PEDOLOGICAL DATA

If agronomists, geneticists and breeders possess field records about the sites of collection, they are better able to predict how the material collected will perform in new environments. The topographical and pedological data recorded during the collecting missions are summarized here.

### Latitude

Samples were collected through a range of 34degrees (7-42°N) (*see* Table 15.3). The entire collection from 7° to 14°N is from Ethiopia, while that from 21° to 42°N was collected from the Mediterranean region. The area from 15°to 20°N was not explored at all. In Ethiopia, most of the samples from 7° to 8°N were hexaploid wheats; those from 9° to 14°N were mixed, but from 9° to 12°N were mainly tetraploids; those from 13° to 14°N were mainly hexaploids.

Moving north from Ethiopia towards the Mediterranean, at a latitude of 21-22°N, on the Hoggar and Tassili plateaux of Algeria, hexaploid wheats only were found. Above 22°N, mixed populations were again found to be the rule. Among the Mediterranean countries, Algeria was sampled through nearly 15degrees (21-36°N), Italy through 11 degrees (21-32°N), Libya and Egypt through 8 degrees (23-32°N), Greece through 7 degrees (23-26°N and 37-41°N), Spain through 7 degrees (35°- 42°N), Morocco through 5 degrees (29-34°N) and Tunisia through 3 degrees (33-36°N).

More than one-third of the collection was sampled between 35° and 40°N (that is, mainly from southern Spain and Italy, northern Algeria, Tunisia and Greece). To establish whether any wheat is cultivated between 0° and 7°N and between 15° and 20°N, further sampling is needed. Sampling is also needed between 11° and 22°N and between 27° and 30°N, assuming that the low number of samples was not because of the lack of wheat cultivation or the presence of modern varieties. In any case, before taking any action the passport data from Bari should be integrated with those of the IBPGR, ICARDA and United States Department of Agriculture (USDA) collections.

### Altitude

Samples were collected from sea level to 3600 m, but mostly between sea level and 1400 m (*see* Table 15.4). Nearly all the samples from Ethiopia were collected between

**Table 15.3** Distribution of wheat accessions collected (°N)

| | 7-8° | 9-10° | 11-12° | 13-14° | 21-22° | 23-24° | 25-26° | 27-28° | 29-30° | 31-32° | 33-34° | 35-36° | 37-38° | 39-40° | 41-42° | NR[a] | Total |
|---|---|---|---|---|---|---|---|---|---|---|---|---|---|---|---|---|---|
| Algeria | — | — | — | — | 21 | 56 | 4 | 2 | 4 | 2 | 43 | 177 | — | 1 | — | 69 | 379 |
| Egypt | — | — | — | — | — | 13 | 66 | 14 | 7 | 5 | — | — | — | — | — | 20 | 125 |
| Greece | — | — | — | — | — | 19 | 31 | — | — | — | — | — | 19 | 69 | 6 | 87 | 231 |
| Italy | — | — | — | — | — | — | — | — | — | 1 | — | 5 | 130 | 95 | 83 | 88 | 402 |
| Libya | — | — | — | — | — | 5 | 35 | 7 | 4 | 45 | — | — | — | — | — | — | 96 |
| Morocco | — | — | — | — | — | — | — | — | 15 | 33 | 5 | — | — | — | — | 2 | 55 |
| Spain | — | — | — | — | — | — | — | — | — | — | — | 27 | 109 | 62 | 4 | 206 | 408 |
| Tunisia | — | — | — | — | — | — | — | — | — | — | 41 | 78 | — | — | — | — | 119 |
| Ethiopia | 95 | 108 | 26 | 23 | 21 | 93 | 136 | 23 | 30 | — | — | — | — | — | — | 104 | 356 |
| Total | 95 | 108 | 26 | 23 | 21 | 93 | 136 | 23 | 30 | 86 | 89 | 287 | 258 | 227 | 93 | 576 | 2171 |

**Table 15.4** Distribution of wheat accessions collected by altitude (m above sea level)

| | 0-200 | 400 | 600 | 800 | 1000 | 1200 | 1400 | 1600 | 1800 | 2000 | 2200 | 2400 | 2600 | 2800 | 3000 | 3200 | 3400 | 3600 | NR[a] |
|---|---|---|---|---|---|---|---|---|---|---|---|---|---|---|---|---|---|---|---|
| Algeria | 67 | 24 | 37 | 51 | 112 | 20 | 47 | 5 | 6 | — | — | — | — | — | — | — | — | — | 3 |
| Egypt | 32 | 2 | — | — | — | — | — | — | — | — | — | — | — | — | — | — | — | — | 91 |
| Greece | 53 | 51 | 38 | 27 | 19 | 2 | — | — | — | — | — | — | — | — | — | — | — | | 41 |
| Italy | 24 | 48 | 67 | 85 | 50 | 13 | 9 | — | — | — | — | — | — | — | — | — | — | — | 106 |
| Libya | 11 | 22 | 55 | 6 | 1 | — | — | — | — | — | — | — | — | — | — | — | — | — | 1 |
| Morocco | 15 | 14 | 4 | 9 | 5 | 3 | 2 | — | — | — | — | — | — | — | — | — | — | — | 3 |
| Spain | 16 | 51 | 68 | 57 | 47 | 12 | 3 | — | — | — | — | — | — | — | — | — | — | — | 154 |
| Tunisia | 48 | 2 | 208 | 20 | 14 | — | — | — | — | — | — | — | — | — | — | — | — | — | 7 |
| Ethiopa | — | — | — | — | — | — | 7 | 3 | 20 | 37 | 43 | 48 | 47 | 56 | 27 | 2 | 3 | 1 | 62 |
| Total | 266 | 214 | 289 | 243 | 254 | 64 | 68 | 8 | 26 | 37 | 43 | 48 | 47 | 56 | 27 | 2 | 3 | 1 | 475 |

a No record

1800 and 3000 m, while those from below 1800 m were collected in the oases and in the Mediterranean region. Further sampling is needed from the 1000-3600 m range.

## Landform

Samples were collected from nine different kinds of landform (*see* Table 15.5). Most came from undulating land (444 samples), followed by plains (312), mountains (257) and rolling land (152). Samples were also collected from hilly land (67), flood plains (54) and dissected hilly land (35). Very few samples were taken from steeply dissected (6) and swampy (5) areas.

## Site

Although the collecting team had eight choices of site (*see* Table 15.6), in 862 cases no site was recorded. This was because of difficulties in deciding the exact position of a site. However, most samples were collected from plains (506) and valley sides (271). A fair number of samples came from depressions (155), convex slopes (138), valley bottoms (117) and terraces (83), and a few came from crests (24) and summits (15).

## Soils

Samples were collected from nine types of soil, ranging from sandy to calcareous (*see* Table 15.7 *overleaf*). Most samples were collected from clay soils (518). Some were collected from sand (162), sandy clay (153), silty clay (108), sandy silt (104), silt (94), and loam (93), and a few from calcareous (68) and highly organic (28) soils. The low number of samples from good soils is not surprising since exploration and collection usually took place in remote and marginal areas where good, fertile soils are rare.

The classification of soil for stoniness ranged from 'no stones at all' to 'virtually paved' (*see* Table 15.8 *overleaf*). As expected, most of the wheat fields had no stones (608), and thus tillage was unaffected (324). But samples were also taken from soils with affected (259) and badly affected (141) tillage. In only a few cases was the soil impossible to till (17), and in fewer still impossible to till because it was 'virtually paved' (4). Tillage was affected in Greece (41), Egypt (35), Italy (33), Tunisia (25) and Algeria (24), and badly affected in Italy (30), Algeria (29) and Greece (20).

Four categories of soil depth were used (*see* Table 15.9 *overleaf*). Most of the wheats were found on fields with a normal ploughing depth (591), with greater than normal depth (358) and with very deep soils (269); only a few samples (49) were collected from thin soils, mainly in Italy (25) and Algeria (10).

Soils were classified according to drainage conditions: excessively drained, well drained, moderately drained and inadequately drained (*see* Table 15.10 *overleaf*). In

**Table 15.5** Distribution of wheat accessions collected according to landform

| | Swamps | Steeply dissected | Hilly dissected | Flood plains | Hilly | Rolling | Mountains | Plains | Undulating | No record | Total |
|---|---|---|---|---|---|---|---|---|---|---|---|
| Algeria | 2 | — | 1 | 13 | 7 | 51 | 28 | 91 | 121 | 65 | 379 |
| Egypt | — | 1 | — | 8 | — | — | — | 28 | 2 | 86 | 125 |
| Greece | — | 2 | 6 | 9 | 17 | 29 | 48 | 7 | 20 | 93 | 231 |
| Italy | — | 3 | 26 | — | 20 | 20 | 113 | 16 | 60 | 144 | 402 |
| Libya | — | — | — | 7 | 4 | 3 | 1 | 27 | 24 | 30 | 96 |
| Morocco | — | — | — | 7 | 5 | 6 | 5 | 15 | 7 | 10 | 55 |
| Spain | — | — | 1 | — | 7 | 10 | 15 | 20 | 54 | 301 | 408 |
| Tunisia | — | — | — | 8 | — | 11 | 7 | 58 | 18 | 17 | 119 |
| Ethiopia | 3 | — | 1 | 2 | 7 | 22 | 40 | 50 | 138 | 93 | 356 |
| Total | 5 | 6 | 35 | 54 | 67 | 152 | 257 | 312 | 444 | | |

**Table 15.6** Distribution of wheat accessions collected according to site

| | Summits | Crests | Terraces | Valley bottoms | Convex slopes | Depressions | Valley sides | Plains | No record | Total |
|---|---|---|---|---|---|---|---|---|---|---|
| Algeria | 4 | 1 | 11 | 24 | 22 | 30 | 77 | 138 | 72 | 379 |
| Egypt | — | — | 1 | 19 | — | 1 | 1 | 16 | 87 | 125 |
| Greece | — | 9 | 10 | 22 | 21 | 18 | 27 | 26 | 98 | 231 |
| Italy | 3 | 4 | 40 | 21 | 52 | 33 | 60 | 38 | 151 | 402 |
| Libya | 1 | 3 | — | 3 | 3 | 20 | 2 | 34 | 30 | 96 |
| Morocco | — | 2 | — | 2 | 12 | 10 | — | 19 | 10 | 55 |
| Spain | — | 2 | 10 | 13 | 3 | 13 | 30 | 29 | 308 | 408 |
| Tunisia | — | — | 1 | — | 12 | 5 | 10 | 71 | 20 | 119 |
| Ethiopia | 7 | 3 | 10 | 13 | 13 | 25 | 64 | 135 | 86 | 356 |
| Total | 15 | 24 | 83 | 117 | 138 | 155 | 271 | 506 | 862 | 2171 |

**Table 15.7** Distribution of wheat accessions collected according to soil texture

| | Highly organic | Calcareous | Loam | Silt | Sandy silt | Silty clay | Sandy clay | Sand | Clay | No record | Total |
|---|---|---|---|---|---|---|---|---|---|---|---|
| Algeria | — | 16 | 2 | 57 | 41 | 1 | 51 | 30 | 113 | 68 | 379 |
| Egypt | — | — | 15 | 1 | 12 | 74 | 1 | 15 | 70 | 125 | |
| Greece | — | — | 2 | 25 | 22 | 17 | 23 | 26 | 8 | 108 | 231 |
| Italy | 28 | 5 | 47 | 10 | 16 | 46 | 16 | 7 | 83 | 144 | 402 |
| Libya | — | — | — | — | 1 | 7 | 3 | 35 | 1 | 49 | 96 |
| Morocco | — | — | — | — | 1 | 14 | 15 | 11 | 9 | 5 | 55 |
| Spain | — | 29 | 5 | — | 5 | 12 | 13 | — | 38 | 306 | 408 |
| Tunisia | — | 18 | — | — | — | 3 | 8 | 51 | 26 | 13 | 119 |
| Ethiopia | — | — | 22 | 1 | 6 | 1 | 20 | 1 | 225 | 80 | 356 |
| Total | 28 | 68 | 93 | 94 | 104 | 108 | 153 | 162 | 518 | 843 | 2171 |

**Table 15.8** Distribution of wheat accessions collected according to soil stoniness

| | None | Tillage unaffected | Tillage affected | Tillage difficult | Tillage impossible | Virtually paved | No record | Total |
|---|---|---|---|---|---|---|---|---|
| Algeria | 205 | 54 | 24 | 29 | 1 | — | 66 | 379 |
| Egypt | 8 | 4 | 35 | — | — | — | 78 | 125 |
| Greece | 23 | 39 | 41 | 20 | — | 2 | 106 | 231 |
| Italy | 151 | 44 | 33 | 30 | 3 | 1 | 140 | 402 |
| Libya | 27 | 38 | 5 | 5 | — | — | 21 | 96 |
| Morocco | 20 | 17 | 5 | 6 | 1 | 1 | 5 | 55 |
| Spain | 66 | 29 | 14 | — | 1 | — | 298 | 408 |
| Tunisia | 21 | 58 | 25 | 4 | — | — | 11 | 119 |
| Ethiopia | 87 | 41 | 77 | 47 | 11 | — | 93 | 356 |
| Total | 608 | 324 | 259 | 141 | 17 | 4 | 818 | 2171 |

**Table 15.9** Distribution of wheat accessions collected according to soil depth

| | <Plough depth | Very deep | >Plough depth | Normal ploughing depth | No record | Total |
|---|---|---|---|---|---|---|
| Algeria | 10 | 54 | 70 | 170 | 75 | 379 |
| Egypt | 1 | 5 | 41 | 1 | 77 | 125 |
| Greece | 5 | 42 | 23 | 50 | 111 | 231 |
| Italy | 25 | 33 | 91 | 107 | 146 | 402 |
| Libya | 2 | 24 | 5 | 44 | 21 | 96 |
| Morocco | — | 19 | 11 | 20 | 5 | 55 |
| Spain | 1 | 15 | 33 | 59 | 300 | 408 |
| Tunisia | 2 | 34 | 19 | 53 | 11 | 119 |
| Ethiopia | 3 | 43 | 65 | 87 | 158 | 356 |
| Total | 49 | 269 | 358 | 591 | 904 | 2171 |

a No reaction

**Table 15.10** Distribution of wheat accessions collected according to drainage

| | Inadequately drained | Moderately drained | Well drained | Excessively drained | No record | Total |
|---|---|---|---|---|---|---|
| Algeria | 39 | 73 | 164 | 33 | 70 | 379 |
| Egypt | — | 11 | 32 | — | 82 | 125 |
| Greece | 18 | 18 | 65 | 15 | 115 | 231 |
| Italy | 53 | 40 | 120 | 40 | 149 | 402 |
| Libya | 4 | 17 | 40 | 8 | 27 | 96 |
| Morocco | 15 | 2 | 23 | 8 | 7 | 55 |
| Spain | 1 | 5 | 87 | 7 | 308 | 408 |
| Tunisia | 9 | 11 | 76 | 11 | 12 | 119 |
| Ethiopia | 38 | 168 | 42 | 19 | 89 | 356 |
| Total | 177 | 345 | 649 | 141 | 859 | 2171 |

a No reaction

most cases (859), data on drainage were not recorded at all. From the data available, it seems that most samples were collected from well-drained soils (649), fewer from moderately drained soils (345), and fewer still from inadequately drained soils (177) and excessively drained soils (141). In nearly all countries, wheat was collected from all four different drainage conditions. This means that at any latitude and longitude between Ethiopia and the Mediterranean, breeders can select material adapted to these different degrees of drainage.

## SYSTEMATICS OF COLLECTED WHEATS

From Ethiopia to the Mediterranean region, most of the wheats were tetraploids and hexaploids and, in general, quite different in morphology and genetic composition. As regards morphology, spikes from Ethiopia, Hoggar and Tassili (in Algeria), and from the Mediterranean countries showed different shapes as well as different patterns of variation.

At species level, *Triticum durum* and *T. aestivum* was widespread (*see* Table 15.11). *T. polonicum* was found mainly in Ethiopia (Porceddu et al., 1973) and occasionally in Morocco (Perrino et al., 1986); *T. monococcum*, in southern Spain and Italy (Perrino and Hammer, 1982); *T. dicoccum* in Ethiopia, Italy (Perrino and Hammer, 1982) and northern Spain; and *T. spelta* in northern Spain and Italy.

In Italy, *T. spelta* and *T. dicoccum* have recently been reintroduced and are now cultivated in some marginal areas. This is largely because restaurants in northern Italy and France are using the flour of these two wheats to make bread and pasta.

As expected, once introduced into Italy the special forms of *T. aestivum* found in the Hoggar and Tassili and other oases proved highly susceptible to mildew and other diseases because of the higher relative humidity during the growing period.

Natural hybrids between *Aegilops ventricosa* and *Triticum durum* were found in northern Algeria and were more common there than expected.

## GENETIC VARIABILITY

### Agronomic traits

An analysis of variability (Porceddu, 1979a) for eight characters, carried out on 15 populations of wheat collected in 1972-73 from Ethiopia, Algeria and Italy (*see* Table 15.12), showed significant differences among origins for all characters except for flag leaf and culm senescence. The Ethiopian wheats had the highest frequency for early ripening, shortest-stemmed plants, and lowest length of upper internode and leaf sheath. For the same characters, the Italian material showed the highest variation and the highest values. Since the five populations from each country were selected for their high variability and, at least in the case of Ethiopia and Algeria, for the very distant sites from which they were collected, we conclude that, at least for the characters analyzed, diversity was greater for Mediterranean than for Ethiopian wheats.

### Quantitative spike characters

A comprehensive study on the geographical diversity of eight quantitative spike characters in a world collection consisting of more than 3000 entries of durum wheat (Spagnoletti-Zeuli and Qualset, 1987) showed that in general the diversity of the

**Table 15.11** Distribution of *Triticum* species collected in the Mediterranean region and Ethiopia

| | *T. monococcum* | *T. dicoccum* | *T. durum* | *T. turgidum* | *T. polonicum* | *T. aestivum* | *T. spelta* |
|---|---|---|---|---|---|---|---|
| Algeria | | | + | + | | + | |
| Egypt | | | + | + | | + | |
| Greece | | | + | + | | + | |
| Italy | + | + | + | + | | + | + |
| Libya | | | + | + | | + | |
| Morocco | | | + | + | + | + | |
| Spain | + | + | + | + | | + | + |
| Tunisia | | | + | + | | + | |
| Ethiopia | | + | + | + | + | + | |

**Table 15.12** Variability for eight agronomic characters observed on 15 selected wheat populations from three different origins [a]

| Characters | Ethiopia [b] | | | Algeria [b] | | | Italy [b] | | | Significance among origins |
|---|---|---|---|---|---|---|---|---|---|---|
| | min | max | diff | min | max | diff | min | max | diff | |
| Heading time in days from May | 10 | 18 | 8 [f] | 19 | 23 | 4 [e] | 8 | 24 | 16 [e] | f |
| Flag leaf senesence from May 1st | 34 | 38 | 5 [e] | 33 | 41 | 8 [e] | 33 | 37 | 4 [f] | n.s |
| Culm senesence from May 1st | 13 | 16 | 3 [f] | 14 | 17 | 3 [f] | 13 | 17 | 4 [f] | n.s. |
| Plant height (cm) [c] | 73 | 80 | 7 [e] | 92 | 114 | 22 [e] | 86 | 121 | 35 [e] | e |
| Upper internode length (cm) [c] | 34 | 37 | 3 [f] | 43 | 49 | 6 [e] | 43 | 55 | 12 [e] | e |
| Leaf sheath length (cm) [c] | 18 | 20 | 2 [f] | 23 | 25 | 2n.s. | 21 | 27 | 6 [f] | e |
| Flag leaf length (cm) [d] | 17 | 20 | 3 [f] | 20 | 21 | 1n.s. | 17 | 24 | 7 [e] | f |
| Flag leaf breadth (cm) [d] | 10 | 14 | 4 [f] | 10 | 14 | 4 [f] | 16 | 18 | 2 [f] | e |

a Five populations from Ethiopia, five from Algeria and five from Italy (Sicily); b Each value is referred to the population mean value, that is, the average of 500 plants (five plants selected at random from each progeny for 100 progenies present in each population); c At maturity; d at 30 days after heading; e $P = 0.05$; f $P = 0.01$

material from the Mediterranean was as great as that from West Asia (*see* Figure 15.2). The wheats from the two subregions were very different from each other. Wheats from France clustered with USSR and USA materials; those from Ethiopia clustered with those from India, while the Afghanistan material formed a very distinct group on its own. The Ethiopian wheats were more similar to those of West Asia than were the Mediterranean ones. In other words, at least for the characters considered in this study, the Mediterranean wheats have diversified from those in the primary centre of diversity more than the Ethiopian ones have.

## Resistance to leaf and stem rusts

Studies of resistance to stem and leaf rust were carried out in collaboration with the Agriculture Canada Research Station, Winnipeg, Manitoba (Jedel et al., 1988. In 1985, a total of 494 accessions of tetraploid and hexaploid wheats from Bari were planted in the rust nursery near Winnipeg. Spreader rows were inoculated with a composite of leaf rust races prevalent in western Canada, and a mixture of 10 stem rust races.

Most of the tetraploid accessions from all countries were highly resistant to leaf rust, while most of the hexaploid accessions from Algeria and Egypt were susceptible or segregating for leaf rust reaction (*see* Table 15.13 *overleaf*). Sources of resistance to leaf rust were found in the accessions from Ethiopia, Greece, Italy and Spain. Resistance to stem rust was generally greater in tetraploid than in hexaploid accessions. The hexaploid accessions from Egypt all showed resistance to stem rust, while those from Algeria showed very little resistance. Sources of resistance to stem rust were also found in accessions from Ethiopia, Greece, Italy and Spain. Only 15 accessions were found to be resistant to all races of leaf rust, and only 10 to all races of stem rust.

## Resistance to rusts and powdery mildew in primitive wheats

The aim of research carried out with the Cereal and Plant Pathology Institutes in Rome (Pasquini et al., in press) was to identify genotypes resistant to Italian isolates of leaf rust, stem rust and powdery mildew in a collection of *T. monococcum* and *T. dicoccum* accessions from different ecogeographical areas.

The results show that *T. monococcum* may be a good source of resistance to rusts and powdery mildew (*see* Table 15.14 *overleaf*). Almost all the accessions tested were resistant to biotypes of leaf rust and a good number showed low infection when inoculated with the isolates of powdery mildew and stem rust. Less resistance to rust and powdery mildew was observed in *T. dicoccum* than in *T. monococcum*. Nevertheless, many of the *T. dicoccum* accessions tested proved to be resistant or moderately resistant to one or more mildew biotypes. Many *T. dicoccum* accessions showed moderate susceptibility to leaf rust.

**Figure 15.2** Cluster analysis performed for eight quantitative spike characters: spike length, awn length, number of nodes per spike, spike internode length, awn/spike ratio, kernel number/spike, kernel weight/spike and weight/kernel

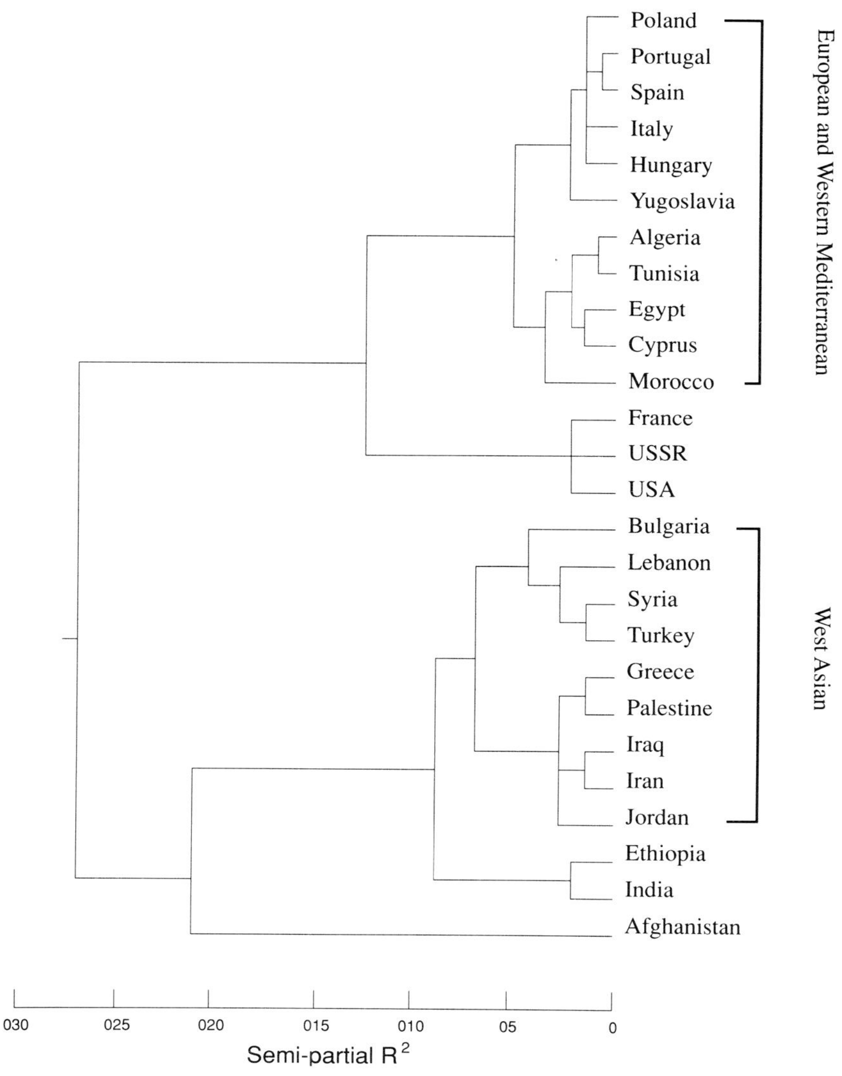

**Table 15.13** Adult plant resistance in a field nursery to leaf rust and stem rust in *Triticum* accessions in the Germplasm Institute, Bari

| Country of origin | Ploidy level | Leaf rust No. of accessions | | | Stem rust No. of accessions | | |
|---|---|---|---|---|---|---|---|
| | | R[a] | M | S | R | M | S |
| Algeria | Hesistant | 3 | 2 | 39 | 1 | 1 | 41 |
| | Composite | 1 | 0 | 2 | 2 | 0 | 3 |
| | Tetraploids | 20 | 0 | 2 | 11 | 1 | 8 |
| Egypt | Hesistant | 2 | 5 | 5 | 10 | 0 | 0 |
| | Composite | 2 | 0 | 0 | 4 | 0 | 0 |
| | Tetraploids | 12 | 0 | 0 | 5 | 1 | 4 |
| Ethiopia | Hesistant | 12 | 2 | 14 | 9 | 4 | 12 |
| | Composite | 3 | 0 | 2 | 2 | 1 | 5 |
| | Tetraploids | 24 | 1 | 1 | 13 | 3 | 8 |
| Greece | Hesistant | 13 | 9 | 17 | 16 | 2 | 15 |
| | Composite | 1 | 2 | 1 | 7 | 0 | 4 |
| | Tetraploids | 43 | 3 | 1 | 29 | 1 | 8 |
| Italy | Hesistant | 9 | 12 | 16 | 5 | 12 | 5 |
| | Composite | 0 | 1 | 1 | 7 | 1 | 1 |
| | Tetraploids | 57 | 1 | 0 | 26 | 18 | 10 |
| Spain | Hesistant | 64 | 26 | 96 | 44 | 2 | 137 |
| | Composite | 12 | 1 | 3 | 6 | 0 | 11 |
| | Tetraploids | 31 | 1 | 3 | 25 | 1 | 5 |
| Tunisia | Hesistant | 1 | 0 | 0 | 1 | 0 | 0 |
| | Composite | 0 | 0 | 0 | 0 | 0 | 0 |
| | Tetraploids | 1 | 0 | 0 | 0 | 0 | 1 |

a Reactions: R = resistant; M = mixture of susceptible and resistant plants; S = susceptible
Source: Jedel et al. (1988)

## Storage proteins

It has been established that pasta- or bread-making quality is related to the nature of gliadins and glutenins. Analysis of durum wheat accessions collected in several Mediterranean countries revealed the presence of additional variants for these proteins. Thus, the limited genetic diversity of commercial varieties can be broadened by using the variability found in primitive cultivars and landraces.

New allelic variants were also detected for the high-molecular-weight (HMW) glutenin subunits, which play a major role in determining bread-making quality.

Extensive electrophoretical analyses have revealed the occurrence, in bread and durum wheats, of genotypes lacking some groups of gliadins or glutenins (Lafiandra

**Table 15.14** Resistance of two primitive wheats to mildew and rust

| | N[a] | R-S[b] | Mildew | | | Leaf rust | | | Stem rust | | |
|---|---|---|---|---|---|---|---|---|---|---|---|
| | | | 1B[c] | 2B | 3B | 1B | 2B | 3B | 1B | 2B | 3B |
| *Triticum monococcum* | 12 | R | 2 | 2 | 2 | 1 | 2 | 9 | 0 | 1 | 7 |
| | | MR | 3 | 3 | 1 | 0 | 0 | 0 | 0 | 0 | 0 |
| | | MS | 4 | 2 | 0 | 0 | 0 | 0 | 1 | 0 | 0 |
| | | S | 0 | 1 | 0 | 0 | 0 | 0 | 1 | 1 | 2 |
| *T. dicoccum* | 33 | R | 10 | 4 | 0 | 2 | 1 | 0 | 4 | 1 | 1 |
| | | MR | 10 | 7 | 0 | 4 | 0 | 0 | 1 | 1 | 0 |
| | | MS | 8 | 10 | 2 | 3 | 8 | 4 | 7 | 1 | 0 |
| | | S | 4 | 2 | 4 | 11 | 5 | 8 | 1 | 4 | 23 |

a No. of accessions; b R = resistant; MR = medium resistant; MS = medium susceptible; S = susceptible; c B = biotype
Source: Modified from Pasquini et al. (in press)

et al., 1985, 1988). These mutations, easily detected by protein electrophoresis, may affect the processing characteristics of flour and semolina as well as their nutritional value (for example, amino acid composition). Further studies are therefore needed.

## GENETIC EROSION

Considerable genetic erosion had taken place in Sicily (Perrino and Martignano, 1973; Perrino and Hammer, 1983), where only 21 traditional varieties out of 45 recognized by De Cillis (1942) were found. Thus, over 30 years about half traditional varieties had been lost. Similar erosion was found in Cyprus and Greece (Papadakis, 1929; Porceddu, 1979b), Algeria and Tunisia (Porceddu and Olita, 1973; Perrino et al., 1976a, 1976b), Libya, Egypt and Spain (Al Alazzeh et al., 1982, Perrino et al., 1984c). In the North African countries in general, and especially in Morocco (Perrino et al., 1986), more severe genetic erosion was noted than in Italy, Spain and Greece. The greatest variability was found in mountainous areas of Algeria and Sicily.

Of more than 400 wheat varieties cultivated in Italy after the first World War (De Cillis, 1927) only about 1 or 2 per cent have been collected over the past 20 years. The number of new or relatively new wheat varieties cultivated in Italy today is not only lower (about 150) but intrinsically less diverse.

## DISTRIBUTION OF WHEAT GERMPLASM

Soon after it was founded, the Germplasm Institute started to distribute duplicated samples from both its own and the USDA's collections to institutions operating in at least 35 countries (*see* Table 15.15). In all, over 45 000 samples have been distributed to date, mostly to breeding programmes in Europe (21 815) and Asia (16 807), but also

**Table 15.15** Wheat germplasm distributed throughout the world by the Germplasm Institute, Bari since 1971

| | | | |
|---|---|---|---|
| France | 7 | | |
| Czechoslovakia | 11 | | |
| Poland | 20 | | |
| West Germany | 25 | | |
| East Germany | 191 | | |
| England | 246 | | |
| USSR | 371 | | |
| Bulgaria | 518 | | |
| Portugal | 733 | | |
| Greece | 880 | | |
| Spain | 1064 | | |
| Sweden | 1367 | | |
| Italy | 16 382 | Europe | 21 815 |
| China | 49 | | |
| Japan | 50 | | |
| Lebanon | 50 | | |
| Israel | 57 | | |
| India | 705 | | |
| Turkey | 6 691 | | |
| Syria | 9205 | Asia | 16 807 |
| Nigeria | 50 | | |
| Morocco | 55 | | |
| Libya | 94 | | |
| Tunisia | 117 | | |
| Egypt | 205 | | |
| Ethiopia | 349 | | |
| Algeria | 891 | Africa | 1761 |
| Canada | 673 | | |
| Mexico | 709 | | |
| USA | 2582 | North America | 3964 |
| Venezuela | 8 | | |
| Argentina | 10 | | |
| Bolivia | 873 | | |
| Australia | 291 | Australia | 291 |
| Total | 45 529 | | 45 529 |

to programmes in North America (3964), Africa (1761), South America (891) and Australia (291). By stimulating breeders to enlarge the genetic base of cultivars or increase the number of cultivars available, distributing germplasm in this way may, in the long run, create new centres of diversity. It seems that this has already happened in barley (Peeters, 1988). Thus plant breeding has an opportunity to refute the accusation that it leads to genetic erosion.

## CONCLUSION

Sixty collecting missions in the Mediterranean and Ethiopian centres of genetic diversity for wheat have yielded 2171 populations which, distributed throughout the world, have stimulated wheat breeders to incorporate new sources of variability for agronomic traits, resistance to diseases, and protein, nutritional and technological qualities into new wheat lines and varieties.

The field data recorded during collection have not been used extensively for selecting samples to distribute to breeders, but they could be used for studies of evolution or co-evolution with wild species and of the ecogeographical distribution of important traits.

The characterization and evaluation of wheat germplasm needs the involvement of breeders, pathologists, geneticists and statisticians to obtain good data. Germplasm can be fully used only if large collections are screened in different environments. To date, no other cheaper and more efficient methods of evaluation have been developed.

The distribution of new varieties increases genetic erosion, but in some cases, because of the effect of climatic conditions and infestation by new biotypes or races of parasites on new varieties, genetic resources will be needed for the transfer of genes from wild species to cultivated wheats. Collecting is a slow process, but it will continue to play an important role in plant breeding activities.

## *Discussion*

J. P. Srivastava: I would like to have your views regarding loss of genetic variability. Is it real or imaginary?

P. Perrino: It is very real if you compare the number of varieties cultivated today with those grown in the past in countries such as Greece and Italy or those in the centres of diversity.

S. Jana: Why should there be less variability within populations in more recently collected landraces than those that were collected many years ago?

P. Perrino: Because today more varieties are found by collectors in place of 'true landraces' than in the past.

S. JANA: Several years ago I did an experiment on this issue of loss of genetic diversity. I took Suneson's composite crosses, particularly the one which involved the US world collection, some landraces from the Middle East, ICARDA's preliminary observation nurseries, and co-operative trials from Canada, which includes advanced material from the breeders before they are considered for release. I ran standard statistical diversity analysis and got expected results. Suneson's crosses had the highest amount of diversity, followed by the landraces from the Middle East, then the preliminary observation nurseries and finally the co-op trials from Canada, which had the lowest phenotypic diversity for a range of characters. We also compared these preliminary observation nurseries each year from 1978 to 1984 and found that phenotypic diversity increased slightly at the end of six years.

P. PERRINO: Yes, but in Italy even 20 years ago it was futile to collect barley landraces because there was no variability left as a result of genetic erosion. If you had compared true landrace diversity from a really variable population sample, your results would have been different.

S. JANA: Are you referring to variability among landraces of barley or within a landrace? I believe it would make no difference if a landrace was 20 or even 200 years old if one is considering variability within landraces. But if you are considering the differences among landraces that is a different matter. I suspect that genebank curators are more interested in the latter type of landrace variability. I studied within-sample variability with expected results. I do not see any contradictions.

A. DAALOUL: I would think that in the selected germplasm nurseries there is considerable genetic diversity, but if we look at the landrace diversity utilized in the breeding programmes at present, I would agree with Dr Perrino that it is far less than the diversity existing in bygone years. The genebase of released varieties is very narrow in the countries of this region.

M. OBANNI: I would also agree with Dr Perrino. Take the case of Morocco. The durum wheat collection which was assembled in the 1920s has been lost. We have made other collections to make up for this loss but I am sure that the diversity is much less than that which existed earlier. In 1965 a French plant breeder reported finding two durum landraces BD027 and BD01554 and *Triticum timooheevi* in Morocco which were resistant to the Hessian fly. Unfortunately, today we cannot find this material.

J. P. SRIVASTAVA: I appreciate this stimulating discussion. But we also have to look at the economic realities of life. We cannot tell the farmers to continue to grow their old varieties and ignore the newly introduced higher yielding material. My objective in raising this issue was that we should also not go to the other extreme and destroy all genetic variability in our cultivated crops and go for a single dominant variety. We have to develop new varieties targetted to different ecological zones in a country. As long as the national programmes realise this and continue to utilize diverse material for varying environments, there is little danger of genetic erosion but we just cannot put the clock back and continue to rely on obsolete landraces which may no longer exist.

Wheat Genetic Resources: Meeting Diverse Needs
Edited by J. P. Srivastava and A. B. Damania

Published by John Wiley & Sons

# 16

# *Evaluation and Utilization of Ethiopian Wheat Germplasm*

H. MEKBIB and G. HAILE MARIAM

In terms of hectarage and production, wheat is Ethiopia's fifth most important crop, after teff, barley, maize and sorghum. It covers 13 per cent of the 6 million ha or so under major crops (cereals, pulses and oil crops) in the country, with a total production of about 774 000 tonnes (CSA, 1987). Some two-thirds of the wheat grown in Ethiopia consists of tetraploid forms and one-third of hexaploid forms (Tessema, 1987). Germplasm evaluation data indicate that more than 70 per cent of the wheat collection consists of tetraploid wheats, mainly *durum* and *aethiopicum* convarieties and some intermediate forms of these; the rest consists of traditional cultivars of hexaploid wheat, mainly *T. aestivum* conv. *vulgare* (*see* Table 16.1 *overleaf*). Both convarieties are landrace populations which have evolved under local conditions in farmers' fields from time immemorial (Worede, 1974), and they show a wealth of variability.

In this paper we describe the agromorphological characteristics of Ethiopian wheat germplasm and their importance for wheat breeding.

## USES OF WHEAT IN ETHIOPIA

In urban areas and in large villages, *T. aestivum* is used for making common bread, and the two major tetraploids, *T. turgidum* conv. *durum* and *aethiopicum*, for macaroni/spaghetti and pastries. In both rural and urban areas, these two main tetraploids are used, either separately or mixed with bread wheat, to make home-made local breads (*dabo*, *ambasha*, *kitta*, *dabo-kolo* and, whenever teff is not available, *injera*), boiled or roasted whole grains (*nifro*, *kolo*), crushed grains cooked in milk/water, with butter and spices added (*kinche*), and porridge (*genfo*).

**Table 16.1** Wheat germplasm grouped taxonomically and in terms of maturity duration

| *Triticum* species | Kernel colour | | | | | | | | | Total | % |
|---|---|---|---|---|---|---|---|---|---|---|---|
| | White to yellow | | | Red to brown | | | Purple to black | | | | |
| *T. durum* | 89 | 238 | 9 | 44 | 93 | 8 | 78 | 104 | 6 | 670 | 58.2 |
| *T. turgidum* | 17 | 107 | 1 | 5 | 16 | 1 | 3 | 17 | 1 | 168 | 14.6 |
| *T. dicoccum* | — | — | — | 33 | 24 | 3 | — | — | — | 60 | 5.3 |
| *T. polonicum* | — | 11 | 1 | — | — | — | — | — | — | 12 | 1 |
| *T. pyramidale* | — | 1 | — | — | — | — | — | — | — | 1 | 0.1 |
| *T. vulgare* | 52 | 123 | 7 | 25 | 26 | 4 | — | — | — | 237 | 20.6 |
| *T. compactum* | 2 | — | — | — | — | — | — | — | — | 2 | 0.2 |
| Total | 160 | 480 | 18 | 107 | 159 | 16 | 82 | 121 | 7 | 1150 | |

*T. turgidum* ssp. *dicoccon* is the most important minor wheat in Ethiopia. Its flour is mixed with boiling water and butter to make a soup (*aja*), consumed mostly by women during pregnancy and nursing. It is also consumed as *kinche* and sometimes as *genfo* and *kitta*.

Farmers occasionally cultivate *T. turgidum* conv. *polonicum* in pure stands (Habtemariam and Mekbib, 1988) for *kitta*, *genfo* and *kolo*.

## GENETIC DIVERSITY IN ETHIOPIAN WHEAT

Wheat in Ethiopia is grown in the plateau areas at different elevations, but mainly between 2100 and 2900 m (Demissie, 1986). The bulk of the collection at the Plant Genetic Resources Centre/Ethiopia (PGRC/E) consists of the main species and subspecies described by Vavilov (1929), Ciferri and Giglioli (1939) and more recently by Sakamoto and Fukui (1972) and Porceddu et al. (1973). These are: *T. turgidum* ssp. *turgidum* conv. *durum*; *T. turgidum* ssp. *turgidum* conv. *aethiopicum*; *T. turgidum* ssp. *turgidum* conv. *polonicum*; *T. turgidum* ssp. *dicoccon*; *T. aestivum* ssp. *vulgare*; and *T. aestivum* ssp. *compactum*.

Intermediate forms not easily identifiable are often encountered. During the multiplication, characterization and preliminary evaluation work of the past 7 years, it has been noted that accessions collected in the same locality are a mixture of two, three or sometimes more species with different botanical forms. In some farmers' fields, 8-10 different botanical forms were observed in a single sampling.

The greatest variation is in the two major convarieties *durum* and *aethiopicum*. This led Vavilov (1951) to list tetraploid wheats among the other crops with Ethiopia as their centre of diversity. *T. turgidum* ssp. *dicoccon* and conv. *polonicum* are poorer in variation, although particular agromorphological characteristics still differentiate them from those of other countries.

The tetraploid wheats have been grown long enough in the Ethiopian highlands to have evolved diversified forms and unique characteristics not known elsewhere (Harlan, 1974). This considerable diversity makes them interesting from both a taxonomic and a breeding point of view.

The hexaploid wheats are, by contrast, of more recent introduction. They were probably introduced by Portuguese explorers in the 18th century (Ciferri and Giglioli, 1939), and are of more limited diversity.

According to Nastasi (1964), Ethiopian wheat types are believed to have changed little in the past few centuries. However, the traditional cultivars have been almost completely replaced by improved varieties in some of the more developed areas of Arsi and Shewa provinces (Worede, 1983). Fortunately, the danger of genetic erosion was considerably reduced by the establishment of the PGRC/E in 1976, which has so far collected 5017 accessions for conservation and utilization. A total of 10 825 accessions (including donations, selections and repatriated materials) has been assembled over the years from a wide range of areas in Ethiopia (*see* Figure 16.1).

**Figure 16.1** Altitudinal ranges and frequency of occurrence of wheat germplasm collection in Ethiopia

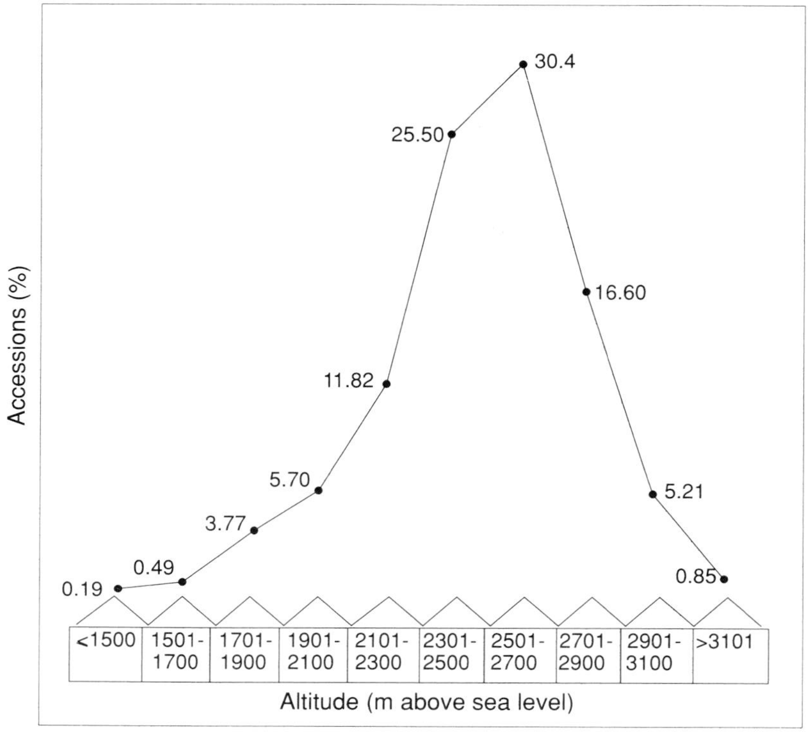

## CHARACTERISTICS OF ETHIOPIAN WHEATS

The majority of the crop germplasm accessions characterized and multiplied so far are landraces. These populations can pose difficulties in characterization and evaluation. Genetically homogeneous accessions of self- pollinating crops are easy to handle, whereas heterogeneous accessions cause complications for several reasons. First, it is difficult to record variable characters adequately. Second, it is not easy to regenerate variable populations without losing rare types. Last, but not least, these problems slow down germplasm utilization. To overcome these complications as far as possible, the PGRC/E has started to separate heterogeneous accessions of self-pollinating crops, such as wheat and barley, into agromorphological categories. Each category has some obvious and important agronomic and taxonomic characters in common, although individuals may vary in other characters. Although it makes the task of evaluation easier, splitting these populations into different categories is not recommended. To overcome the difficulties of handling a large number of accessions, the PGRC/E is considering developing a scoring system that allows accurate data recording and processing for highly heterogeneous populations.

The characters of greatest interest to the plant breeders (yield, seed quality, environmental stress tolerance, disease and pest reactions) are generally less heritable and/or their determination needs special experimental designs or sophisticated equipment. The task of recording such characters is regarded as further, rather than initial, evaluation and is generally considered to be the main responsibility of plant breeders. However, the PGRC/E has initiated several further evaluation activities, in close cooperation with the breeders concerned.

A good example is the ETHIO-Swedish Academy for Research Cooperation with Developing Countries (SAREC) project. By 1988, more than 60 per cent of the total 5017 PGRC/E collection of wheat accessions had been characterized. From the observations made during characterization and initial evaluation, Ethiopian wheats are mostly early types with small vegetative and floral organs (*see* Tables 16.2 and 16.3).

Apart from quantitative characters influenced by growing conditions (such as plant height and days to maturity), Ethiopian wheats possess characteristics which distinguish them from those of other countries. This has already been indicated by several scientists, including Percival (1927), Vavilov (1929), Ciferri and Giglioli (1939), and Harlan (1969). The most important of these characteristics is the exceptional diversity of forms existing in the two major tetraploid wheats, *T. turgidum* conv. *durum* and *aethiopicum*. Among other typical characters of the Ethiopian tetraploid wheats are the presence of varieties with purple kernels, anthocyanin pigmentation in the vegetative organs of the young plant, leaf pubescence and awnless forms in the convariety *durum*. Remarkable also is the fact that the morphological characters of the Ethiopian river or cone wheats differ considerably from those of the Mediterranean types.

The Ethiopian hexaploid wheats are, as previously mentioned, of more recent introduction. They are of more limited diversity, and taxonomically not as interesting as the main tetraploid wheats. Most of the types known are *T. aestivum* conv. *vulgare*;

**Table 16.2** Characterization results of Ethiopian wheat germplasm accessions for some quantitative characters

| Character | No. of accessions | Min | Max | Mean | SD | CV |
|---|---|---|---|---|---|---|
| Days to 50% flowering | 5886 | 41 | 120 | 77.6 | 10.6 | 13.7 |
| Plant height in cm | 5880 | 15 | 195 | 95.9 | 19.4 | 20.2 |
| Days to maturity | 5874 | 78 | 198 | 117 | 14.7 | 12.5 |
| Number of spikelets per spike | 5702 | 2.6 | 40 | 18.2 | 3.8 | 21 |
| Number of kernels per ear | 5553 | 12 | 78 | 29.5 | 7.9 | 26.8 |

**Table 16.3** Characterization results of Ethiopian wheat germplasm accessions for some qualitative characters

| Character | State | Frequency | Percent | Cumulative frequency |
|---|---|---|---|---|
| Awnedness | Awned | 4053 | 94 | 4053 |
| | Awnless | 246 | 5.7 | 4299 |
| | Mixed | 12 | 0.3 | 4311 |
| Spike density | Lax | 1208 | 28.63 | 1208 |
| | Intermediate | 2284 | 54.13 | 3492 |
| | Dense | 727 | 17.23 | 4219 |
| Glume colour | White to yellow | 3191 | 75.63 | 3191 |
| | Red to brown | 1003 | 23.77 | 4194 |
| | Purple to black | 25 | 0.59 | 4219 |
| Glume hairiness | Absent | 3982 | 81.7 | 3982 |
| | WW | 126 | 2.9 | 4108 |
| | High | 236 | 5.4 | 4344 |
| Vitreousness | Soft and starchy | 877 | 20.91 | 877 |
| | Partly vitreous | 327 | 7.79 | 1204 |
| | Vitreous | 2989 | 71.28 | 4193 |
| Size of kernels | < 6 mm | 1912 | 46.87 | 3876 |
| | 6-8 mm | 1964 | 46.87 | 3876 |
| | > 8 mm | 314 | 7.44 | 4190 |

the others are *T. aestivum* conv. *compactum* and some intermediate forms. In both cases, anthocyanin pigmentation is absent. According to Ciferri and Giglioli (1939), the earliest introductions of bread wheat varieties are similar to the Indian types, while Ethiopian club wheats do not differ from the Eurasian forms.

## UTILIZATION OF WHEAT GERMPLASM

The genebank is responsible for providing the germplasm material needed by plant scientists to develop improved crop varieties. The PGRC/E has done its best to bring its germplasm activities to the attention of breeders and actively involve them in germplasm screening and utilization.

During the course of the PGRC/E's development, the priority has shifted slowly from collection and conservation to evaluation and utilization. This shift has reflected the growing interest of crop breeders in the available indigenous landraces, as these provide the basic source of genetic variation and breeding material. Several PGRC/E wheat lines have already been incorporated into the crop improvement programmes. The ETHIO-SAREC project has used indigenous landraces to improve durum wheat. In this approach, it is assumed that a genetic line is resistant to at least some of the pathogens or pests prevailing in its particular area (Tessema, 1986). When the seeds of identical lines from different areas, each possessing different genes for resistance, are combined, the resulting mixture will provide partial protection against a broad spectrum of pathogens or pests. This approach also takes various desirable agronomic characters into consideration.

The PGRC/E, in collaboration with the Scientific Phytopathological Laboratory (SPL) at Ambo, has evaluated 502 accessions for resistance to stem, stripe and leaf rust (*see* Figure 16.2). Resistance to various rusts has also been observed in about 510 wheat lines in work conducted by the Ministry of Agriculture.

PGRC/E is participating in the durum wheat germplasm evaluation and documentation project being undertaken by the International Center for Agricultural Research in the Dry Areas (ICARDA). During the 1988 dry season, 200 accessions of durum

**Figure 16.2** Resistance found in 502 PGRC wheat accessions in cooperation with SPL-Ambo

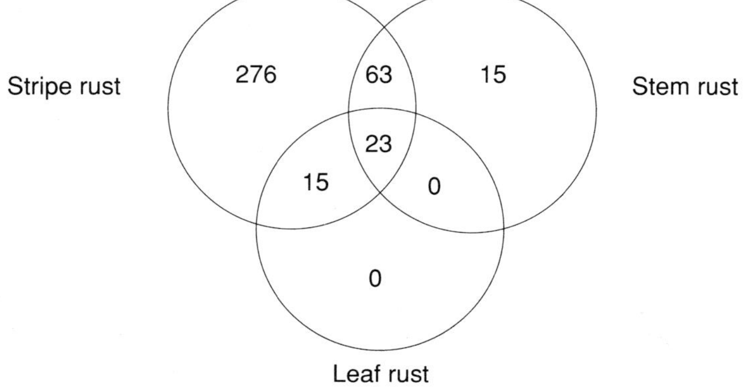

wheat from different countries were evaluated for different characters by PGRC/E in cooperation with durum wheat breeders. The accessions were grown at Debre Zeit Agricultural Research Station. Eleven accessions were selected for use in the national durum wheat breeding programme. The same group of accessions were also planted in a low-rainfall area at Koka during the main rainy season, and the results are now awaited.

## *Discussion*

S. JANA: You mentioned that you have been engaged in landrace improvement. What is the purpose of this work and how do you improve a landrace? And when you have created an 'improved' landrace, what do you do with the original landrace? What is the consequence of landrace improvement on variability within a population?

H. MEKBIB: This work was carried out in collaboration with the breeders and not just by personnel of the PGRC/E. The landrace improvement project was started on durum wheat in 1986. Three landrace populations were collected from the wheat-growing areas in the high, intermediate and the low-lying areas. The landraces were subdivided at the genebank. Half were stored for long-term conservation and the other half were given to the breeders as working material. In this process, the populations were evaluated and lines were selected on the basis of their different agromorphological traits. These were planted again and further selections were made, with priority being given to higher-yielding ones. These were also sometimes crossed with better material coming from the international agricultural research centres.

M. N. SANKARY: I have not observed *Aegilops* species in North or South Yemen. Have you seen any in Ethiopia? If so, do you think that Ethiopia is a secondary or primary centre for wheat evolution?

H. MEKBIB: So far we have not observed *Aegilops* in Ethiopia. If you think the existence of *Aegilops* indicates a primary centre for wheat evolution, then Ethiopia is not that centre, but it could be a secondary centre.

M. N. SANKARY: Do you think the landraces you mentioned are relics from Roman times?

H. MEKBIB: I do not know, but they are certainly very old. It has been said that the hexaploid landraces were introduced to Ethiopia by the Portuguese only a few hundred years ago, but the tetraploid landraces are far older than that.

Wheat Genetic Resources: Meeting Diverse Needs
Edited by J. P. Srivastava and A. B. Damania

Published by John Wiley & Sons

# 17

# *Evaluation of Durum Wheat Lines for Yield, Drought Tolerance and Septoria Resistance in Tunisia*

A. DAALOUL, M. HARRABI, K. AMMAR and M. ABDENNEDHAR

Durum wheat is the most important cereal crop in Tunisia. It covers over 60 per cent of the total cereal growing area of about 1 500 000 ha, and it is the major staple food crop, used to make pasta, couscous, bourghul and rural bread. Efforts to improve this crop began early this century, and fall into three periods. During the first period (1918-38), landraces of durum wheat were classified and then subjected to mass selection and pure line selection. The second period (1939-69) was characterized by an extensive breeding programme in which local lines were crossed with each other and with exotic germplasm; selection from these crosses gave lines with earliness, better leaf and stem rust resistance, and good grain quality. During the third period (1969 to date), high-yielding varieties were developed from segregating germplasm introduced from the Centro Internacional de Mejoramiento de Maíz y Trigo (CIMMYT) or from material crossed in Tunisia.

The use of these high-yielding varieties (Karim, Ben Bechir and Razak) has increased cereal production substantially. Self-sufficiency was achieved in 1985 and 1987, when 2.1 million tonnes and 1.9 million tonnes were produced, respectively. However, production fell drastically to 0.7 million tonnes in 1986 and 0.3 million tonnes in 1988, when rainfall was low and poorly distributed. To reduce these fluctuations, varieties adapted to semi-arid regions need to be developed. The national breeding programme therefore has two components:

1. for favourable areas (average annual rainfall 450-800 mm), breeding specially adapted high-yielding varieties, with good resistance to Septoria and leaf rust; for semi-arid regions, developing lines with good adaptation to harsh conditions (water, cold and heat stress), in order to stabilize grain yields;

2. evaluation of local landraces and of germplasm from the world collection for tolerance to stress in semi-arid regions; hybridization, followed by selection among local lines and high-yielding varieties, in order to combine adaptation with high yield potential.

The Institute of Agronomy of Tunisia (INAT) has been collaborating with the International Center for Agricultural Research in the Dry Areas (ICARDA) since 1987 as a part of the regional network to evaluate durum wheat lines. Evaluation has dealt with agronomic traits (plant height and yield components) and with Septoria resistance. In 1988, when rainfall was only 55 per cent of the average and was poorly distributed, lines were also evaluated for drought tolerance.

## MATERIALS AND METHODS

A total of 5790 lines from the world collection held at ICARDA's genebank were evaluated for agronomic traits and Septoria resistance over 2 years, 1987 and 1988 (*see* Table 17.1).

**Table 17.1** Number of lines and traits evaluated in Tunisia during two years

| Years | Rainfall (mm) (percentage of average) | No. lines | Traits evaluated | |
|---|---|---|---|---|
| | | | Agronomic | Disease resistance |
| 1987 | 660 (165) | 3685 | Plant height<br>Resistance to lodging<br>1000 kernel weight | Septoria |
| | | 200 (elite) | Plant weight<br>Thousand-kernel-weight<br>Drought tolerance | — |
| 1988 | 216 (54) | 1905 | — | Septoria |
| Total | | 5790 | | |

### Agronomic traits

During 1987, the rainfall was higher than normal and was well distributed, allowing good plant development. The 3685 lines planted in hills were evaluated for:

- plant height, measured in cm from the base of the plant to the tip of the head;
- lodging resistance, where plants bent at an angle of over 45° were considered susceptible;

- thousand-kernel-weight (TKW), determined from a sample of 10 heads from each line, for which the total number of kernels (N) and the weight in grammes (W) were measured.

The TKW was deduced as follows:

$$TKW = \frac{W \times 1000\ g}{N}$$

In 1988, 200 elite lines were evaluated. To ensure the survival of at least a few of the plants, a single irrigation of 30 mm was given during late March. The lines were evaluated for:

- drought tolerance, for which two readings were made, the first before the 30 mm irrigation (23 March) and the second 1 month later (28 April). Lines were classified as susceptible (S), intermediate (I) and resistant (R); the difference between the first and second readings revealed the lines which had recovered from drought stress;
- plant height and TKW, determined as in 1987.

## Septoria resistance

A total of 5590 lines were tested in growth chambers for Septoria resistance at the seedling stage.

The inoculum was isolated from sick leaves and multiplied in liquid media containing 3 g of yeast extract (Bacto), 3 g of glucose and 300 ml of distilled water. This inoculum was incubated, with agitation, for 1 week at room temperature. It was then diluted to a concentration of 10 spores per ml. The lines were seeded on laboratory soil in batches of 104 squares at a rate of three kernels per square. The seedlings were grown in growth chambers at a temperature of between 20 and 23°C and a photoperiod of 12 hours, and they were watered every 2 days. At the 3-leaf stage they were inoculated, placed in dew chambers for 72 hours and then returned to the growth chambers.

The readings were made 14 days after inoculation. The lines were classified as above (S, I and R). The degree of attack was recorded as percentage of leaf surface covered with lesions.

# RESULTS

Detailed reports on these evaluations were submitted to ICARDA for inclusion in the evaluation databank and the overall analysis of the world durum wheat collection. The results presented in this paper summarize the main features of importance to durum wheat breeders.

## Agronomic traits

### *Plant height*

The results for 1987 and 1988 (*see* Figures 17.1a and 17.1b) showed that:

- in both years there was a normal distribution for plant height, despite the difference in the number of lines observed;
- the plants were shorter during the dry year, 1988;
- plant height in 1987 was higher and distributed over a fewer number of lines.

This indicates that plant height is a good criterion for measuring diversity in dry, rainfed environments.

### *Resistance to lodging*

This trait was recorded only during the good year, 1987, when the 3685 lines were classified into two groups, one consisting of susceptible and the other of resistant lines. The results showed that 49.6 per cent of the lines were susceptible to lodging and 50.4 per cent were resistant (*see* Figure 17.2).

### *Thousand-kernel-weight*

During 1987, out of the 3685 lines tested, 427 were selected as having better TKW than the national check, Karim (60 g). In 1988, 16 lines were selected. The code number and passport data of these lines are available at the ICARDA databank.

### *Drought tolerance*

This trait was measured in 1988. The first reading for drought tolerance enabled the 200 elite lines to be classified as S, I or R: 11.76 per cent of the lines were classified as R; 54.3 per cent as I; and 33.94 per cent as S (*see* Figure 17.3). The second reading was made after irrigation and enabled a further classification (*see* Figure 17.4): 19.46 per cent were classified as R; 72.40 per cent as I; and 8.14 per cent as S. This showed that:

- the elite lines have a good level of drought tolerance: 66.06 per cent of plants did not suffer from severe drought and, following irrigation, this increased to 91.86 per cent;
- the percentage of recovery was high: 7.7 per cent of the genotypes were reclassified from I to R and 18.1 per cent recovered from S to I (*see* Table 17.2 *overleaf*).

**Figure 17.1(a)** Distribution of plant height in 1987
**Figure 17.1(b)** Distribution of plant height in 1988
**Figure 17.2** Distribution of lines for resistance to lodging measured on 3685 lines in 1987
**Figure 17.3** Distribution of lines in relation to their resistance to water stress before irrigation
**Figure 17.4** Distribution of lines in relation to their resistance to water stress after irrigation followed by stress
**Figure 17.5** Distribution of 3684 lines for their Septoria resistance in 1987
**Figure 17.6** Distribution of 1905 lines for their Septoria resistance in 1988

**Figure 17.1(a)**

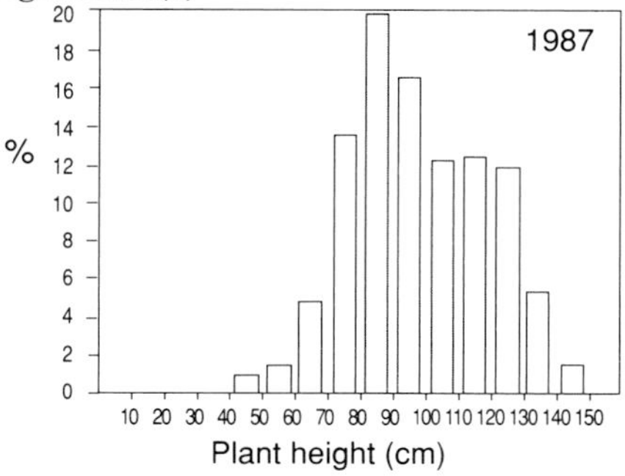


**Figure 17.1(b)**

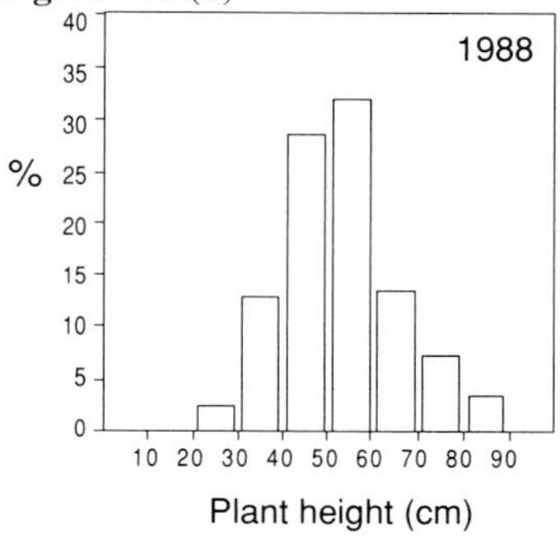


**Figure 17.2**

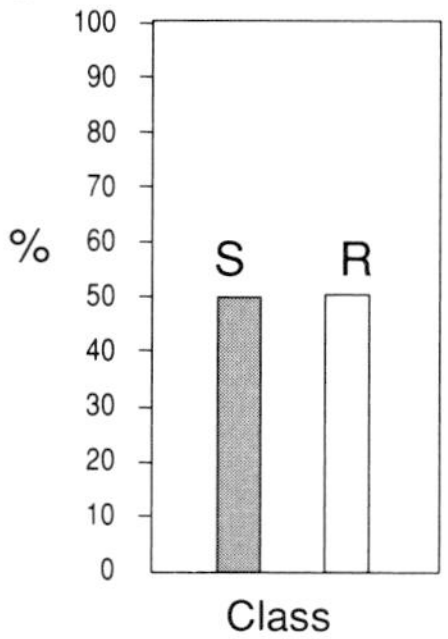


**Figure 17.3**

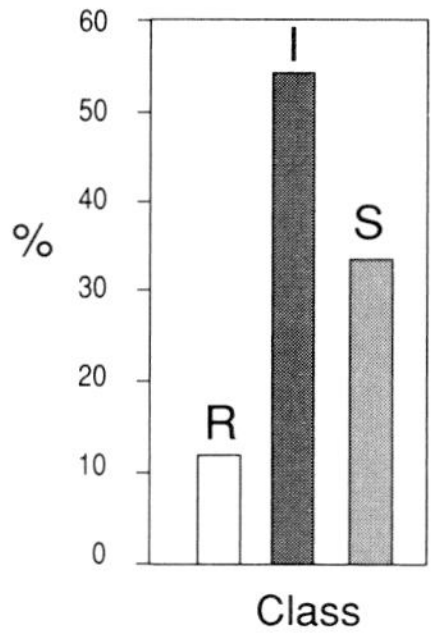


**Figure 17.4**

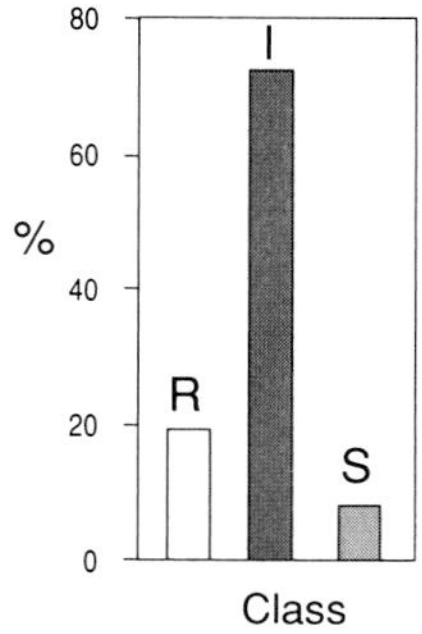


**Figure 17.5**

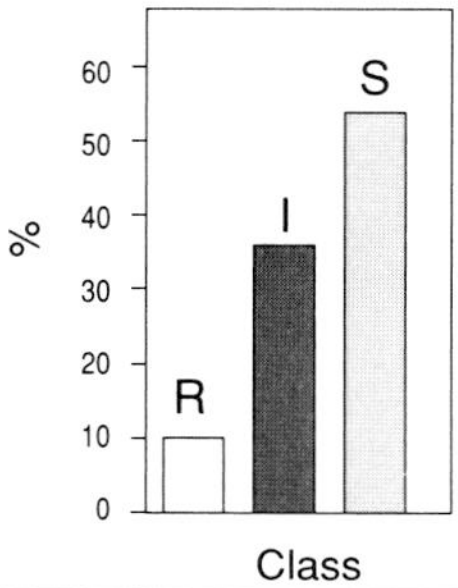


**Figure 17.6**

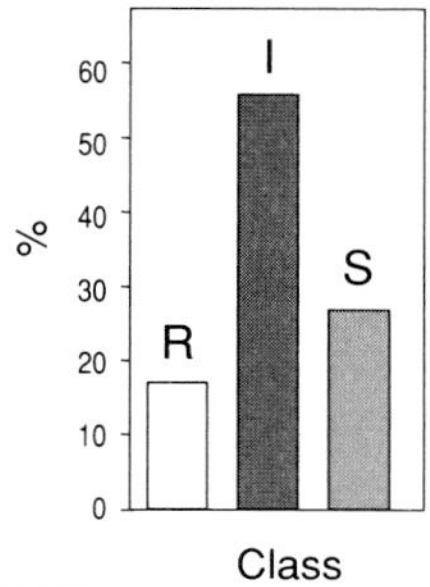


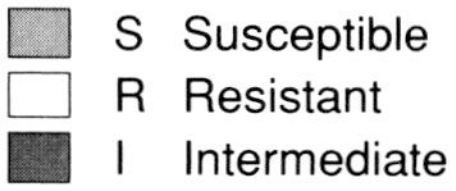

The screening of 200 elite lines under dry conditions in 1988 at INAT's experiment station in Mornag enabled a further 16 lines with intermediate or good drought tolerance and better TKW than the national check (Karim) to be selected (*see* Table 17.3).

**Table 17.2** Percentage of recovery of lines tested for a severe drought followed by 30 mm irrigation

| **Readings** | **Class** [a] | | |
|---|---|---|---|
| | **R** | **I** | **S** |
| Before irrigation | 11.76 | 54.3 | 33.94 |
| After 30 days following 30 mm irrigation | 19.46 | 72.4 | 8.14 |
| Recovery | +7.7 | +18.1 | -25.8 |

a R = resistant; I = intermediate; S = susceptible

**Table 17.3** Performance of 16 selected lines from 200 elite lines from ICARDA, 1988

| **No. of elite accessions** | **1st reading of drought stress (%)** [b] | **2nd reading of drought stress following 30 mm irrigation (%)** [b] | **TKW**[a] **(g)** |
|---|---|---|---|
| 10 | S | I | 62.78 |
| 43 | I | R | 60.56 |
| 44 | I | I | 70.80 |
| 48 | I | I | 63.26 |
| 49 | R | R | 60.19 |
| 50 | I | I | 60.00 |
| 119 | I | I | 61.96 |
| 120 | I | I | 50.29 |
| 124 | I | I | 59.60 |
| 128 | I | I | 63.40 |
| 135 | I | I | 73.05 |
| 138 | I | I | 63.04 |
| 155 | I | I | 68.23 |
| 167 | I | I | 64.19 |
| 177 | R | R | 60.92 |
| 198 | S | I | 65.09 |

a TKW = thousand-kernel-weight; b R = resistant; I = intermediate; S = susceptible

### Septoria resistance

In 1987, the 3684 lines tested for Septoria resistance at the seedling stage were classified into three groups: 10 per cent of lines were classified as R; 36 per cent as I; and 54 per cent as S (*see* Figure 17. 5). In 1988, the 1905 lines tested: 17 per cent were classified as R; 56 per cent as I; and 27 per cent as S (*see* Figure 17. 6). Detailed data on the reaction type and the degree of attack are on file in the ICARDA databank.

## CONCLUSION

These evaluations, carried out on a large number of lines from the ICARDA world collection of durum wheat, have enabled a few lines with good drought resistance, lodging resistance, TKW, and Septoria resistance to be selected and made available to the national durum wheat breeding programme. They have also provided the ICARDA network with substantial data that have enabled lines with specific traits to be selected and made available to regional breeding programmes. These activities should be complemented by continued testing of other lines, the creation of special nurseries with elite lines for specific traits, and research on heritability and the ability to combine multiple traits.

## *Discussion*

A. MADDUR: Can you elaborate on the criteria used for classifying your material for drought resistance in the field?

A. DAALOUL: We used a three-level score: R for resistant plants which did not show wilting or damage and continued to grow normally; I for intermediate, where leaves were curled with very little further growth and 50 per cent of the leaves were still green; and S for susceptible, where all leaves were yellow and growth stopped completely.

A. MADDUR: Why did you not include the components of yield — spikes per plant and mean kernels per spike — which are as important as kernel weight in determining yield in wheat?

A. DAALOUL: These yield components are very important but because of the large number of accessions involved in the study we had to reduce the number of characters to be observed. Also, in 1987, the drought in Tunisia did not permit good spike development so that these data would be misleading and correlation studies could not be made.

C. JOSEPHIDES: Did you notice if the lines which were tall and resistant to lodging had solid stems?

A. Daaloul: Yes, some of the tall lines resistant to lodging had solid stems. However, several others did not possess solid stems and they should be investigated from the anatomy point of view.

O. F. Mamluk: When you are evaluating such large numbers of accessions of durum wheat, I would suggest you record not only the reaction of plants to Septoria, but also the formation of pectnidia. This is important because some plants that show intermediate reaction still do not allow the fungus to produce pectnidia. Secondly, in Septoria screening, plant maturity is of some importance, so I would suggest you could also include data on early-maturing types.

A. Daaloul: This work in Tunisia was carried out in collaboration with Dr Harrabi, who is a pathologist. We tested plants in the seedling stage only, but if we find it necessary we could record formation of pectnidia also. As regards your second suggestion, we did not evaluate for Septoria in the field but in the growth chamber because of the large numbers involved (about 5000 accessions) and so were unable to record early- or late-maturing types.

Wheat Genetic Resources: Meeting Diverse Needs
Edited by J. P. Srivastava and A. B. Damania

Published by John Wiley & Sons

# 18

# *Effects of Sowing Season on Yield and Quality of Iraqi and Hungarian Wheat Varieties*

J. A. SHAMKHE, L. CSEUZ and J. MATUZ

Wheat growing is very different in Iraq and Hungary. In Iraq, most wheats grown are spring-sown or facultative types; fertilizer application is low (80 kg N/ha and 50 kg $P_2O_5$/ha; no potassium is applied to wheat) under both rainfed and irrigated conditions; and the average yield is about 1 t/ha. In Hungary, all the wheat cultivars are winter types; they are grown at high rates of fertilizer application (250 kg/ha of N, $P_2O_5$ and $K_2O$ in the ratio of 1 : 1 : 1), with other chemicals, and mainly without irrigation; and the average yield is about 5 t/ha.

Despite these differences, the two countries share a number of common problems. For example, each experiences a marked dry season. The basis for improving wheat production in both countries is breeding for high yield, good grain quality, disease resistance, drought tolerance and resistance to lodging. To meet these objectives, we initiated a breeding programme, the first stage of which was to study the productivity and quality of some of the varieties which are popular in Iraq. This paper summarizes the results of this study.

## MATERIALS AND METHODS

There were six varieties in the experiment (*see* Table 18.1 *overleaf*). The trial was sown at the Kiszombor Station of the Cereals Research Institute in the autumn (22 October 1987) and the spring (28 March 1988). The 6.5 $m^2$ plots were planted in four replications (complete randomized block design.). Sowing density was 500 seeds/$m^2$. The trial was planted with an Oyjord eight-row planter. The previous crop had been green pea. Before planting, fertilizer was applied at a rate of 60 kg N/ha, 100 kg $P_2O_5$/ha and

**Table 18.1** The origin and genealogies of the parental varieties

| Name | Origin | Genealogy |
|---|---|---|
| Sabir-beg [a] | Iraq | Local variety |
| Abu-graib [a] | Iraq | Ajeeba x Inia x Mexico 24 |
| Noori [b] | Iraq | Sabir-beg x Inia R66 x Mexico 2 x Mexico 10 |
| Mexipak [b] | Pakistan | (Frontana x Kenya 58 - Newthach/Nori 10 - Baant) Grabo 55 |
| GK Kincsö [c] | Hungary | Arthur x Sava |
| GK Ságvári [c] | Hungary | Aurora - GT76.150 |

a Facultative; b Spring-grown; c Winter-type

100 kg $K_2O$/ha. The weather was favourable at both plantings. The spring-type varieties were protected from frost by thick snow cover, so no frost damage was observed. During the growing season (1 September 1987 to 30 June 1988) 390 mm of rainfall was recorded, 70 mm less than the long-term average. The autumn-planted trial was harvested on 8 July 1988. Winter-type wheats sown in the spring did not head completely. Their heading occurred late in June, and because of the heat in July all the spikes became sterile. The spring-planted trial was harvested on 8 August 1988. The plot yields were measured and converted to t/ha.

Grain quality was studied by means of a modified Zeleny test (Lelley, 1973) and a modified sodium dodecyl sulphate (SDS) test (Matuz et al., 1986). In addition, quality was determined by the usual Hungarian methods, such as valorigraph value and loaf volume (Karácsonyi, 1970). (A valorigraph is a modified farinograph used to characterize the bread-making quality of flour.) Protein content was determined by the Kjeldahl method.

## RESULTS

The results of the autumn- and spring-sown trials showed that, in Hungary, autumn sowing gives better results (provided there is no frost damage) than spring sowing, even for the spring-type wheats. Under autumn conditions, the Iraqi varieties ripened very early (they headed 8-10 days earlier than the Hungarian varieties) and were approximately 20 cm taller than when spring sown.

The winter-type varieties headed incompletely and all the heads were found to be sterile. As the data show, autumn-sown Abu-graib, Noori and Mexipak achieved the yield levels of the Hungarian varieties (*see* Figure 18.1). In the spring-sown trial, Sabir-beg yielded 21 per cent, Abu-graib 37 per cent, Noori 41 per cent and Mexipak 49 per cent less than in the autumn-sown trial.

The highest yielding variety in the autumn-sown trial was GK Ságvári; next was Mexipak, followed by GK Kincsö, Abu-graib and Noori; the lowest yielding was

**Figure 18.1** Grain yield of wheats

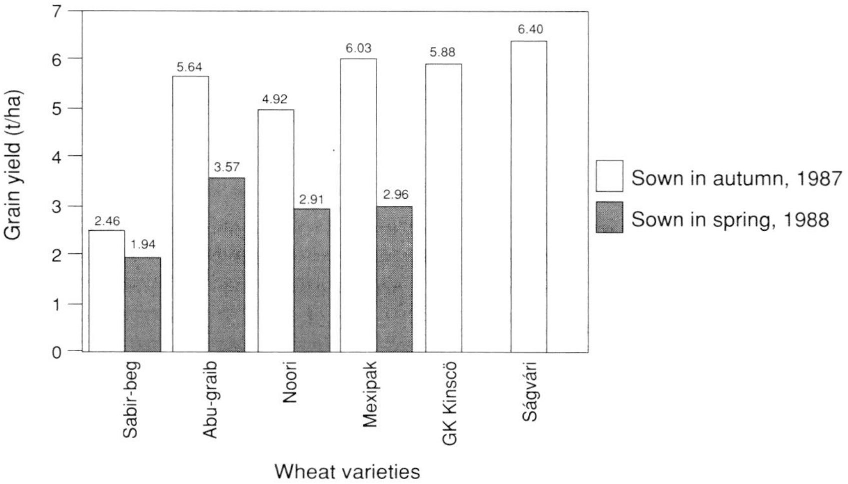


Sabir-beg. When spring sown, the order changed, the highest yielding variety being Abu-graib, followed by Mexipak, Noori and Sabir-beg.

Planting season also affected grain protein content and quality. Protein content was 1 to 3 per cent higher in the autumn-sown trial than in the spring-sown trial. The greatest difference was found in Sabir-beg, the protein content of which was 15.1 per cent when autumn sown but only 11.7 per cent when spring sown (*see* Figure 18.2 *overleaf*). Iraqi varieties generally had higher protein content and better grain quality than Hungarian varieties (GK Kincsö and GK Ságvári are medium-quality wheats).

The Zeleny value of Sabir-beg and Noori was outstanding when these varieties were autumn sown, but under spring sowing conditions the Zeleny value was lower (*see* Figure 18.3 *overleaf*). The Iraqi varieties' SDS values were similar; planting season did not significantly affect this character (*see* Figure 18.4 *overleaf*). In terms of the valorigraph values, Noori and Abu-graib performed the best. Planting season had more effect on the valorigraphic value of Sabir-beg and Mexipak than on that of other varieties (*see* Figure 18.5 *overleaf*).

The results of the baking test showed that Noori produced the largest loaf volume. The effect of planting season on this character was ambiguous: Sabir-beg and Mexipak gave larger loaf volumes when autumn sown, whereas Noori and Abu-graib gave large volumes when spring sown (*see* Figure 18.6 *page 200*).

No previous trials studying winter- and spring-type wheats under autumn and spring conditions have been carried out in Hungary. In our trial, we found autumn sowing significantly more favourable than spring sowing for all the varieties studied. In the

**Figure 18.2** Protein content of wheats

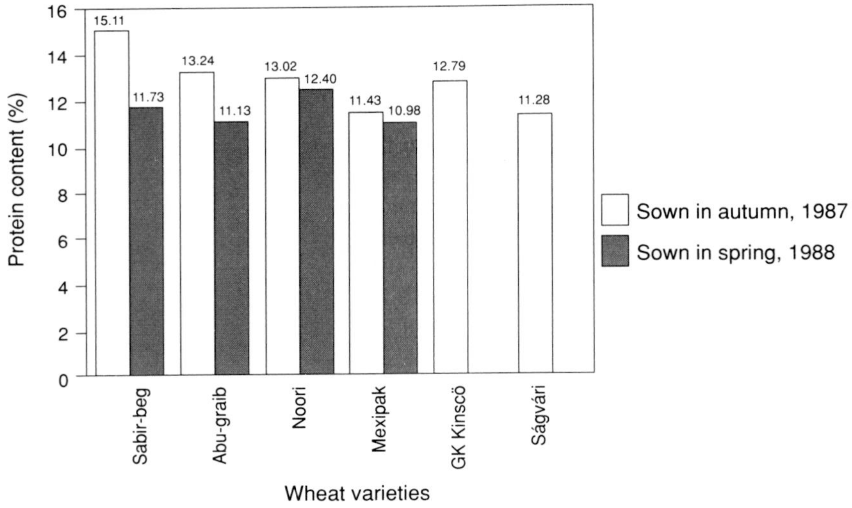


**Figure 18.3** Zeleny value of wheats

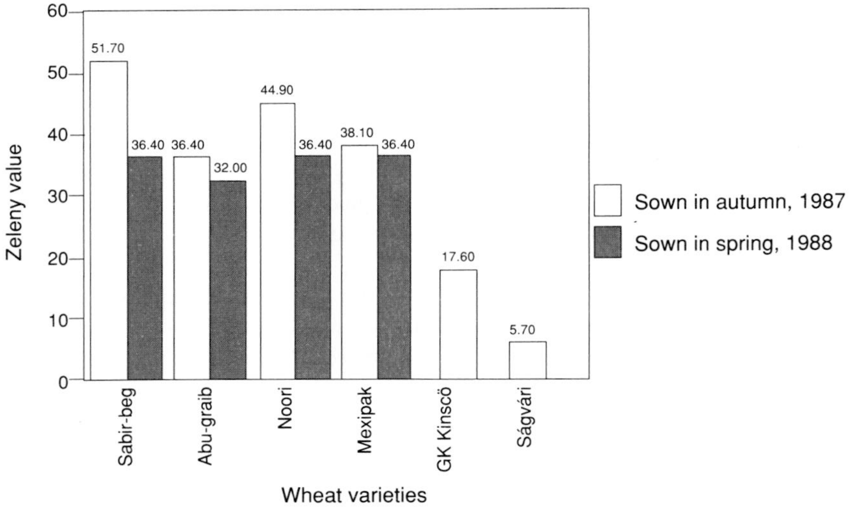

**Figure 18.4** Sodium dodecyl sulphate (SDS) value of wheats

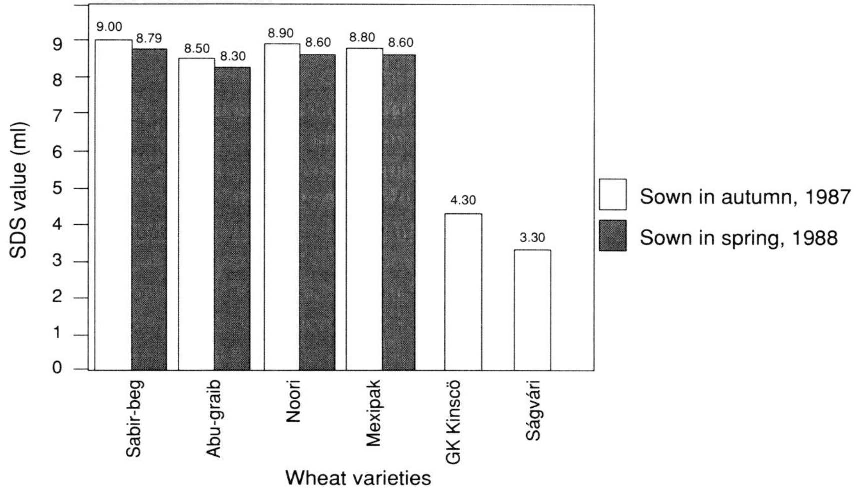


**Figure 18.5** Valorigraphic value of wheats

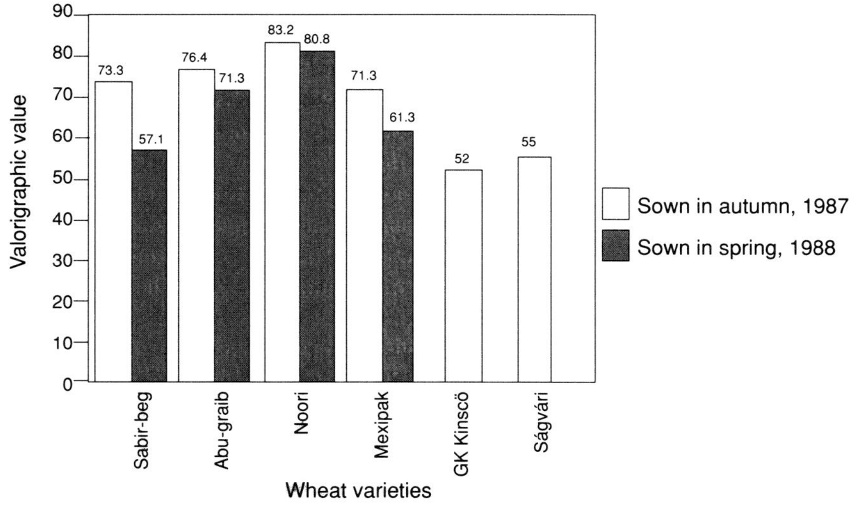

**Figure 18.6** Loaf volume

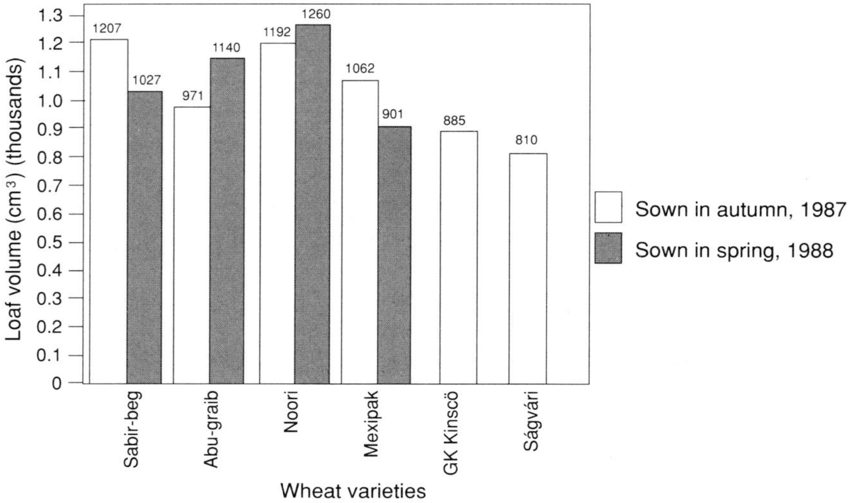


autumn-sown trial, the wheats yielded 21-51 per cent more and their grain quality was better. Daaloul's (1981) results from planting season trials, carried out using winter x spring origin lines, agree with ours.

Sowing winter- and facultative-type durum wheat lines in both planting seasons has also shown the superiority of autumn planting (Beke and Matuz, 1982).

Quality tests show that the protein content and bread-making quality of Hungarian varieties could be improved by crossing them with Iraqi varieties. To achieve this goal, we have included these varieties in our breeding programme.

Wheat Genetic Resources: Meeting Diverse Needs
Edited by J. P. Srivastava and A. B. Damania

Published by John Wiley & Sons

# 19

# *Evaluation and Conservation of Wheat Genetic Resources in India*

S. K. Mithal and M. N. Kopper

Wheat is one of India's most important cereal crops. Archaeological evidence has revealed its long history of cultivation in this subcontinent. Kernels of dwarf wheat, *Triticum sphaerococcum*, dating back 5000 years have been found at Mohanjo-Daro in India. Pal (1966) reported that the north-western sector of the Indian subcontinent, between the Himalayan range and the Hindukush mountains, is the original home of the common bread wheat, *Triticum aestivum*. According to Howard (1916), *T. sphaerococcum* (Indian dwarf wheat) is also supposed to have originated in the north-western area of the Indian subcontinent. *T. durum* (macaroni wheat) was apparently introduced into India by pilgrims returning from Mecca. *T. dicoccum* (emmer or Khapli wheat) may have entered via the west coast, brought by Arab traders.

These species became highly diversified in the course of time but, because of substantial genetic erosion over the past 25 years, their present status is precarious. Many landraces and earlier varieties have been rapidly replaced by modern ones, particularly in the northern and north-western plains. Industrialization has led to further irreparable losses. Against this background, efforts have also been made to collect, evaluate and conserve this valuable genetic resource to the maximum extent possible.

Three species — *T. aestivum*, *T. durum* and *T. dicoccum* — are grown in India as commercial crops. *T. aestivum* is the most important in all the wheat-growing states, covering about 86 per cent of the country's total wheat area. *T. durum* occupies slightly more than 12 per cent of the total wheat area; it is confined to the central and southern states. The cultivation of *T. dicoccum* is confined largely to southern states, but a small amount is grown in Gujarat. It is being gradually replaced by high-yielding *aestivum* and *durum* types. *T. sphaerococcum* is no longer cultivated in India (Agarwal, 1986).

## EXPLORATION AND COLLECTION

Since the creation of the Plant Introduction Division at the Indian Agricultural Research Institute (IARI) in New Delhi in 1961, efforts have been made to collect, evaluate and conserve indigenous wheat germplasm. The National Bureau of Plant Genetic Resources (NBPGR) started making extensive efforts to collect wheat genetic resources in India in 1977.

Table 19.1 gives the regions/states explored for wheat species between 1976 and 1988. The states of Gujarat, Rajasthan, Madhya Pradesh, Maharashtra, Karnataka, Jammu and Kashmir, and Sikkim, and the hilly areas of Uttar Pradesh and Himachal Pradesh were extensively covered. The NBPGR currently maintains indigenous collections of *T. aestivum* (1475 accessions), *T. durum* (500 accessions) and *T. dicoccum* (37 accessions).

Besides the NBPGR, there are a number of other organizations which maintain wheat germplasm collections in India (*see* Table 19.2).

The emphasis in collection efforts was put on the collection of random population samples from farmers' fields, but some material was also collected from threshing floors and local markets. Sites were selected on the basis of factors such as local diversity, agroecological variation, variation in farming systems and variations in local preferences. The wheat germplasm which was collected from different states included variations in plant type, plant height, straw colour, colour and size of ear-heads, colour and size of awns, tillering capacity, and other morphological and agronomic characteristics.

Drought-tolerant *T. aestivum* and *T. durum* wheats which exhibited considerable morpho-agronomic variability were collected from the Saurashtra region of western Gujarat. *T. durum* wheats were also collected from unirrigated pockets in Madhya Pradesh, Rajasthan and Karnataka. Samples of *T. durum* which had adapted to saline soils were collected from the *bhal* tracts in Limri and the adjoining belt of Surendernagar in Gujarat; *aestivum* materials showing tolerance to soil salinity were collected from well-irrigated sites in the drier areas of south-western, central and northern Rajasthan. Collections of *T. aestivum* which had extreme tolerance to cold, arid conditions were made from the north-western Himalayas, Lahul and Spiti (Himachal Pradesh).

## EVALUATION OF INDIGENOUS WHEAT GERMPLASM

### Morphological and agronomic characters

The evaluation of indigenous wheat germplasm collections has yielded some interesting results. The assessment of *T. aestivum*, *T. durum* and *T. dicoccum* cultivars collected from Gujarat, Himachal Pradesh, Karnataka, Madhya Pradesh, Rajasthan

**Table 19.1** Indigenous wheat germplasm collected from different states during the period 1976-88

| States | No. of accessions | | | Total |
|---|---|---|---|---|
| | *Triticum aestivum* | *T. durum* | *T. dicoccum* | |
| Gujarat | 208 | 248 | 17 | 473 |
| Rajasthan | 561 | 88 | — | 649 |
| Madhya Pradesh | 126 | 279 | — | 405 |
| Karnataka | 16 | 169 | 20 | 205 |
| Himachal Pradesh | 235 | — | — | 235 |
| Sikkim | 37 | — | — | 37 |
| Uttar Pradesh | 148 | 31 | — | 179 |
| Jammu and Kashmir | 117 | — | — | 117 |
| Maharashtra | 51 | 215 | 19 | 285 |
| Total | 1499 | 1030 | 56 | 2585 |

**Table 19.2** Wheat germplasm collecting in India

| Centre | Indigenous collections | Total collections |
|---|---|---|
| National Bureau of Plant Genetic Resources, New Delhi | 2012 | 11 947 |
| Indian Agricultural Research Institute, New Delhi | 1437 | 17 000 |
| Punjab Agricultural University, Ludhiana | — | 4000 |
| Haryana Agricultural University, Hisar | 65 | 1500 |
| Vivekanand Parvatiya Krishi Anusandhan Shala, Almora | 250 | 3884 |
| G.B. Plant University of Agriculture and Technology, Pantnagar | 48 | 4432 |
| Gujarat Agricultural University, Junagarh | 88 | 3738 |
| Mahatma Phule Agricultural University, Niphad | 345 | 930 |
| Zonal Agricultural Research Station. Powarkheda | — | 1866 |
| Indian Agricultural Research Institute, Regional Station, Simla | 10 | 1260 |
| Agricultural Research Station (U.A.S.), Dharwad | — | 1500 |
| Bihar Agricultural College, Sabour | 277 | 277 |
| Total | 4532 | 52 334 |

and Sikkim according to 18 morphological and agronomic descriptors has been made. The $X^2$ values estimated for each descriptor in *T. aestivum* and *T. durum* are shown in Table 19.3 (*overleaf*).

The *aestivum* cultivars were collected from Gujarat (68), Himachal Pradesh (86), Karnataka (18), Madhya Pradesh (14), Rajasthan (14) and Sikkim (22). The *durum* cultivars were collected from Gujarat (24), Rajasthan (23), Madhya Pradesh (21) and

**Table 19.3** $Chi^2$ values of *Triticum aestivum* and *T. durum* for relative proportions of cultivars from different states and for different characters

| **Characters** | **Species** | |
|---|---|---|
| | ***T. aestivum*** | ***T. durum*** |
| Initial growth habit | 66.8 [a] | 46.6 [a] |
| Leaf colour at five leaf stage | 22.5 [a] | 48.2 [a] |
| Days to 50% spike emergence | 123.5 [a] | 230.4 [a] |
| Days to 50% spike maturity | 153.0 [a] | 163.4 [a] |
| Plant height (cm) | 278.2 [a] | 83.7 [a] |
| Combined spike-awn length (cm) | 59.8 [a] | 46.8 [a] |
| Spike length (cm) | 127.8 [a] | 11.7ns |
| Awn-length (cm) | 177.7 [a] | 64.6 [a] |
| Spike-awn ratio | 187.4 [a] | 67.6 [a] |
| Spike density | 52.9 [a] | 56.5 [a] |
| Grain yield per plant (g) | 63.6 [a] | 78.7 [a] |
| Tillers per plant | 458.2 [a] | 154.1 [a] |
| Grain weight per spike (g) | 179.4 [a] | 116.3 [a] |
| Grains per spike | 230.2 [a] | 60.9 [a] |
| Thousand-grain-weight (g) | 93.5 [a] | 88.0 [a] |
| Grain colour | 68.7 [a] | 23.5 [a] |
| Grain uniformity | 18.4 [b] | 61.6 [a] |
| Degree of seed shrivelling | 20.5 [b] | 32.8 [a] |

a, b P = 0.01 and 0.05 respectively, ns = not significant

**Table 19.4** Range for 12 quantitative characters studied in indigenous wheat germplasm

| **Characters** | ***Triticum aestivum*** | | ***T. durum*** | | ***T. dicoccum*** | | **Control** [a] | |
|---|---|---|---|---|---|---|---|---|
| | **min** | **max** | **min** | **max** | **min** | **max** | **min** | **max** |
| Days to 50% spike emergence | 81 | 124 | 84 | 122 | 86 | 108 | 81 | 87 |
| Days to 50% spike maturity | 115 | 149 | 118 | 150 | 123 | 132 | 133 | 134 |
| Plant height (cm) | 45 | 116 | 40 | 112 | 61 | 103 | 100 | 112 |
| Combined spike-awn length (cm) | 8 | 23 | 8 | 25 | 11 | 21 | 14.3 | 18.3 |
| Spike length (cm) | 6 | 14 | 5 | 14 | 7 | 9 | 9 | 10.7 |
| Awn-length (cm) | — | 13 | 1 | 14 | — | 12 | 4.6 | 7.6 |
| Spike-awn ratio | 1 | 2.8 | 1 | 3 | 1 | 2.4 | 1.5 | 1.8 |
| Grain yield per plant (g) | 1 | 18 | 2 | 17 | 2 | 11 | 14 | 22.5 |
| Tillers per plant | 2 | 18 | 2 | 17 | 4 | 16 | 6 | 9 |
| Grain weight per spike (g) | 0.2 | 2.5 | 0.2 | 2.2 | 0.1 | 1.3 | 1.1 | 1.4 |
| Grains per spike | 5 | 83 | 6. | 51 | 7 | 52 | 37 | 43 |
| Thousand-grain-weight (g) | 1 | 5 | 1.3 | 5.2 | 1.4 | 3.5 | 2.6 | 3.6 |

a Control varieties included HD 2204 and Sonalika

**Table 19.5** Variability in indigenous wheat collections for different characters in different states

| States | Characters |
|---|---|
| *Triticum aestivum* | |
| Gujarat | Leaf colour at five-leaf stage, plant height, grain yield, tiller number, grain uniformity, degree of seed shrivelling |
| Himachal Pradesh | Initial growth habit, spike emergence, combined spike-awn length, spike length, awn length, grain number per spike, spike-awn ratio, grain uniformity |
| Karnataka | Grain weight per spike, thousand-grain-weight, grain colour |
| Madhya Pradesh | Spike density |
| Rajasthan | Spike maturity, spike density, tiller number |
| Sikkim | Spike-awn ratio, grain colour, degree of seed shrivelling |
| *Triticum durum* | |
| Gujarat | Spike emergence, spike maturity, combined spike-awn length, spike length, awn-length, spike-awn ratio, spike density, tiller number, grain number per spike, thousand-grain-weight, grain uniformity |
| Rajasthan | Initial growth habit, spike emergence, spike maturity, plant height, thousand-grain-weight, grain yield, grain colour |
| Madya Pradesh | Spike emergence, plant height, awn-length, spike-awn ratio, thousand-grain-weight, tiller number, grain yield, degree of seed shrivelling |
| Karnataka | Leaf colour at five-leaf stage, combined spike-awn length, grain weight per spike, grain number per spike, thousand-grain-weight |

Karnataka (117). The *dicoccum* group comprised 14 cultivars from Gujarat and 12 from Karnataka.

The $X^2$ tests indicated significant differences at the 1 per cent level of significance among the materials from different states for all the descriptors except grain uniformity and degree of seed shrivelling in the *aestivum* cultivars (significant only at the 5 per cent level) and spike length in *durum* cultivars (not significant, even at the 5 per cent level) (*see* Tables 19.4 and 19.5).

The *aestivum* group from Himachal Pradesh was comparatively diverse for initial growth habit, days to spike emergence, combined spike and awn length, spike length, awn length, grain number per spike and grain uniformity. The materials from both Himachal Pradesh and Sikkim showed relatively high diversity for spike-awn ratio. Rajasthan material was more diverse in terms of days to spike maturity, spike density and number of tillers. Gujarat material was more diverse in terms of leaf colour at five-leaf stage, plant height, grain yield per plant, grain uniformity and degree of seed shrivelling. However, the materials from both Rajasthan and Gujarat states were more diverse for number of tillers per plant. The material from Karnataka was more diverse for grain weight per spike, thousand-kernel-weight (TKW) and grain colour. The material from Sikkim showed high diversity for grain colour and degree of seed shrivelling.

The *durum* group from Gujarat exhibited relatively high diversity for combined spike and awn length, spike length, awn length, spike-awn ratio, spike density, number of tillers per plant, number of grains per spike and grain uniformity; materials from Gujarat, together with Rajasthan and Madhya Pradesh, showed high diversity for days to spike emergence; and Gujarat and Rajasthan accessions showed high diversity for spike maturity. Rajasthan material showed more diversity in terms of initial growth habit, plant height, grain yield per plant and grain colour, whereas Karnataka material was more diverse in terms of grain weight per spike and leaf colour at the five-leaf stage. For TKW, the materials from all four states were equally highly diverse. The material from Madhya Pradesh also exhibited high diversity for plant height, awn length, spike-awn ratio, number of tillers per plant, grain yield per plant, and degree of seed shrivelling. Karnataka material also showed high diversity for combined spike and awn length and number of grains per spike.

In the *dicoccum* group, the material from Gujarat was observed to be more variable for most of the characters studied.

A wide range of variation was observed for a number of morphological characters in *T. aestivum* and *T. durum* types, including size and arrangement of leaves on the stem, leaf colour and pubescence, colour of the awns (from light green to black), angle of arrangement of the awns (from acute to widely spreading), colour of the ear-heads (from yellowish to green and brown), and colour of the grains (from white to amber and from light red to dark red). Awned, hooded or awnless ear-heads were observed in *T. aestivum* types.

## Disease resistance

The indigenous materials were screened at Punjab Agricultural University, Ludhiana for several diseases, and a number of types resistant to one or more diseases have been identified. Table 19.6 shows the accession numbers and states of origin of the types resistant to black, brown or yellow rusts, or to two or all three of these rusts (Gupta et al., 1983). IC 35119 and IC 35127 from Karnataka in *T. durum*, IC 36706 and IC 36729 from Himachal Pradesh and IC 47490 from Karnataka in *T. aestivum*, and IC 47453 from Karnataka in *T. dicoccum* all showed high levels of resistance to rust under epihytolic conditions. IC 28594, a *durum* accession from Gujarat, was observed to be highly resistant to brown and yellow rusts, and also showed resistance to loose smut and powdery mildew.

The Khapli or emmer wheat of India possesses a high degree of rust resistance. In Punjab, where yellow rust is widespread, the local wheat had a measure of tolerance (Pal, 1966).

**Table 19.6** Resistance in wheat germplasm to black, brown and yellow rust

| State | *Triticum aestivum* | | *T. durum* | | *T. dicoccum* | |
|---|---|---|---|---|---|---|
| Gujarat | IC — | — | — | 28 594 [a] | — | — |
| Himachal Pradesh | IC 36 705 | 36 706 | — | — | — | — |
| | IC 36 723 | 36 729 | — | — | — | — |
| Karnataka | IC 35 162 | 47 490 | 35 119 | 35 127 | 35 173 | 47 543 |
| | IC 47 497 | 47 515 | — | — | — | — |
| | IC 47 536 | — | — | — | — | — |
| Madhya Pradesh | IC 31 482 | — | — | — | — | — |
| Rajasthan | — | — | 29 052 | — | — | — |
| Uttar Pradesh | IC 47 335 | — | — | — | — | — |

a Also resistant to loose smut and powdery mildew (Gupta et al., 1983)

## Salinity tolerance

The materials were screened for salinity tolerance at the Central Soil Salinity Research Institute, Karnal, with very promising results. Table 19.7 shows the accession numbers and states of origin of accessions observed to be tolerant to saline conditions (Singh, 1983). Among the *T. aestivum* types, IC 26727 from Rajasthan, IC 26729, IC 26734 and IC 26740 from Himachal Pradesh, and IC 28609 and IC 28614 from Gujarat showed tolerance, as did *T. durum* types from Gujarat, Rajasthan, Madhya Pradesh and Karnataka. In *T. dicoccum*, IC 28596 from Gujarat was the only salt-tolerant type.

A number of other sources of salt-tolerant materials have been reported. According to Rana (1977), a local wheat, Kharchia, found in Rajasthan, possessed a high degree of tolerance to saline conditions. Rana (1977) also reported that Kharchia 65 and HD

**Table 19.7** Tolerance in wheat germplasm to salinity

| State | *Triticum aestivum* | | *T. durum* | | | *T. dicoccum* |
|---|---|---|---|---|---|---|
| Gujarat | IC 28 609 | 28 614 | 28 515 | 28 517 | 28 518 | 28 596 |
| | IC — | — | 28585-A | 28526 | — | — |
| Himachal Pradesh | IC 28 729 | 28 734 | — | — | — | — |
| | IC 28 740 | — | — | — | — | — |
| Karnataka | IC — | — | 35 043 | 35 111 | 35 118 | — |
| | IC — | — | 35125 | — | — | — |
| Madhya Pradesh | IC — | — | 32 021 | 32 029 | 32 041 | |
| Rajasthan | IC 26 727 | — | 29 075 | 29 079 | — | — |

Source: Singh, 1983

1982 tolerated soil salinity. Two varieties, K 7435 and HD 2177, were reported by Kumar et al. (1983) as salt-tolerant types. Kumar and Yadav (1983) developed salt-tolerant mutants BHP 10 and BHP 31 from a moderately salt-tolerant variety, HD 1553. The variety WH 157 released from Haryana Agricultural University in 1976 was salt tolerant (Rana and Singh, 1977). CSW 538, CSW 540 and Rata wheat have also been reported as tolerant to saline/alkaline soils (Rana, 1986).

## Protein content

Austin et al. (1982) reported that a number of *aestivum* and *durum* wheats possess high protein and high lysine contents (*see* Table 19.8). The protein and lysine content of these materials ranged from 16.06 to 18.82 per cent and from 2.52 to 2.85 per cent, respectively.

**Table 19.8** Sources of high protein and high lysine contents in wheat material

| Variety | Protein percentage | Lysine percentage |
|---|---|---|
| *Triticum aestivum* | | |
| HD (M) 1404 | 16.59 | 2.80 |
| HW106 | 16.76 | 2.76 |
| HW112 | 16.15 | 2.55 |
| NP 787 | 16.06 | 2.73 |
| NP 790 | 16.21 | 2.76 |
| NP 791 | 16.71 | 2.70 |
| NP 794 | 16.11 | 2.53 |
| NP 796 | 16.63 | 2.73 |
| A-2-1-8 | 16.42 | 2.84 |
| A-375 | 16.50 | 2.67 |
| A-475 | 16.17 | 2.85 |
| *T. durum* | | |
| IRM 4971 | 17.21 | 2.55 |
| IRN 4975 | 16.94 | 2.62 |
| IRN 4979 | 16.22 | 2.70 |
| IRN 4982 | 18.18 | 2.57 |
| IRN 4991 | 17.21 | 2.58 |
| IRN 5002 | 18.22 | 2.69 |
| IRN 5003 | 18.82 | 2.62 |
| IRN 5007 | 18.36 | 2.58 |
| IRN 5027 | 18.64 | 2.52 |

Source: Austin et al., 1982

## Exotic germplasm

Exotic wheat germplasm has been introduced continuously since 1936, initially through the IARI. With the establishment of the NBPGR in New Delhi in 1977, introduction activities intensified. Since 1982, exotic materials have been grown in the post-entry quarantine isolated nursery at the NBPGR for the purposes of health inspection, seed multiplication and distribution. Most of these materials are from the International Center for Agricultural Research in the Dry Areas (ICARDA), the Centro Internacional de Mejoramiento de Maíz y Trigo (CIMMYT) and the United States Department of Agriculture (USDA). Every year, 4000 to 8000 new seed materials are introduced in the form of various international test nurseries and trials.

A wheat field day is organized each year during March. Wheat breeders and other scientists from research institutes and university departments throughout the country participate and select materials for use in their breeding programmes. The materials selected are supplied before the next cropping season begins. In 1988 alone, 10 183 materials were supplied to researchers. Evaluation data on these materials are published in the form of a catalogue each year and distributed to wheat breeders.

## Future programmes

The indigenous germplasm collection is still in the process of being characterized and documented. A comprehensive catalogue will be brought out for breeders' use. Purnia District and the adjoining areas of Bihar State will be explored to add to the collection.

Wheat Genetic Resources: Meeting Diverse Needs
Edited by J. P. Srivastava and A. B. Damania

Published by John Wiley & Sons

# 20

# *Evaluation of Local Wheat Landraces in Breeding Programmes in Syria*

K. OBARI

The evaluation of local landraces of wheat plays an important role in providing useful germplasm for national breeding programmes. It is important to use some of the more eroded landraces in breeding programmes and to conserve these genetic resources, maintaining their diversity under storage.

Studies on local wheat landraces were carried out at the International Center for Agricultural Research in the Dry Areas (ICARDA) near Aleppo and at the Agricultural Research Council (ARC) near Douma in 1984. Between 1986 and 1988, some 346 local landraces (bread and durum) and a few Italian varieties were evaluated at various national research stations in Syria. This paper presents some of the results of this evaluation.

## MATERIALS, METHODS AND RESULTS

Local durum wheat landraces have many names, including Haurani, Gezia, Itali, Senn El-Jamal, Biaddi, Hridieh, Jori and Baladi Ahmar. The names for local bread wheat landraces include Qaraa, Mexipak, Salamoni and Mexiky. The experiments were conducted in wide range of locations in Syria (*see* Table 20.1 *overleaf*)

Seed (20 g) from each sample was planted in two 2.5 m-long rows with a distance between them of 25 cm. A randomized block design was used. Fertilizer application consisted of 100 kg of $P_2O_5$/ha and 25 kg N/ha during seedbed preparation. A further 25 kg N/ha was applied just before heading. Weeding and irrigation were carried out as and when required. The results were entered into a computerized database for analysis and utilization.

**Table 20.1** Year, research station and number of cultivars tested 1986-88 in Syria

| Year | Research station and cultivar nos. | Total |
|---|---|---|
| 1986 | Qarahta Dam (30), Deir Elzzor (40), Aleppo (40), Homs (40) | 150 |
| 1987 | Deir Elzzor (30), Aleppo (30), Homs (20), Hassakeh (20) | 100 |
| 1988 | Aleppo (30), Hassakeh (25), Qarahta (21), Deir Elzzor (20) | 96 |
| Total | | 346 |

The main phenological characters studied were germination, days to heading, days to flowering and maturity, and growth habit. The main morphological characters recorded were plant type, spike density, lodging, glume hair, plant height, spikelets per spike and awn characteristics. Grain yield, reaction to diseases and pests, and seed colour and size were also recorded.

The 24 lines designated as promising are listed in Table 20.2. Some of the results are given in Table 20.3.

**Table 20.2** Promising lines identified after evaluation studies in Syria, 1986-88

| Year | Number and name of promising accessions | Centre name |
|---|---|---|
| 1986 | Haurani (336), Mobalad (103 ) | Qarahta (Damascus) |
| 1986 | Haurani (152), Hammari (116 ) | Deir Elzzor |
| 1986 | Haurani (464), Senator Capelli (343) | Aleppo |
| 1986 | Haurani (69), Jori Mobalad (104) | Homs |
| 1987 | Senator Capelli (83), Tunsieh (75) | Deir Elzzor |
| 1987 | Haurani (14), Haurani (13) | Homs |
| 1987 | Haurani Beladi (105), Jori Mobalad (104) | Hassakeh |
| 1987 | Siedieh (44), Senn El Jamal (29) | Aleppo |
| 1988 | Strampelli (186), Gemelli (187 ) | Aleppo |
| 1988 | Gerardo (223), Sansone (225) | Deir Elzzor |
| 1988 | Jazira 17 (170), Haurani (167) | Hassakeh |
| 1988 | Hammari (148), Haurani (530 ) | Qarahta (Damascus) |

**Table 20. 3** Results of 24 local landraces and introduced cultivars evaluated at various agricultural research stations, 1986-88

| Year | Station name | Acc. No and name | Planting date | Dates of germination | Dates of maturity | Dates of harvest | Grain harvest | Plant type | Spike density | Disease [a] resistance | Lodging [b] | Glume hair | Plant height | Spikelets per spike | Yield [c] per plot | Plant colour |
|---|---|---|---|---|---|---|---|---|---|---|---|---|---|---|---|---|
| 1986 | Qarahta | 336 Haurani | 22.11.85 | 19.12.85 | 20.5.86 | 4.6.86 | Medium | Erect | Dense | High | P | Absent | 121 | 20 | 820 | White |
| 1986 | Qarahta | 103 Mobalad | 22.11.85 | 20.12.85 | 6.5.86 | 1.6.86 | Early | Erect | Dense | High | P | Absent | 90 | 19 | 711 | White |
| 1986 | Deir Elzzor | 152 Haurani | 26.11.85 | 5.12.85 | 15.5.86 | 24.5.86 | Medium | Semi erect | Dense | Medium | I | Absent | 100 | — | 1201 | Red |
| 1986 | Deir Elzzor | 116 Hammari | 26.11.85 | 6.12.85 | 18.5.86 | 24.5.86 | Medium | Semi erect | Dense | High | I | Low | 115 | — | 1175 | Red |
| 1986 | Aleppo | 464 Haurani | 30.11.85 | 23.12.85 | 10.6.86 | 16.6.86 | Early | Semi erect | Dense | Medium | — | — | 105 | 17 | 1116 | Brown |
| 1986 | Aleppo | 343 S. Capelli | 30.11.85 | 23.12.85 | 11.6.86 | 16.6.86 | Early | Erect | Dense | — | — | — | 90 | 19 | 891 | Brown |
| 1986 | Homs | 69 Haurani | 23.11.85 | 6.12.85 | 23.5.86 | 3.6.86 | Medium | Prostrate | Inter-mediate | — | — | — | 185 | 12 | 397 | — |
| 1986 | Homs | 107 Jori | 23.11.85 | 7.12.85 | 22.5.86 | 3.6.86 | Early | Erect | — | — | — | — | 120 | 12 | 391 | — |
| 1987 | Deir Elizzor | 83 S. Capelli | 25.11.86 | 15.12.87 | 30.5.87 | 10.6.87 | Very late | Semi erect | Dense | Low | — | Absent | 152 | 23 | 807 | Red |
| 1987 | Deir Elizzor | 75 Tunsirh | 25.11.86 | 16.6.86 | 3.6.87 | 10.6.87 | Late | Semi erect | Dense | Medium. | — | Low | 135 | 23 | 785 | Red |
| 1987 | Homs | 14 Haurani | 30.11.86 | 4.1.87 | — | 20.6.87 | Medium | Erect | Dense | Medium | I | Low | 101 | 22 | 771 | Red |
| 1987 | Homs | 107 Jori Mobalad | 30.11.86 | 18.1.87 | — | 20.6.87 | Medium | — | Dense | Low | P | Low | 100 | 40 | 544 | Red |

**Table 20. 3** (contd)

| Year | Station name | Acc. No. and name | Planting date | Dates of germination | maturity | harvest | Grain harvest | Plant type | Spike density | Disease [a] resistance | Lodging [b] | Glume hair | Plant height | Spiklets per spike | Yield [c] per plot | Plant colour |
|---|---|---|---|---|---|---|---|---|---|---|---|---|---|---|---|---|
| 1987 | Aleppo | 44 Siedieh (D) | 24.11.86 | 8.12.86 | 2.6.87 | 20.6.87 | Medium | Semi erect? | Inter-mediate | Medium | P | Low | 100 | 20 | 1078 | White |
| 1987 | Aleppo | 29 Senn El Jamal | 24.11.86 | 10.12.86 | 4.6.87 | 20.6.87 | Late | — | Dense | High | I | Low | 107 | 22 | 928 | White |
| 1988 | Aleppo | 186 Strampelli (IS) | 20.11.78 | 23.12.78 | 27.5.88 | 20.6.88 | Medium | Semi erect | Very dense | — | I | — | 105 | — | 1329 | Purple |
| 1988 | Aleppo | 187 Gemelli SI | 20.11.78 | 24.12.87 | 1.6.88 | 20.6.88 | Medium | Semi erect | Very dense | — | S | — | 100 | — | 1143 | Brown |
| 1988 | Deir Elzzor | 223 Gerardo DI | 30.11.87 | 13.12.87 | 25.5.88 | 9.6.88 | Medium | Semi erect | Inter-mediate | Low | I | Absent | 125 | — | 1171 | Light brown |
| 1988 | Deir Elzzor | 225 Sansone DI | 30.11.87 | 15.12.78 | 25.5.88 | 9.6.88 | Medium | Semi erect | Dense | Medium | — | Absent | 65 | — | 961 | Light brown |
| 1988 | Hassakeh | 170 Jazira 17 | 1.12.87 | 16.12.87 | 8.6.88 | 19.6.88 | Early | — | — | High | — | — | 85 | — | 717 | Light brown |
| 1988 | Hassakeh | 167 Horani | 1.12.87 | 16.12.87 | 5.6.99 | 9.6.88 | Early | — | — | — | — | — | 110 | — | 587 | Light brown |
| 1988 | Qarahta | 148 Hammari | 24.11.87 | 9.12.87 | 24.5.88 | 7.6.88 | Early | Erect | Inter-mediate | Medium | P | Low | — | — | 561 | Light brown |
| 1988 | Qarahta | 530 Haurani | 24.11.87 | 9.12.87 | 25.5.88 | 7.6.88 | Early | — | Dense | — | — | Absent | — | — | 486 | Dark brown |

a Early stage evaluations of reaction to disease are not included; b For lodging: P = prostrate, I = intermediate, S = susceptible; c Plot acreage equals = 1.25 $m^2$

Wheat Genetic Resources: Meeting Diverse Needs
Edited by J. P. Srivastava and A. B. Damania

Published by John Wiley & Sons

# 21

# *Morphological Variation in* Triticum dicoccoides *from Jordan*

A. A. JARADAT and B. O. HUMEID

The presence of wild emmer wheat, *Triticum dicoccoides*, in Jordan is not mentioned in any Flora of the West Asian region (Post and Dinsmore, 1933; Mouterde, 1966; Bor, 1968) and collection missions carried out as recently as the early 1980s (Witcombe et al., 1981) failed to locate it. In a review of the species (Poiarkova, 1988) the catchment area of the Upper Jordan was considered as the centre of distribution for this species, but it has since has been found at more than 30 locations in the mesic and xeric regions of Jordan (Jaradat and Jana, 1987; Jaradat et al., 1988). A large number of accessions have been collected and are now maintained at several genebanks.

A few of the collected samples were found to be associated with the *Quercus* shrub. This association is reported to be the natural habitat of the species (Bor, 1968). However, most accessions were collected in or near relatively undisturbed sites, and in cultivated fields. In 1988, sympatrically distributed accessions of *T. dicoccoides* and *T. durum* were collected from 12 sites, all in northern Jordan. These collections are of special importance because of the possibility of introgression between the two species.

The reduced genetic variability of cultivated wheat (Frankel and Soulé, 1981) makes this major food crop increasingly vulnerable to biotic and abiotic stresses. Wild emmer wheat has been shown to harbour considerable genetic diversity, with many types adapted to different climates and soil types (Srivastava et al., 1988), and to include diverse sources of disease resistance (Moseman et al., 1984), drought tolerance (Blum et al., 1983), cold tolerance and other desirable agronomic traits (Nevo et al., 1984; Poiarkova, 1988). In view of the generally higher homogeneity of the newly bred wheat cultivars (Plucknett et al., 1983), the genetic diversity of wild emmer should prove extremely useful in wheat improvement.

In this paper we report on the substantial variation among accessions of wild emmer from Jordan for several qualitative and agronomic traits.

## MATERIALS AND METHODS

The wild emmer collections used in this study were assembled over the period 1983-86 from 10 locations in Jordan (*see* Figure 21.1). These locations represent the mesic and xeric Mediterranean environments in the country. The collection sites extended from 10 km south-west of Irbid in northern Jordan to 14 km north-west of Madaba in central Jordan.

**Figure 21.1** Map of Jordan showing collection sites of wild emmer

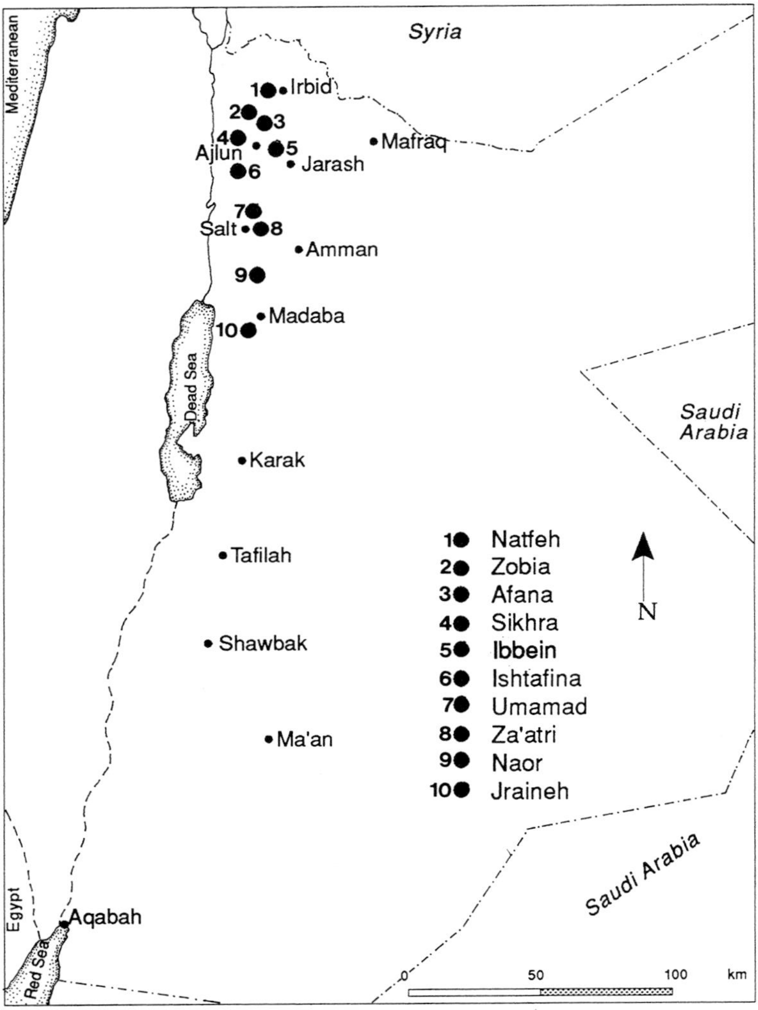

A field experiment consisting of 42 entries and three replicates was established. For such a large field experiment, we used the 'nearest neighbour analysis' as described by Wilkinson et al. (1983). Three improved wheat varieties (Bezostaya, Sonalika and Sham 1) and a durum wheat landrace (AJ 84027) were included in the experiment as check varieties.

The field experiment was carried out at the main research station of the International Center for Agricultural Research in the Dry Areas (ICARDA) during the 1986-87 growing season. Twenty young plants, from each of the 42 accessions, were scored for early vigour, juvenile growth habit, waxiness, penultimate leaf attitude, flag leaf attitude, reaction to yellow rust and fertility of the basal floret in the spike. The average diversity index (H') was estimated for each of the qualitative traits in each of the collection sites, following Jain et al. (1975).

The following traits were recorded for 15 mature plants per accession: plant height, peduncle length, spike length, number of spikelet groups per spike, number of productive tillers and leaf area. The ratio of number of spikelet groups per spike to flag leaf area was calculated and is considered as a crude measure of source-sink relationship.

## RESULTS

### Qualitative traits

The wild emmer collection displayed a remarkable level of diversity for all the qualitative traits under study. The H' estimates ranged from 0.23 for fertility of basal floret to 0.89 for waxiness (*see* Table 21.1 *overleaf*). Two qualitative traits are of particular importance — fertility of the basal floret and reaction to yellow rust. A small proportion of the whole collection (20 per cent) had fertile basal florets. However, in some locations, such as Umamad and Ishtafina, this proportion was as high as 60 per cent. In these locations, wild emmer was collected in or near durum wheat fields.

Reaction to yellow rust displayed a high level of diversity (H' = 0.89). A high proportion (54 per cent) of the material was resistant, 41 per cent was moderately resistant and only 5 per cent was susceptible.

### Quantitative traits

Summary statistics for days to heading, days to maturity and filling period are presented in Table 21.2 (*overleaf*). Days to heading and filling period showed relatively high variability. Accessions with early heading dates and short filling periods came from locations at low to medium elevations (< 700 m), while those with late heading dates and long filling periods came from mountainous areas with elevations over 900 m.

**Table 21.1** Average diversity indices estimates (H's) and phenotypic classes for several qualitative traits in a wild emmer collection from Jordan

| Trait | H' | Percent phenotype |
|---|---|---|
| Early vigour | 0.69 | 54 healthy, 46 medium |
| Juvenile growth habit | 0.69 | 49 medium, 91 prostrate |
| Growth class | 0.84 | 63 spring, 30 facultative, 7 winter |
| Penultimate leaf attitude | 0.64 | 67 erect, 33 medium |
| Flag leaf attitude | 0.29 | 8.5 medium, 91.5 horizontal |
| Waxiness | 0.89 | 8.5 strong, 57 medium, 34.5 absent |
| Yellow rust reaction | 0.84 | 54 resistance, 41 medium, 5 susceptible |
| Fertility of basal floret | 0.23 | 20 fertile, 80 sterile |

**Table 21.2** Summary statistics for days to heading, days to maturity and filling period, for wild emmer, improved wheat varieties and a durum landrace collection from Jordan

| Trait | Wild emmer | | | Improved cultivars | | | Landrace | | |
|---|---|---|---|---|---|---|---|---|---|
| | Min | Mean | Max | Min | Mean | Max | Min | Mean | Max |
| Heading | 115 | 135 | 148 | 125 | 130 | 135 | 130 | 131 | 133 |
| Maturity | 156 | 160 | 162 | 159 | 161 | 162 | 162 | 162 | 162 |
| Filling | 13 | 25 | 36 | 26 | 33 | 44 | 29 | 31 | 33 |

Comparisons between the check varieties and the wild emmer collection for six quantitative traits are given in Table 21.3.

**Table 21.3** Summary statistics and mean comparison for several agronomic traits between accessions of wild emmer, improved wheat varieties and a durum landrace from Jordan

| Traits | Wild emmer Range | Improved varieties Range | Landrace Mean |
|---|---|---|---|
| Plant height | 72.0 - 87.6[a] | 71.0 - 80.7 | 86.12 |
| Peduncle length | 32.3 - 42.6 | 35.3 - 36.0 | 53.30[a] |
| Spike length | 10.7 - 14.6[a] | 8.3 - 12.4 | 7.26 |
| Productive tillers | 20.6 - 30.1[b] | 16.5 - 18.9 | 19.13 |
| Spikelet groups | 17.9 - 23.4 | 17.9 - 24.0 | 24.13[b] |
| Flag leaf area | 17.0 - 30.0 | 23.3 - 40.7 | 48.87[b] |
| Spikelet groups/flag leaf area | 0.64 - 1.25[a] | 0.59 - 0.768 | 0.494 |

a P = 0.05; b P = 0.01

Plant height (*see* Figure 21.2) and peduncle length (*see* Figure 21.3a *overleaf*) displayed similar geographic distribution. They were positively and significantly correlated (r = 0.597, P = 0.019). Generally, short, slim wild emmer plants came from sites at high elevation, while stout, robust plants came from sites at low elevation. The low yet highly significant correlation coefficient indicates that both types are present in more than one location.

**Figure 21.2** Minimum, mean and maximum values for plant height in a collection of wild emmer from different locations in Jordan

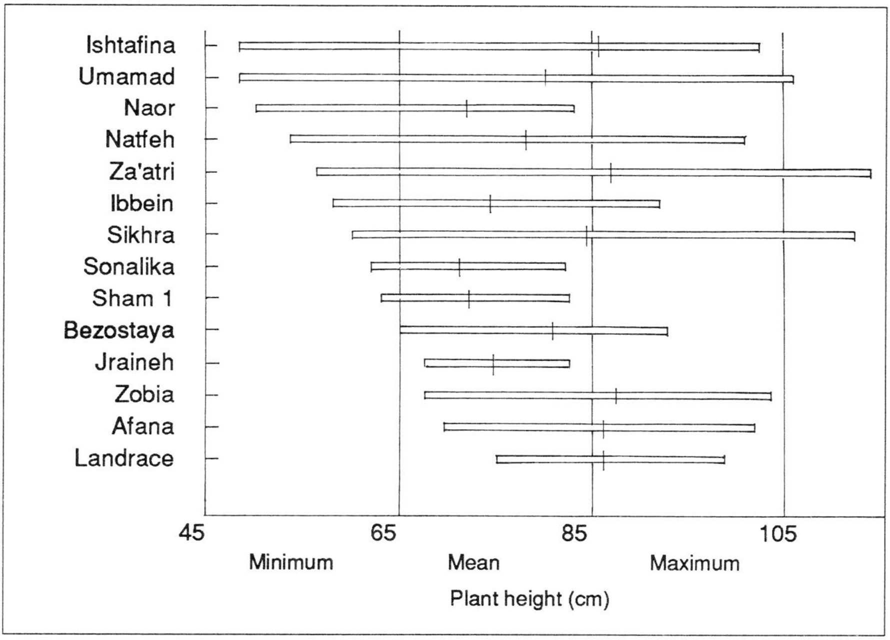


Variability in spike length is shown in Figure 21.3b (*overleaf*). Except for Bezostaya, the cultivated wheats had shorter than average spikes, while wild emmer accessions showed a wide range of spike lengths, above and below the mean. Spike length was negatively (r = -0.494) but not significantly (P = 0.061) correlated with flag leaf area.

Deviation from the mean for spikelet groups per spike is presented in Figure 21.3c (*overleaf*). Wild emmer accessions from Ishtafina had the largest number among the wild emmer collection. This important agronomic trait was positively and significantly correlated (r = 0.62, P = 0.014) with flag leaf area. The association between the two traits is clear when comparing Figures 21.3c and 21.3d (*overleaf*).

**Figure 21.3** Deviations from the mean for (a) number of spikelet groups/spike, (b) peduncle length, (c) spike length, (d) flag leaf area, (e) productive tillering capacity and (f) the ratio of a/d (the number of spikelet groups/spike/flag leaf area) in a collection of wild emmer from different locations in Jordan

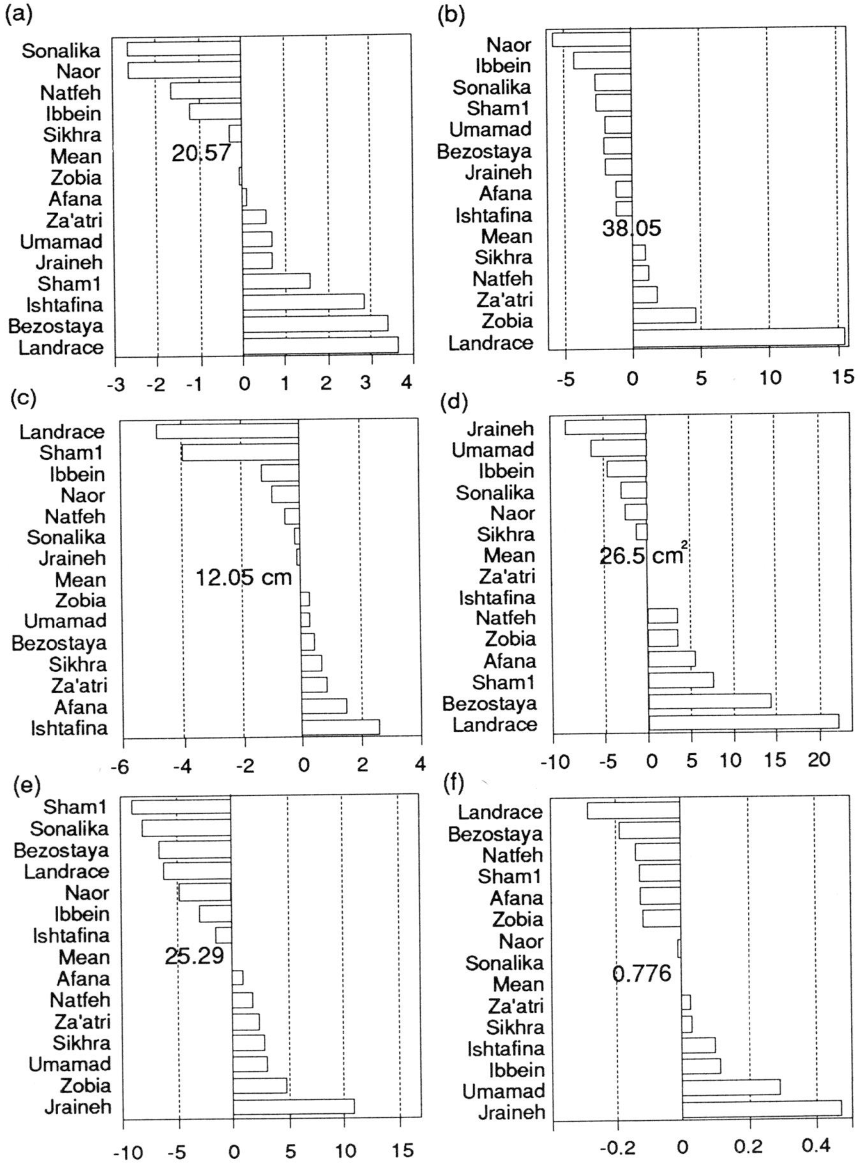

The productive tillering capacity of wild emmer was negatively and significantly correlated with flag leaf area (r = -0.517, P = 0.048). Wild emmer genotypes with small flag leaves have the capacity to produce more fertile tillers (*see* Figure 21.3e). Accessions from two dry locations (Umamad and Jraineh) are characterized by their high ratio of spikelet groups to flag leaf area (*see* Figure 21.3f) and high productive tillering capacity values. These accessions might harbour genes for drought tolerance and, possibly, for higher photosynthetic capability.

## CONCLUSION

The results of this study both confirmed and added to earlier findings (Jaradat et al., 1988) concerning the variability of wild emmer from Jordan. However, the high H' estimate for yellow rust (0.84) and the high proportion of accessions resistant to this pathogen (54 per cent) add to the importance of the wild emmer genepool in Jordan. Rust resistance must have been acquired over a long period of host-pathogen co-evolution, and should be used for breeding purposes.

The geographical variation for the quantitative traits was dramatically displayed when the accessions were grown in a relatively standardized environment. This confirms earlier findings by Frankel and Soulé (1981) and Nevo et al. (1984). Quantitative traits, which differed significantly among accessions from different sites, originated by natural selection as adaptations to the specific environment in which each accession evolved. A similar conclusion was reported for wild and domesticated barley species in Jordan (Jaradat, 1989). Remarkable variability was found, for example, for days to heading and for filling period. As expected, early-heading genotypes with short filling periods came from warmer sites at low elevations, while genotypes from cooler sites at higher elevations were characterized by late heading and longer filling periods. Early heading and a short filling period are important adaptive traits for drought tolerance (Blum et al., 1983).

Two races of wild emmer wheat are present in this collection from Jordan. The first is the robust race (Kushner et al., 1984; Poiarkova, 1988). This race is characterized by tall, early-maturing plants which have large spikes, a large number of spikelets per spike and a large number of tillers per plant. In a recent review of the morphology, geography and intraspecific taxonomics of *T. dicoccoides*, Poiarkova (1988) stated that: 'To date, the existence of the robust wild emmer is known only from the catchment of the Sea of Galilee'. However, the race has now been found in relatively dry areas of Jordan, in or near durum wheat fields, at elevations of between 500 and 700 m.

The second race is the 'grassy', slender type, which was considered by Kushner and Halloran (1982) to be the primary form of *T. dicoccoides*. It is characterized by short, slender plants with short spikes and fewer tillers than the robust type. It was collected from mountainous areas of Jordan, at elevations between 900 and 1350 m.

Our results indirectly indicated that some of the tested genotypes are drought tolerant and/or of higher photosynthetic capability than the check varieties. This was expressed as a higher ratio of number of spikelet groups to unit flag leaf area.

## *Discussion*

S. Jana: Why do you have to utilize *dicoccoides* to lengthen or shorten grain filling period or such other characters in durum wheat?

A. A. Jaradat: We are looking at the variability for whatever traits we can observe in the *dicoccoides* collection; grain filling period was one such trait. Whether the source of genes for this trait comes from *dicoccoides* or the durum landraces is for the breeder to decide. But we found a much broader variation for this trait in *dicoccoides* than in the landraces.

A. B. Damania: In your estimation, how many morphological forms of *T. dicoccoides* exist in Jordan or elsewhere?

A. A. Jaradat: I would like to refer you to the papers by Kushnir and Halloran (1982) in which the primitive grassy form of *dicoccoides* is described as a morphological variant, and by Poiarkova (1988) which discusses morphology, geography and intraspecific taxonomy of this wild species. I believe there are essentially two races — a grassy type which was considered as the primitive form of *T. dicoccoides* and a robust, more advanced type. In my data on plant height combined with spike length and number of spikelets per spike, the two extremes represent the grassy form and the robust form and any intermediaries are natural variants across the scale. But I do not know if there is a genetical basis for these differences.

A. B. Damania: Electrophoretical studies on storage proteins of single seeds from both types might indicate some genetical links and/or differences in the banding pattern.

A. A. Jaradat: Yes. We plan to carry out electrophoretical studies.

J. G. Waines: I believe electrophoresis will not be able to distinguish between the two types. It can only tell you if there is any correlation between another enzymatic character or not. I think you have to cross the two and analyse the genetics of the differences. There are similar forms of morphologically different diploid wheats in Turkey. There is a large and a small form and they are distinct, although growing in the same population. Also, you have a tetraploid which is very similar to the diploid and it is difficult to tell them apart. You can usually only do that if you look at the chromosome number.

S. Jana: Dr Helena Poiarkova spent her sabbatical at the University of Saskatchewan and brought some of the seeds from crosses between the two *T. dicoccoides* forms. The segregation was not very clear but there seem to be two genes involved, with one being dominant.

Wheat Genetic Resources: Meeting Diverse Needs
Edited by J. P. Srivastava and A. B. Damania

Published by John Wiley & Sons

# 22

# *Characterization and Utilization of Sicilian Landraces of Durum Wheat in Breeding Programmes*

G. Boggini, M. Palumbo and F. Calcagno

In the first half of this century, durum wheat landraces were the only cereal crops cultivated in Sicily (Perini and Verona, 1954) (*see* Table 22.1 *overleaf*). The large number of landraces cultivated in that period should not be attributed to inefficient farming but to the need to find the most suitable landraces for specific environmental conditions. Each landrace has particular characteristics suited to certain climatic conditions (De Cillis, 1942). It is because of their high adaptability that some landraces are still cultivated today, particularly in the mountainous inland areas of the island.

The Germplasm Institute of Bari collected Sicilian landraces of durum wheat (Porceddu and Bennett, 1971; Perrino and Martignano, 1974), studied their morphological and genetic characteristics (Porceddu, 1979; Bogyo et al., 1980; De Pace et al., 1983; Spagnoletti-Zeuli et al., 1983, 1984, 1985), and used the material in breeding programmes. These programmes led to the development of the variety Norba, which, in trials carried out by the Experimental Research Institute for Cereal Crops, proved to have excellent grain yield and adaptability to the different areas of cereal cultivation in Italy (Spagnoletti-Zeuli et al., 1988). The Experimental Research Institute for Cereal Crops (Catania Section) and the Wheat Research Station of Sicily, at Caltagirone, have also made considerable use of these landraces in breeding programmes. Preliminary morphological, physiological, agronomical and technological studies were carried out on some landraces (Boggini et al., 1985, 1987) to identify the best parents.

In this paper we present the results of agronomic trials and technological evaluation of some landraces; agronomic and technological comparisons between lines within landraces; and the utilization of Sicilian landraces in breeding programmes.

## MATERIALS AND METHODS

Agronomic trials and technological evaluations were carried out in Catania in the 1984-85 and 1985-86 seasons on 15 landraces, which were compared with five check varieties commonly grown in Sicily. The characteristics studied are given in Table 22.2. The trials were conducted using a randomized block design with three replications and plots of 10 $m^2$. The landraces were grown according to the agronomic practices most suitable for the improved cultivars (seeding rate of 450 germination

**Table 22.1** Landraces grown in the Sicilian provinces in 1950-51

| Sicilian provinces | Landraces | Wheat area% |
|---|---|---|
| Agrigento | Russello, Timilia, Sammartinara, Maiorcone | 89 |
| Trapani | Biancuccia, Duro di Crotone, Timilia, Ruscia | 60 |
| Enna | Russello, Bufala, Timilia | 52 |
| Siracusa | Timilia, Ruscia, Farro, Maiorca | 50 |
| Ragusa | Regina, Ruscia, Trentino, Urria | 37 |
| Catania | Regina, Ruscia, Trentino | 20 |
| Palermo | Timilia, Sammartinara, Biancuccia, Realforte, Maiorca, Maiorcone | <10 |
| Caltanissetta | Timilia, Sammartinara, Biancuccia, Realforte, Maiorca, Maiorcone | <10 |

Sources: Perrino and Verona, 1954

**Table 22.2** Procedures used for agronomic and quality measurements

| Measurements | Method |
|---|---|
| 1 Grain yield (t/ha) | Agronomic trial |
| 2 Heading time | Days from 1 January |
| 3 Plant height (cm) | Agronomic trial |
| 4 Grain test weight (kg/hl) | Agronomic trial |
| 5 Thousand grain weight (g) | Agronomic trial |
| 6 Number of kernels/spike | Agronomic trial |
| 7 Number of spikes per square metre | Agronomic trial |
| 8 Harvest index | Biological yield/grain yield |
| 9 Protein concentration (% d.m.) | Micro-Kjeldahl (N x 5.7) |
| 10 Wet gluten content (% d.m.) | Glutomatic system |
| 11 Ash content (% d.m.) | 600°C for 12 hours |
| 12 Organolectic pasta judgement (0 - 10) | D'Egidio et al., 1982 |
| 13 Total organic matter (g starch/100g pasta) | D'Egidio et al., 1982 |

seeds/$m^2$, nitrogen fertilizer application of 90 kg N/ha, and anticryptogamic treatment). The objective was to evaluate the yield potential of the landraces, despite the risk of lodging.

In 1983, to identify variability within each landrace, electrophoretic analyses of the gliadins (Dal Belin Peruffo et al., 1980) and of the glutenins (Payne et al., 1980) were carried out on a sample of 100 seeds from 24 landraces, using half kernel. Eighteen landraces were heterogeneous — that is, composed of different genotypes. These genotypes were then grown using the other half kernel, and then morphologically and agronomically characterized.

The trials began in the 1984-85 season and were conducted for 3 years. A total of 75 lines were tested in 7.65$m^2$ plots using a randomized block design, with two replications in the first 2 years and three in the third year. The characters observed are those described under 1 to 5 in Table 22.2. Data for protein content and sedimentation value in sodium dodecyl sulphate (SDS) are available for only the second and third years (Axford et al., 1979).

In 1979, Regina and Vallelunga were crossed with Creso to attenuate the undesirable morphological characteristics of the two landraces (very tall plants, very long cycle and susceptibility to leaf rust). The two landraces were chosen for their excellent adaptability to the Sicilian environment, as indicated by their widespread cultivation in the past.

Starting with the $F_2$ generation, selection was carried out on the basis of plant height, earliness and resistance to leaf rust, while attempting to retain the structure and vigour of the Sicilian landraces. The lines that were morphologically uniform in the $F_5$ generation were evaluated in preliminary agronomic trials in the 1985-86 growing season. During this season, the rainfall (162.6 mm between sowing and maturity) and temperatures (above 30°C from the second half of May onwards) allowed evaluation for performance under drought conditions. Some 54 lines with good yield capacity were selected for further evaluation in comparison with the check cultivars Appulo, Capeiti, Creso and Duilio.

These trials were carried out in the 1986-87 and 1987-88 seasons. Fertilizer was applied at a rate of 90 kg $P_20_5$/ha and 80 kg N/ha, according to common practice for durum wheat in the region. Randomized block design was used, with three replications and plots of 10 $m^2$.

## RESULTS

### Agronomic trials and technological evaluation of 15 landraces

In 1984-85, the climatic conditions were particularly favourable to the crop, whereas in 1985-86 high temperatures during the heading/flowering period resulted in reduced yields. Significant differences between the two germplasm groups (landraces and, as checks, improved cultivars) were observed for the following characteristics: time to

**Table 22.3** Agronomic and quality measurements of landraces and varieties grown in Catania 1984-85 and 1985-86

| | Grain yield (t/ha) | Heading time | Plant height (cm) | Grain test weight (kg/hl) | TKW (g) | Kernels spike (n°) | Spike/MQ (n°) | Harvest index | Protein conc. (% s.s) | Wet gluten content (%) | Ash content (%) | Organolectic pasta | |
|---|---|---|---|---|---|---|---|---|---|---|---|---|---|
| | | | | | | | | | | | | Judgement (0 - 10) | Total organic matter g starch/100g pasta |
| **Landraces** | | | | | | | | | | | | | |
| Castiglione Gl. | 1.91 | 110.3 | 130.2 | 80.9 | 31.4 | 39.3 | 423.3 | 0.31 | 11.70 | 30.0 | 1.04 | 6.3 | 1.41 |
| Crotone | 1.63 | 109.0 | 132.3 | 80.5 | 33.0 | 38.0 | 419.7 | 0.26 | 11.82 | 27.5 | 0.93 | 6.7 | 1.66 |
| Farro Lungo | 2.09 | 110.2 | 129.2 | 79.9 | 35.0 | 33.0 | 400.0 | 0.34 | 11.92 | 30.5 | 0.97 | 6.0 | 1.59 |
| Gioia | 1.89 | 107.2 | 123.2 | 80.3 | 35.9 | 37.0 | 478.7 | 0.30 | 11.95 | 30.5 | 0.99 | 4.0 | 1.98 |
| Lina | 1.94 | 108.0 | 126.8 | 80.4 | 32.7 | 38.7 | 508.0 | 0.29 | 12.15 | 31.5 | 0.94 | 5.2 | 1.63 |
| Margherito | 1.84 | 110.3 | 137.5 | 81.0 | 35.7 | 39.7 | 421.7 | 0.26 | 11.73 | 34.5 | 1.00 | 6.0 | 1.54 |
| Martinella | 1.82 | 109.2 | 123.7 | 78.8 | 31.9 | 28.0 | 466.7 | 0.33 | 11.84 | 29.0 | 1.01 | 5.3 | 1.86 |
| Russello | 2.31 | 110.8 | 136.8 | 80.5 | 35.2 | 45.0 | 382.7 | 0.37 | 12.17 | 31.0 | 1.01 | 6.0 | 1.44 |
| Scorsonera | 2.10 | 109.3 | 130.7 | 81.5 | 35.6 | 37.7 | 533.3 | 0.27 | 13.19 | 31.0 | 1.02 | 6.5 | 2.02 |
| Sicilia Lutri | 2.72 | 111.0 | 109.0 | 81.8 | 28.9 | 29.7 | 625.7 | 0.37 | 14.45 | 42.0 | 1.01 | 5.3 | 1.59 |
| Sicilia Reste Nere | 2.41 | 108.2 | 104.0 | 81.1 | 33.0 | 32.3 | 520.0 | 0.44 | 12.09 | 33.5 | 0.94 | 5.7 | 1.95 |
| Timilia | 2.08 | 114.3 | 114.8 | 80.9 | 28.3 | 26.3 | 615.7 | 0.29 | 13.57 | 34.0 | 0.98 | 5.1 | 1.88 |
| Tripolino | 2.63 | 108.3 | 137.3 | 81.6 | 42.3 | 45.3 | 367.7 | 0.33 | 12.28 | 32.0 | 1.02 | 3.5 | 2.66 |
| Urria | 2.48 | 108.2 | 121.8 | 80.9 | 31.2 | 47.0 | 482.3 | 0.41 | 11.87 | 30.0 | 0.97 | 6.7 | 2.12 |
| Vallelunga Gl | 2.47 | 108.2 | 133.0 | 81.5 | 34.1 | 35.0 | 425.7 | 0.41 | 11.59 | 31.0 | 1.05 | 4.7 | 2.06 |
| Landraces mean | 2.16 | 109.5 | 126.6 | 80.7 | 33.6 | 36.8 | 471.4 | 0.33 | 12.29 | 31.87 | 0.99 | 5.53 | 1.80 |
| ± SE | ± 0.12 | ± 0.6 | ± 2.5 | ± 0.8 | ± 1.9 | ± 0.4 | ± 10.2 | ± 0.01 | ± 0.21 | ± 0.87 | ± 0.07 | ± 0.25 | ± 0.07 |
| **Varieties test** | | | | | | | | | | | | | |
| Capeiti | 2.19 | 93.3 | 105.8 | 81.9 | 32.6 | 43.0 | 456.3 | 0.53 | 13.06 | 37.5 | 0.86 | 6.7 | 1.47 |
| Creso | 3.17 | 110.0 | 83.7 | 81.0 | 37.2 | 45.7 | 379.7 | 0.55 | 12.24 | 35.0 | 1.01 | 6.0 | 1.42 |
| Karel | 2.69 | 94.3 | 83.5 | 78.2 | 32.5 | 58.7 | 490.0 | 0.61 | 11.95 | 30.5 | 0.88 | 6.3 | 1.69 |
| Produra | 2.16 | 90.2 | 75.3 | 78.8 | 32.3 | 22.7 | 729.0 | 0.45 | 14.45 | 34.5 | 0.98 | 6.0 | 1.65 |
| Trinakria | 2.34 | 93.8 | 109.7 | 80.3 | 38.2 | 28.7 | 446.7 | 0.35 | 16.20 | 47.0 | 0.80 | 6.7 | 1.31 |
| Varieties mean | 2.51 | 96.3 | 91.6 | 80.0 | 34.5 | 39.7 | 500.3 | 0.50 | 13.58 | 36.90 | 0.91 | 6.34 | 1.51 |
| ± SE | ± 0.15 | ± 0.6 | ± 2.2 | ± 0.8 | ± 1.6 | ± 1.4 | ± 14.5 | ± 0.02 | ± 0.78 | ± 2.76 | ± 0.04 | ± 0.16 | ± 0.07 |
| t -value | 0.79 [a] | 2.60 [b] | 2.61 [b] | 0.59 [a] | 0.23 [a] | 0.38 [a] | 0.35 [a] | 12.14 [c] | 2.30 [b] | 2.35 [b] | 0.15 [a] | 1.84 [a] | 2.23 [b] |

a Not significant; b $P = 0.05$; c $P = 0.01$

heading, plant height, harvest index, protein and wet gluten content, and surface stickiness in a pasta cooking test (*see* Table 22.3).

No significant difference between landraces and cultivars was observed for grain yield. Creso was the most productive cultivar (3.17 t/ha), followed by the landraces Sicilia Lutri, Tripolino, Urria and Vallelunga Glabra. The yield from Vallelunga Glabra was similar to that from Karel (2.69 t/ha). Good yields were also obtained from the cultivars Sicilia Reste Nere and Russello; these yields were similar to those from Trinakria (2.34 t/ha). The productivity of all the other genotypes in the trials was poor.

Creso and all the landraces headed 1-2 weeks later than the other cultivars. Some landraces (Gioia, Lina, Urria, Vallelunga Glabra, Sicilia Reste Nere and Tripolino) were earlier than Creso. As expected, the landraces were significantly taller than the cultivars. Sicilia Lutri (109.0 cm) and Sicilia Reste Nere (104.0 cm) were the shortest landraces. The cultivars had a slightly higher number of kernels per spike than the Sicilian populations, but some landraces (Urria, Tripolino and Russello) showed a level of head fertility comparable with some of the cultivars. The number of spikes per $m^2$ was similar in the two groups, but the cultivar Produra and the landraces Sicilia Lutri and Timilia tillered more than the other cultivars. There was a significant difference in the harvest indices between the two groups; among the cultivars, the lowest index was obtained in Trinakria and, among the landraces, in Cotrone, Margherito, Scorsonera, Gioia and Castiglione Glabro. The thousand-kernel-weight (TKW) of the two groups was similar; however, the high value of Tripolino deserves attention, as do the good values obtained for Gioia, Margherito, Scorsonera and Russello, which were similar to the values for Creso and Trinakria.

These results indicate that Sicilian durum wheat landraces have good yield potential and that they differ from the improved cultivars only in plant height, harvest index and days to heading. In terms of the first two of these traits, landraces may be considered undesirable in a high-input agriculture system, but the same is not necessarily true in the case of time to heading and maturity. Early heading is, in fact, simply a mechanism to escape drought, not a mechanism of resistance to it. Late-heading cultivars which are able to cope with drought must possess one or more avoidance or resistance mechanisms (water retention, or the capacity to extract water from deeper soil layers).

From a qualitative point of view, the most important differences between the two groups were found in protein and wet gluten contents (*see* Table 22.3). Trinakria confirmed its high capacity to accumulate protein, as did Sicilia Lutri, Timilia and Scorsonera. Sicilia Lutri was also the most productive landrace, in spite of having the lowest TKW. In the pasta-making test, no significant difference between the groups was observed, but the surface stickiness index of cooked pasta was significantly different. All the improved cultivars, particularly Trinakria and Capeiti, were well suited to pasta making (*see* Table 22.3). Good landraces for pasta making were Urria, Cotrone and Scorsonera; the first, however, had a high surface stickness index. Farro Lungo, Castiglione glabro, Russello and Margherito also showed good quality. Some landraces, including Tripolino, Gioia, Vallelunga Glabra, Timilia, Lina, Sicilia Lutri and Maritinella were very poor for pasta making.

**Table 22.4** Range of agronomic and qualitative characters and significance of their variability among lines within landraces

| Landraces | Number of lines | Range (minimum : maximum) and significance | | | | | | |
|---|---|---|---|---|---|---|---|---|
| | | Grain yield (t/ha) | Plant height (cm) | Heading time (dd from 1/4) | TKW (g) | Grain test weight (kg/l) | Protein conc. (% DM) | Sedimentation value (mm) |
| Biancuccia | 4 | (19.9 - 30.4) [a] | (119 - 124) [b] | (33 - 36) [a] | (34.7 - 38.8) [a] | (78.1 - 79.8)[a] | (16.6 - 17.9) [a] | (29.0 - 50.5) [c] |
| Castiglione Gl. | 5 | (24.3 - 34.4) [a] | (121 - 134) [a] | (31 - 36) [a] | (33.1 - 40.5) [a] | (80.9 - 82.1) [a] | (16.5 - 18.9) [a] | (32.0 - 48.5) [a] |
| Farro lungo | 7 | (26.3 - 30.1) [a] | (113 - 126) [a] | (31 - 35) [a] | (35.0 - 38.7) [a] | (78.0 - 80.8) [a] | (16.5 - 17.6) [a] | (25.0 - 47.0) [a] |
| Gioia | 2 | (26.7 - 29.8) [a] | (116 - 124) [a] | (29 - 30) [a] | (32.3 - 33.0) [a] | (79.8 - 80.1) [c] | (17.1 - 18.1) [a] | (34.0 - 35.5) [a] |
| Giustalisa | 3 | (20.5 - 28.5) [a] | (118 - 121) [a] | (29 - 30) [a] | (33.3 - 36.2) [a] | (76.0 - 80.2) [a] | (16.3 - 17.6) [a] | (28.5 - 31.5) [a] |
| Lina | 2 | (23.5 - 27.3) [a] | (129 - 133) [a] | (29 - 29) [a] | (38.6 - 40.8) [a] | (80.1 - 81.5) [a] | (16.5 - 17.5) [a] | (32.0 - 36.5) [a] |
| Martinella | 5 | (26.2 - 33.4) [a] | (110 - 123) [a] | (25 - 31) [a] | (33.6 - 37.8) [a] | (76.2 - 83.8) [c] | (17.0 - 18.4) [a] | (27.0 - 36.0) [b] |
| Pavone | 3 | (28.4 - 39.0) [a] | (117 - 132) [b] | (30 - 32) [a] | (33.1 - 35.6) [a] | (81.7 - 82.7) [a] | (16.9 - 17.5) [a] | (29.0 - 33.0) [a] |
| Realforte | 6 | (13.6 - 29.4) [a] | (101 - 124) [a] | (30 - 39) [a] | (30.1 - 37.6) [a] | (74.3 - 82.3) [c] | (17.1 - 18.7) [c] | (29.5 - 38.5) [c] |
| Ruscia | 5 | (25.1 - 29.4) [a] | (126 - 138) [a] | (27 - 32) [a] | (32.1 - 35.4) [a] | (78.3 - 81.0) [a] | (17.5 - 18.9) [a] | (32.5 - 42.0) [b] |
| Russello | 2 | (26.0 - 29.9) [a] | (125 - 126) [a] | (31 - 33) [a] | (36.1 - 38.3) [a] | (81.9 - 82.5) [c] | (17.3 - 17.7) [a] | (32.5 - 35.5) [a] |
| Scorsonera | 2 | (23.9 - 30.5) [a] | (117 - 131) [c] | (34 - 34) [a] | (36.7 - 39.3) [a] | (80.4 - 83.3) [a] | (17.5 - 18.0) [a] | (31.0 - 35.0) [a] |
| Semenzella | 7 | (23.4 - 32.6) [a] | (108 - 121) [a] | (26 - 36) [a] | (29.5 - 36.1) [a] | (78.8 - 82.5) [a] | (16.7 - 18.2) [a] | (26.0 - 36.5) [a] |
| Sicilia Lutri | 2 | (31.5 - 37.3) [a] | (106 - 112) [a] | (31 - 34) [a] | (30.2 - 32.1) [a] | (81.8 - 82.0) [a] | (17.0 - 18.5) [a] | (34.0 - 34.0) [a] |
| Sicilia R. Nere | 5 | (25.0 - 35.8) [b] | (99 - 120) [a] | (31 - 34) [a] | (30.1 - 40.5) [c] | (80.2 - 81.8) [a] | (16.7 - 18.4) [a] | (29.5 - 37.5) [a] |
| Urria | 3 | (23.3 - 33.8) [a] | (114 - 129) [a] | (30 - 35) a | (34.7 - 36.9) [a] | (77.2 - 79.5) [b] | (15.7 - 17.6) [a] | (26.0 - 35.5) [c] |
| Vallelunga Gl. | 3 | (25.8 - 43.8) [a] | (129 - 139) [a] | (26 - 34) c | (30.8 - 41.0) [b] | (80.2 - 84.2) [c] | (16.4 - 18.1) [a] | (28.5 - 43.5) [b] |
| Vallelunga Pb. | 5 | (20.8 - 35.3) [a] | (122 - 127) [a] | (27 - 34) a | (33.1 - 41.8) [c] | (78.1 - 82.1) [b] | (7.3 - 18.0) [a] | (28.5 - 35.5) [a] |

a Not significant; b $P = 0.05$; c $P = 0.01$

## Agronomic and technological comparison between lines within landraces

Lines in all the landraces except Gioia and Lina differed in one or more of the seven traits considered (*see* Table 22.4.). Vallelunga Glabra and Realforte differed in several traits. The most variable traits were grain test and sedimentation value. In terms of kernel weight, 12 landraces showed variability in one or more qualitative parameters. This is interesting from a breeding perspective because of the need to improve the technological quality of durum wheat using parents adapted to the climate of southern Italy.

Using the data in Table 22.4, several lines were selected on the basis of agronomic and/or qualitative traits as parents for use in breeding programmes (*see* Table 22.5). Scorsonera B and Sicila Lutri B, selected only for their agronomic traits, possess the gliadin component 42, whereas all the others have the component 45, although some of them were not selected for quality. Three selected lines (Ruscia B, Vallelunga Glabra H and Vallelunga Pubescente B) possess the high-molecular-weight (HMW)

**Table 22.5** Lines selected for agronomic and/or qualitative characters, and gliadin and glutenin composition

| | | Selection character | | Storage protein composition | | |
|---|---|---|---|---|---|---|
| **Landraces** | **Lines** | **Agronomic** | **Qualitative** | **Gliadin** | **HMW[a]** | **Glutenin** |
| | | | | | 1A | 1B |
| Biancuccia | C | | + | 45 | — | 6 + 8 |
| Biancuccia | D | + | | 45 | — | 20 |
| Castiglione | A | | + | 45 | — | 13 + 16 |
| Farro Lungo | E | + | | 45 | — | 20 |
| Giustalisa | B | + | + | 45 | — | 13 + 16 |
| Martinella | B nc | | + | 45 | — | 20 |
| Pavone | A | + | | 45 | — | 20 |
| Realforte | A sp | + | + | 45 | — | 13 + 16 |
| Realforte | A p | | + | 45 | — | 13 + 16 |
| Ruscia | B | | + | 45 | 2* | 6 + 8 |
| Russello | A | | + | 45 | — | 13 + 16 |
| Scorsonera | B | + | | 42 | — | 13 + 16 |
| Semenzella | F | + | + | 45 | — | 13 + 16 |
| Semenzella | G | + | + | 45 | — | 13 + 16 |
| Sicilia Lutri | B | + | | 42 | — | 13 + 16 |
| Sicilia Reste Nere | A | + | + | 45 | — | 6 + 8 |
| Urria | B | + | + | 45 | — | 6 + 8 |
| Vallelunga Gl. | D | + | | 45 | — | 13 + 16 |
| Vallelunga Gl. | E | + | + | 45 | — | 7 + 8 |
| Vallelunga Gl. | H | | + | 45 | 2* | 6 + 8 |
| Vallelunga Gl. | B | + | + | 45 | 2* | 20 |

a HMW = high-molecular-weight

glutenin 2*, while Vallelunga E possesses a '7+8' pair. These glutenin compositions give a high technological quality to the storage proteins (Boggini and Pogna, 1989).

The frequent presence of the '13+16' glutenin component is interesting, as it is rare in commonly cultivated Italian varieties (Pogna et al., 1985). Although this composition should give the glutenins a high quality (Payne et al., 1981), this was not confirmed by these results or by earlier results (Boggini et al., 1987; Boggini and Pogna, 1989).

## Utilizing Sicilian landraces in breeding programmes

Total rainfall was markedly different in the two years (*see* Table 22.6). In 1986-87 it was close to the 10-year average, whereas in 1987-88 it was much lower. In particular, spring rainfall in 1987-88 was 65 per cent lower than the 10-year average. The maximum temperatures from May onwards were over 34°C. Thus, 1988 was characterized by drought conditions, whereas the 1987 rainfall and temperatures figures were normal for the area, allowing the genotypes to express their full yield potential. As a result, the yield means were markedly different in the two years: 5.39 t/ha in 1986-87 and 2.24 t/ha in 1987-88 (*see* Table 22.7).

In both years, the productivity of the new lines (means of 5.44 t/ha in 1986-87 and 2.26 t/ha in 1987-88) was higher than that of the checks (means of 4.73 t/ha in 1986-87 and 1.94 t/ha in 1987-88). The new lines suffered less from drought than did the checks, producing 16.5 per cent more than them in 1987-88. But even in 1986-87 the lines yielded 15 per cent more than the checks. On average over the 2-year period, the lines produced 15.3 per cent more than the checks (lines 3.85 t/ha, checks 3.34 t/ha).

In 1986-87 Duilio was the most productive check (5.41 t/ha); 12 lines were more productive than Duilio, and seven were less productive (*see* Table 22.8). In 1987-88 Capeiti was the best check (2.18 t/ha), with 10 new lines having higher yields and only two having lower yields. In both years, Appulo and Creso had lower yields than most of the new lines. On average over the 2-year period, 33 lines were more productive than Duilio and 37 more productive than Capeiti. Table 22.9 shows grain yield, plant height and days to heading for the best new lines in comparison to the checks.

**Table 22.6** Autumn, winter and spring rainfall (mm) recorded during two growing seasons 1987-88 and during the previous 10-year period 1977-86 on the experimental farm

| | Growing season | | |
|---|---|---|---|
| **Period** | **1986-87** | **1987-88** | **1977-86** |
| Autumn | 299.8 | 40.9 | 184.5 |
| Winter | 89.2 | 125.3 | 189.8 |
| Spring | 87.0 | 28.8 | 82.1 |
| Total | 476.0 | 195.0 | 456.4 |

**Table 22.7** Yield (t/ha) of durum wheat lines selected from crosses Creso x landraces and of improved cultivars in 1986-87 and 1987-88

| Material | | Grain yield 1986-87 | 1987-88 | Average of two years |
|---|---|---|---|---|
| | mean | 5.44 | 2.26 | 3.85 |
| Lines | max | 6.65 | 3.27 | 4.96 |
| | min | 3.61 | 1.59 | 2.60 |
| Duilio | | 5.41 | 2.05 | 3.73 |
| Capeiti | | 5.06 | 2.18 | 3.62 |
| Appulo | | 4.63 | 1.93 | 3.28 |
| Creso | | 3.81 | 1.61 | 2.71 |
| Cultivar mean | | 4.73 | 1.94 | 3.34 |
| Grand mean | | 5.39 | 2.24 | 3.82 |
| CV | | 5.32 | 9.49 | 7.21 |
| LSD | (P = 0.01) | 0.62 | 0.46 | 0.10 |

**Table 22.8** Number and grain yield of durum wheat lines selected from crosses landraces x Creso with significantly higher yield than improved cultivars in 1986-87 and 1987-88, (P= 0.1)

| Check cultivars | 1986-87 | | 1987-88 | | Average for two years | |
|---|---|---|---|---|---|---|
| | No. | Grain yield | No. | Grain yield | No. | Grain yield |
| Duilio | 12 | 6.26 | 15 | 2.77 | 33 | 4.08 |
| Capeiti | 21 | 6.08 | 10 | 2.89 | 37 | 4.05 |
| Appulo | 33 | 5.87 | 17 | 2.73 | 48 | 3.94 |
| Creso | 50 | 5.56 | 37 | 2.45 | 54 | 3.85 |

**Table 22.9** Average for two years of grain yield, plant height and heading time of the lines significantly more productive than the best check cultivars

| Cross | No. lines | Range | | |
|---|---|---|---|---|
| | | Grain yield (t/ha) | Plant height (cm) | Days to heading from 1 April |
| Creso x Vallelunga | 22 | 4.40 : 3.83 | 82.5 : 72.1 | 29.0 : 19.6 |
| Creso x Regina | 11 | 4.36 : 3.86 | 80.6 : 73.0 | 31.3 : 22.7 |
| Duilio | — | 3.73 | 77.2 | 16.3 |
| Capeiti | — | 3.62 | 89.2 | 18.7 |
| Appulo | — | 3.28 | 85.0 | 18.7 |
| Creso | — | 2.71 | 71.2 | 29.0 |

Figure 22.1a shows that the best yields were obtained by genotypes with a plant height of between 90 and 100 cm. The performance of the shorter-stemmed lines differed markedly from that of the typical landrace, with its vigorous growth. Even in the unfavourable season there was a positive correlation between plant height and grain yield (*see* Figure 22.1b).

**Figure 22.1** Linear correlation between grain yield and plant height of durum wheat lines selected from crosses Creso x landraces and of improved cultivars in (a) 1986-87 and (b) 1987-88

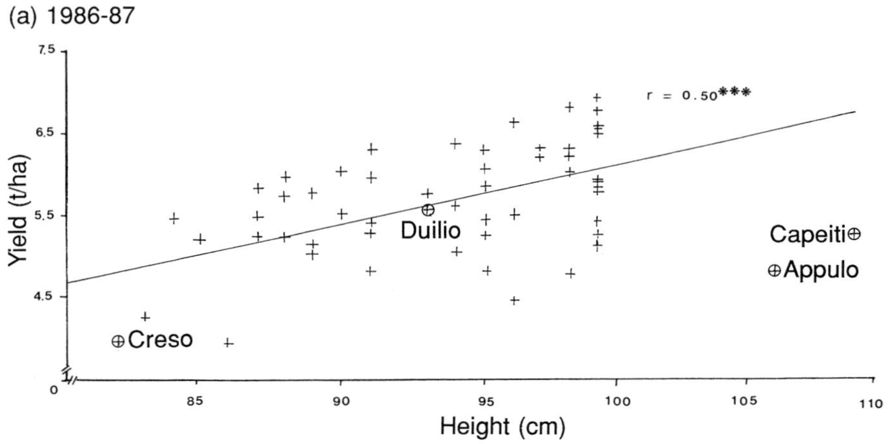


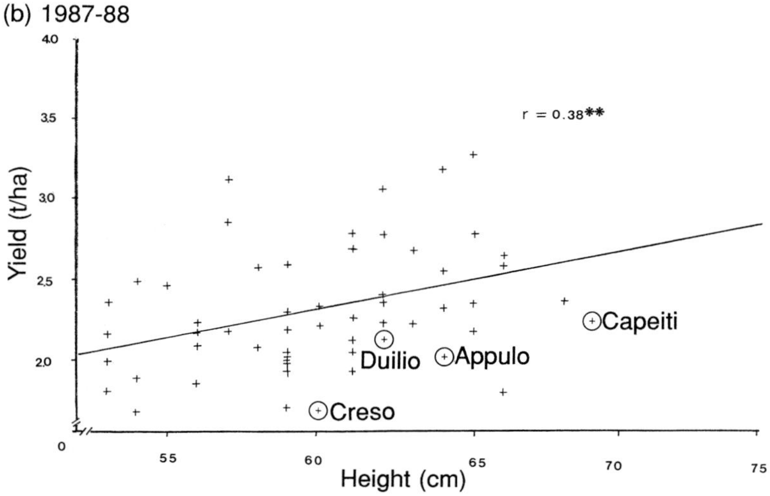

The 1986-87 yields were positively correlated with the length of the vegetative cycle (*see* Figure 22.2a), showing that the new lines had not only inherited a significant increase in this trait, but also an increase in photosynthetic period, a typical trait of the Sicilian landraces. But in the dry 1987-88 period this correlation was inverted, with early heading serving as an effective drought escape mechanism (*see* Figure 22.2b).

**Figure 22.2** Linear correlation between grain yield and earliness of durum wheat lines selected from crosses Creso x landraces and improved cultivars in (a) 1986-87 and (b) 1987-88

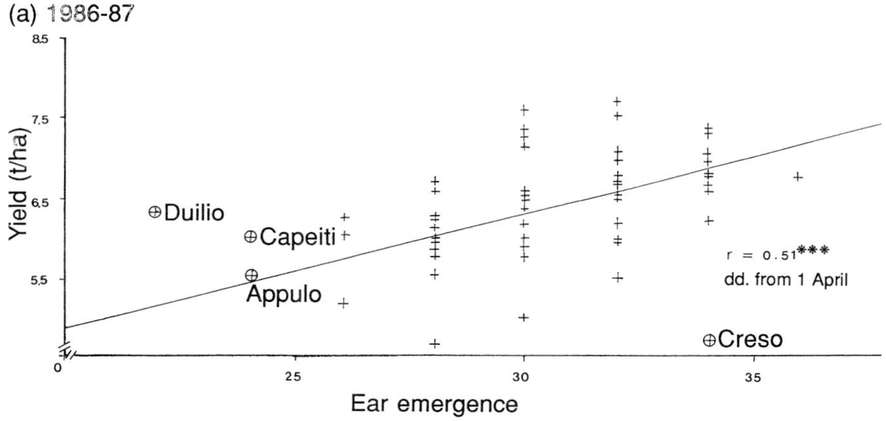


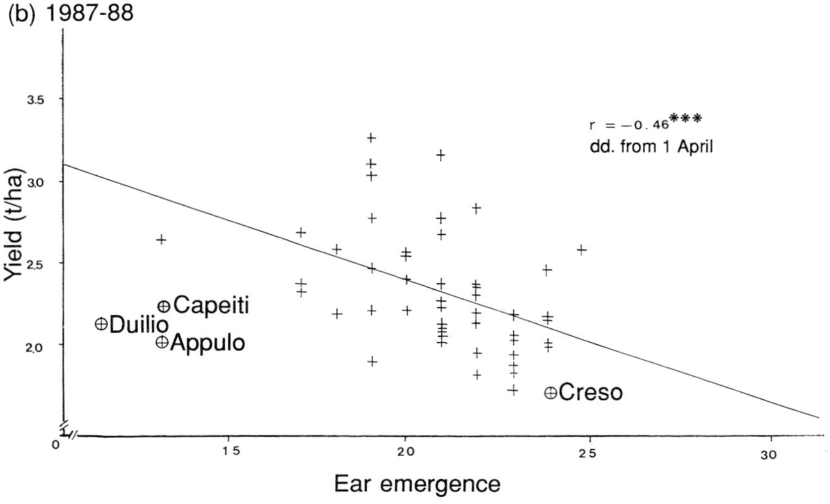

It should be noted, however, that many of the new lines, although later than Duilio, Capeiti and Appulo, produced higher yields. This confirms that the genetic background of Sicilian landraces, conferring adaptation to environmental stresses, has been maintained in spite of the change in morphological traits.

## CONCLUSION

Durum wheat is a major food crop in southern Italy, and genetic improvement represents an important strategy for increasing its yields and yield stability.

The approach taken so far has been to introduce new high-yielding cultivars, even in problematic areas. These genotypes, capable of high yields under favourable conditions, have in most cases proved unsuited to difficult climatic conditions (Duwayri et al., 1987). In contrast, local cultivars adapt better to unfavourable conditions, but show their limitations in good conditions. An alternative approach is to try to combine the high production potential of the new shorter-stemmed cultivars with the stress tolerance of local landraces. Using this approach, highly encouraging results have been obtained by Ceccarelli and Mekni (1985) for barley. As mentioned previously, the Sicilian germplasm is handicapped by its plant height and lateness of cycle but has good tolerance to abiotic stresses and some excellent quality characteristics, which are lacking in most modern Italian cultivars.

After crosses had been carried out between these two groups of germplasm, we were able to identify several improved lines on the basis of agronomic and morphological characteristics. A high leaf area duration (LAD) index and the capacity of the plant to develop the last internode were considered particularly important, as these favour efficient accumulation and transfer of photosynthesized matter to the spike, and hence the production of seeds with a high unit weight.

Little detailed information is available on the biochemical and physiological factors that provide landraces with good tolerance to abiotic stresses. This information would facilitate the selection of more productive and stable lines.

Wheat Genetic Resources: Meeting Diverse Needs
Edited by J. P. Srivastava and A. B. Damania

Published by John Wiley & Sons

# 23

# *Evaluation of Durum Wheat Germplasm in Iran*

A. K. Attary

Located between 25° and 40°N and 44° and 63°E, Iran is 1.64 million km$^2$ in extent and has a population of about 50 million. The country's arable land, however, amounts to only 25 million ha, of which approximately 8.2 million ha are allocated to cereals; of this, 4 million ha are under rainfed wheat, 2.1 million ha are under irrigated wheat and the most of the remaining area is under barley. The total area under durum wheat cultivation is about 330 000 ha. Iran has six agroclimatic regions:

- winter wheat and barley region (rainfed and irrigated mountainous regions with cold temperatures);
- spring wheat and barley region (hot and irrigated);
- spring wheat and barley region (rainfed, with fertile soil);
- facultative region (with a shortage of water);
- spring wheat and barley region (hot, with soil salinity and drought);
- desert (no cultivation).

Most of the crop is rainfed, and consists of landraces of spring or semi-winter types. The grain is used mainy to make flat bread and macaroni. These landraces have adapted to unfavourable growing conditions, but are susceptible to various diseases and have a low yield potential. The Cereal Project of the Seed and Plant Improvement Institute (SPII) in Karadj is planning to improve these landraces with the use of nurseries from the International Center for Agricultural Research in the Dry Areas (ICARDA) and the Centro Internacional de Mejoramiento de Maíz y Trigo (CIMMYT).

## WHEAT GENETIC RESOURCES COLLECTION AND EVALUATION

The cereal genetic resources collection in Iran was started in 1930 by scientists of the Agronomy Department at the University of Tehran. Approximately 8300 samples were collected, of which 230 were durum wheat. Subsequently, each population was subdivided according to morphotypes, following the classification used by Vavilov. The frequency of these morphotypes was determined. Every subspecies and morphotype was maintained in the form of 1 to 3 spikes at room temperature and rejuvenated every 5 years.

Within this collection, over 60 000 lines were described and evaluated for morphological and agronomic characters, and this data was computerized in 1987. New bread wheat varieties, such as Shahpesand, Shahi and Attahe, were selected from these lines and released.

In 1987, the SPII in Karadj started a research programme on durum wheat. Over 250 lines of durum wheat were grown, and were evaluated for their spike characteristics. Five spikes from each line were collected at random and evaluated for spike length, spikelets per spike, number of seeds per spikelet, spike internode length, number of seeds per spike, weight of seeds per spike, and thousand-kernel-weight (TKW). The mean values and ranges for these characters are given in Table 23.1.

**Table 23.1** Mean values and range for characters evaluated in 1987 at Karadj

| Character | Mean | Range |
|---|---|---|
| Spike length (mm) | 72.58 | 60 - 100 |
| Spikelet per spike | 20.45 | 16 - 26 |
| Number of seeds per spikelet | 3.2 | 2 - 5 |
| Spike internode length (mm) | 3.75 | 2.9 - 4.8 |
| Number of seeds per spike | 47.33 | 23 - 70 |
| Weight of seeds per spike (g) | 2.4 | 1 - 4.6 |
| Thousand-kernel-weight (g) | 49.41 | 31 - 71 |

Spike correlation coefficients were also calculated. Grain yield was positively correlated with the number of spikelets per spike, with the weight of seeds per spike, and with the TKW but, as expected, negatively correlated with spike internode length.

The research programme aims to breed improved durum wheat varieties for the rainfed areas, and to increase cultivation under conditions of moisture and temperature stress.

# PART 4

# Utilization for Diverse Needs

Wheat Genetic Resources: Meeting Diverse Needs
Edited by J. P. Srivastava and A. B. Damania

Published by John Wiley & Sons

# 24

# *Evaluation and Utilization of Introduced Germplasm in a Durum Wheat Breeding Programme in Canada*

J. M. CLARKE, J. G. MCLEOD and R. M. DE PAUW

Most of the durum wheat, *Triticum turgidum* L. var. *durum*, in Canada is grown in low-rainfall areas. Strategies to improve the yield potential of both durum and bread wheat, *T. aestivum* L., under dry conditions have been developed at Swift Current (Hurd, 1969; Townley-Smith et al., 1974; Townley-Smith and Hurd, 1979). Crosses involving locally developed and adapted germplasm, with some use of high-yielding introductions, continue to produce new cultivars with higher yields and equal or improved disease resistance and pasta-making qualities compared to earlier cultivars. Similar strategies are widely used in other wheat breeding programmes (Porceddu et al., 1988).

Under drier than average conditions, the old cultivar Pelissier tends to have higher yields than the new cultivars Wakooma and Kyle (*see* Figure 24.1 *overleaf*). Pelissier was introduced into North America from Algeria in 1896 (Joppa and Williams, 1988) and is still grown on about 20 per cent of the durum wheat area in the driest part of south-western Saskatchewan (Prairie Pools Inc., 1988). Pelissier has been used in the production of several widely grown Canadian durum cultivars (Hurd et al., 1972, 1973) and has been the focus of numerous studies of the physiology of adaption to drought (for example, Aamodt and Johnston, 1936; Hurd, 1964; Clarke and McCaig, 1982). The question often arises whether other genotypes with drought adaptation traits equal to, or better than, those found in Pelissier might exist elsewhere.

Srivastava et al. (1988) suggested that landraces and primitive forms of durum wheat from the primary and secondary centres of diversity could provide valuable genes for characters such as drought tolerance. Several examples of the use of landraces or old cultivars in the development of improved cultivars for dryland production can be cited.

**Figure 24.1** Yield response of Kyle (K), Wakooma (W) and Pelissier (P) durum wheat to an environmental index (available water/pan evaporation)

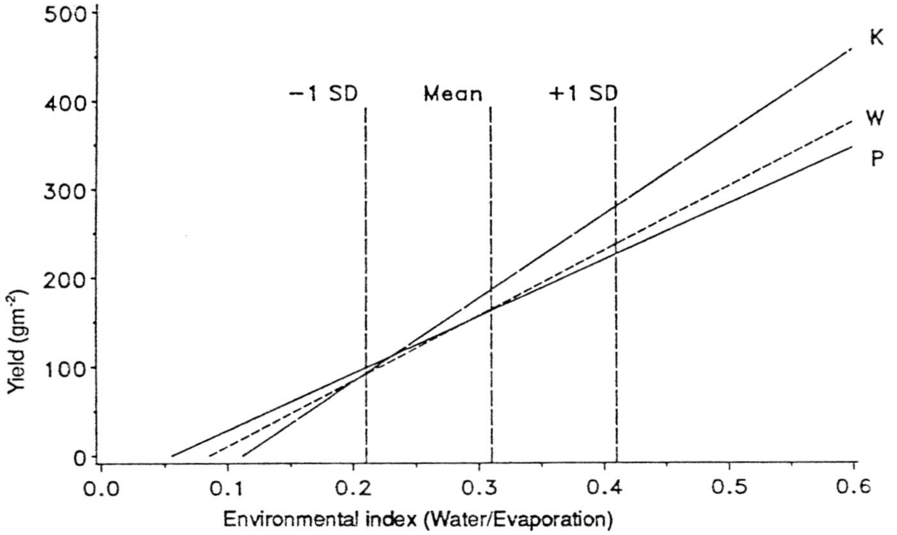


Pelissier, for example, was crossed with the higher-yielding but drought-susceptible cultivar Lakota to produce Wascana (Hurd et al., 1972) and Wakooma (Hurd et al., 1973); both of these cultivars outyield Lakota under dry conditions. Similarly, Duwayri et al. (1987) have produced lines with improved productivity under dry conditions by crossing a cultivar (Haurani) showing good dryland productivity with a cultivar (Stork) showing high yield under optimal growing conditions. These successes seem to have come about by using parental lines about which a great deal was known. Would it be possible to obtain similar results by screening germplasm collections to rapidly identify and use genotypes with superior dryland adaptation?

## SCREENING GERMPLASM FOR DROUGHT TOLERANCE

In 1984 we set out to answer this question by screening about 3600 entries from the durum wheat germplasm collection at the International Center for Agricultural Research in the Dry Areas (ICARDA) (Clarke et al., 1987). Most of the lines screened were also contained in the Small Grains Collection of the United States Department of Agriculture (USDA). Following the initial screening in the 1984 growing season, 660 entries were selected on the basis of visual scoring for agronomic and morphological characters and grown in replicated yield trials in 1985. A subset of 18 apparently high-yielding genotypes were selected for further testing from 1986 to 1988 at two other locations in addition to Swift Current, and for crossing with a high-yielding local line, DT 369 (Wascana/Quilafen).

Precipitation over the testing period ranged from much below to above the long-term average (*see* Table 24.1). Above-normal May and June temperatures were an additional stress factor in 1987 and 1988.

The 1985 yields of the 18 selected genotypes were similar to or higher than that of Pelissier and higher than those of Hercules and Kyle (*see* Table 24.2). The high standard deviations indicate the difficulty of making precise comparisons of genotype yield potentials under very dry conditions. Of the 620 entries tested for yield (40 of the original 660 did not head in 1985), only 25 showed higher yield rank than Pelissier; six of the 18 genotypes selected for further study had higher yield rank than Pelissier.

**Table 24.1** Annual and growing season (May to July) precipitation (mm) at Swift Current, Canada

| | **Growing season** | **Annual** |
|---|---|---|
| 1984 | 100 | 262 |
| 1985 | 73 | 279 |
| 1986 | 205 | 383 |
| 1987 | 129 | 256 |
| 1988 | 143 | 287 |

1896 to 1988 average: growing season 167 mm, annual 358 mm

**Table 24.2** Grain yield per m$^2$ of 620 durum wheat genotypes, 18 selected genotypes and four local checks at Swift Current, Canada in 1985

| | **Yield** | **SD** | **Range** |
|---|---|---|---|
| All genotypes | 58 | 28 | 20 - 131 |
| Selected genotypes | 95 | 17 | 71 - 131 |
| Hercules | 65 | 21 | |
| Kyle | 80 | 20 | |
| Pelissier | 100 | 22 | |
| Wakooma | 85 | 21 | |

Analyses of variance revealed highly significant genotype x environment (G x E) interactions for yield during the 4 years of testing of the 18 selected genotypes and local checks. Such interactions result from different responses to environmental factors, and encompass what physiologists refer to as drought tolerance (Baker, 1987). In our study, the observed interactions involved changes in genotype rank among environments, a situation that Baker (1987) points out as having significant implications for breeding.

Some of the changes in ranking were very pronounced, as in the case of IC 7659 (*see* Table 24.3). In the five environments used as an example, the yield rank of this genotype ranged from highest to lowest. The local checks DT 369, Kyle and Wakooma maintained relatively high ranks throughout the test period, except in 1985. There were few significant differences in yield among genotypes in 1985 and 1988, making the ranking relatively meaningless for these years.

There were changes in yield ranking at Swift Current in 1986 and 1987, although overall test means were nearly identical (*see* Table 24.3). Although rainfall during the growing season was higher in 1986 than in 1987, stored soil water reserves were greater in 1987 than in 1986. The observed changes in yield ranking could have

**Table 24.3** Grain yield per $m^2$ and ranking of selected introductions and local checks in experiments conducted at Swift Current (SC) and Indian Head (IH), Canada

| | | **Yield (rank)** | | | | | |
|---|---|---|---|---|---|---|---|
| **ICARDA No.** | **CI/PI No.** | **SC 1985** | **SC 1986** | **IH 1986** | **SC 1987** | **SC 1988** | **Mean** |
| 7071 | CI 06871 | 77 (20) | 256 (9) | 328 (11) | 238 (9) | 96 (13 ) | 199 (10) |
| 7021 | CI 06878 | 78 (19) | 218 (14) | 297 (16) | 223 (13) | 90 (17) | 181 (17) |
| 7144 | CI 07458 | 85 (14) | 151 (20) | 279 (19) | 223 (14) | 89 (19) | 165 (22) |
| 7659 | PI 11939 | 71 (22) | 118 (21) | 227 (21) | 298 (1) | 113 (5) | 165 (21) |
| 8043 | PI 16694 | 79 (18) | 194 (16) | 215 (22) | 257 (7) | 88 (21) | 167 (19) |
| 8695 | PI 19101 | 86 (12) | — — | 387 (5) | 208 (19) | 100 (10) | 195 (12) |
| 8698 | PI 19102 | 131 (1) | 201 (15) | 284 (18) | 253 (8) | 112 (6) | 196 (11) |
| 9216 | PI 22141 | 85 (13) | 189 (17) | 311 (12) | 213 (16) | 119 (2) | 183 (15) |
| 9369 | PI 26342 | 115 (3) | 279 (4) | 347 (9) | 209 (18) | 95 (16) | 209 (7) |
| 9390 | PI 26495 | 76 (21) | — — | 288 (17) | 193 (21) | 105 (8) | 165 (20) |
| 9430 | PI 26715 | 98 (9) | 185 (19) | 332 (10) | 237 (10) | 101 (9) | 191 (13) |
| 9449 | PI 27189 | 110 (4) | 236 13) | 234 (20) | 188 (22) | 74 (23) | 168 (18) |
| 9544 | PI 27844 | 93 (10) | 243 (12) | 364 (8) | 259 (5) | 89 (20) | 210 (6) |
| 9581 | PI 28315 | 100 (8) | 189 (18) | 203 (23) | 234 (11) | 98 (12) | 165 (23) |
| 9603 | PI 29005 | 109 (5) | 265 (6) | 385 (6) | 224 (12) | 116 (4) | 220 (4) |
| 9659 | PI 29783 | 103 (6) | 277 (5) | 381 (7) | 216 (15) | 98 (12) | 215 (5) |
| 9914 | PI 36721 | 117 (2) | 245 (11) | 297 (15) | 199 (20) | 90 (18) | 189 (14) |
| 9935 | PI 36723 | 88 (11) | 257 (8) | 299 (14) | 188 (23) | 84 (22) | 183 (16) |
| DT 369 | | 80 (17) | 310 (2) | 473 (3) | 262 (4) | 149 (1) | 255 (2) |
| Hercules | | 65 (23) | 265 (7) | 396 (4) | 211 (17) | 95 (15) | 206 (9) |
| Kyle | | 80 (16) | 312 (1) | 505 (1) | 288 (2) | 117 (3) | 261 (1) |
| Pelissier | | 100 (7) | 253 (10) | 303 (13) | 276 (3) | 106 (7) | 207 (8) |
| Wakooma | | 85 (15) | 282 (3) | 500 (2) | 258 (6) | 96 (14) | 244 (3) |
| Mean | | 91 | 235 | 334 | 234 | 101 | 197 |
| LSD = 0.05 | | 10 | 30 | 51 | 20 | 10 | 17 |

resulted from seasonal temperature differences — June and July mean temperatures were greater in 1987 than in 1986 — or from differences in soil water extraction capability. Pelissier, for example, showed improved ranking in 1987, and is known to have an extensive root system (Aamodt and Johnston, 1936; Hurd, 1964).

Averaged over years, experiments and locations, yields of the local checks DT369, Kyle and Wakooma were significantly greater than those of the other cultivars (*see* Table 24.4). Similarly, these three cultivars showed the highest regression slopes, indicating high responsiveness to environmental inputs. According to the interpretation of Finlay and Wilkinson (1963), DT 369, Kyle and Wakooma would have lower yield stability than the other cultivars. Numerous investigators (for example, Lin et al., 1986) have pointed out, however, that genotypes with low regression slopes achieve stability by yielding relatively poorly in favourable environments. Several of the genotypes, such as IC 7017, IC 7144, IC 9430 and IC 9544, showed similar yields and

**Table 24.4** Mean grain yields per $m^2$, yield ranks and slope of the regression of cultivar yield on site mean yield for selected introductions and check cultivars over 13 site-years in Saskatchewan, Canada

| Genotype | Yield | SD | Rank | Slope |
|---|---|---|---|---|
| IC 7017 | 220 | 147 | 5 | 1.17 |
| IC 7021 | 178 | 110 | 19 | 0.86 |
| IC 7144 | 194 | 145 | 12 | 1.15 |
| IC 7659 | 179 | 111 | 18 | 0.85 |
| IC 8043 | 182 | 132 | 16 | 1.03 |
| IC 8695 | 181 | 122 | 17 | 0.92 |
| IC 8698 | 190 | 115 | 14 | 0.92 |
| IC 9216 | 183 | 98 | 15 | 0.77 |
| IC 9369 | 199 | 114 | 10 | 0.90 |
| IC 9390 | 161 | 96 | 23 | 0.69 |
| IC 9430 | 198 | 130 | 11 | 1.05 |
| IC 9449 | 165 | 104 | 22 | 0.81 |
| IC 9544 | 204 | 126 | 8 | 1.01 |
| IC 9581 | 177 | 110 | 21 | 0.85 |
| IC 9603 | 201 | 122 | 9 | 0.97 |
| IC 9659 | 205 | 130 | 7 | 1.03 |
| IC 9914 | 190 | 117 | 13 | 0.94 |
| IC 9935 | 178 | 114 | 20 | 0.90 |
| DT 369 | 249 | 161 | 2 | 1.26 |
| Hercules | 207 | 151 | 6 | 1.21 |
| Kyle | 250 | 175 | 1 | 1.40 |
| Pelissier | 223 | 146 | 4 | 1.15 |
| Wakooma | 239 | 176 | 3 | 1.41 |

regression slopes to Pelissier, and may yet prove useful in the development of cultivars with improved adaptation to dry growing conditions.

It will be several years before this latter point can be clarified, as we are only at the stage of $F_6$ yield testing of the crosses of the selected genotypes with DT 369. Yield testing of $F_2$-derived $F_4$ lines in 1988 indicated some promise of recovering progeny with yields higher than that of DT 369.

The problems of characterizing the mean yield performance of genotypes in a variable environment such as ours make it difficult to rapidly evaluate and incorporate introduced genotypes into a breeding programme. At the same time, there are no physiological screening techniques that can be used to predict precisely the performance of cultivars under dry conditions. As part of our screening of the introduced germplasm, we measured rate of water loss by excised leaves, since low rates of loss have been shown to be positively related to yield under dry conditions (Clarke and Townley-Smith, 1986; Clarke et al., in press). Measurement of this trait is fairly rapid and reasonably precise: only six out of 100 lines appeared to be wrongly classified for leaf water loss in single replicate screening in 1984 (Clarke et al., in press).

In the 1985 yield testing of the 660 genotypes selected from the 3600 grown in 1984, rate of excised-leaf water loss was negatively correlated with yield (*see* Table 24.5). Similarly, mean rate of water loss was lower for the highest-yielding genotypes than for the lowest-yielding ones. However, the interaction of the many traits that determine yield under dry conditions means that single traits such as rate of leaf water loss will not be good predictors of genotype performance. For example, of the 18 genotypes selected for crossing based on 1985 yield performance, two had high rates of water loss, four had intermediate rates, and the remaining 12 had low rates. Only further

**Table 24.5** Correlations of 1984 measurements of morphological and physiological characters with yield of 620 genotypes tested in 1985, and 1984 values of these characters in high (greater than mean +1 SD) and low (less than mean -1 SD) yielding genotypes in 1985

| Character | r | High yield | | Low yield | |
|---|---|---|---|---|---|
| Leaf water loss [a] | -0.18 [h] | 0.53 | 0.04 [g] | 0.77 | 0.05 [g] |
| Days to head | -0.01 | 70.5 | 0.4 | 69.7 | 0.5 |
| Maturity score [b] | 0.22 [h] | 7.3 | 0.2 | 5.5 | 0.3 |
| Agronomic score [c] | -0.39 [h] | 5.2 | 0.2 | 7.7 | 0.2 |
| Glaucousness score [d] | -0.11 [h] | 3.0 | 0.3 | 3.8 | 0.3 |
| Leaf size score [e] | -0.13 [h] | 4.2 | 0.3 | 5.3 | 0.2 |
| Leaf rolling score [f] | -0.02 | 1.7 | 0.1 | 1.8 | 0.1 |

a Leaf water loss expressed as g water lost per g dry weight per hour; b Maturity score: 1 (early), 5, 9; c Agronomic score: 1 (best), 3, 5, 7, 9; d Glaucousness score: 1 (most), 5, 9; e Leaf size score: 1 (large), 5; f Leaf rolling score: 1 (rolled), 2 (not rolled);
g S.D. = standard deviation; h $P < 0.01$

selection and testing of the crosses will reveal the effects of selection for water loss on stress tolerance.

Some of the other characters observed in the 1984 screening, such as agronomic score, glaucousness and leaf size, were significantly correlated with 1985 yields. However, as in the case of rate of excised-leaf water loss, none were precise predictors of yield performance under dry conditions. More precise indicators must be found to facilitate the rapid screening of large germplasm collections for the poorly defined complex of traits which we refer to as drought tolerance.

## USE OF DROUGHT-TOLERANT GENOTYPES IN NORTH AMERICA

Because of the differences in production environment, it may be more difficult to identify drought-tolerant genotypes in germplasm collections for use in North American breeding programmes than for use in programmes near the centres of diversity. Durum wheat is a winter crop in West Asia, but a spring crop in North America. Thus, genotypes with more than a weak vernalization requirement tend not to be adapted to North America. It would be useful to know, therefore, if screening for stress tolerance at ICARDA will identify genotypes of potential use in environments such as ours.

We have data suggesting that the answer to the latter question may be yes. For example, several durum genotypes from the ICARDA programme tested at Swift Current showed high yields relative to local checks under abnormally hot conditions in 1988, but not under the more favourable conditions in 1987 (*see* Table 24.6). In an

**Table 24.6** Grain yields per $m^2$ of recently developed ICARDA durum wheat compared to Haurani and Canadian checks when grown under rainfed and irrigated conditions at Swift Current in 1987 and under rainfed conditions at Swift Current and Indian Head in 1988

| | 1987 | | 1988 | |
|---|---|---|---|---|
| | **Rainfed** | **Irrigated** | **Swift Current** | **Indian Head** |
| Ain Arous | 216 | 325 | 117 | 303 |
| Belikh 2 | 204 | 400 | 114 | 330 |
| Om Rabi | 201 | 251 | 130 | 310 |
| Nile | 199 | 358 | 127 | 322 |
| Sebou | 181 | 437 | 137 | 334 |
| Haurani | 175 | 331 | 102 | 224 |
| Hercules | 148 | 401 | 95 | 274 |
| Kyle | 230 | 634 | 117 | 294 |
| Pelissier | 245 | 584 | 106 | 323 |
| Wakooma | 210 | 559 | 96 | 266 |
| Mean | 201 | 428 | 114 | 298 |
| LSD = 0.05 | 28 | 125 | 14 | 35 |

experiment involving approximately 100 genotypes from the ICARDA collection grown in Canada and Syria (Clarke et al., in press), several cultivars tended to perform well at all locations. For example, 19 of the highest-yielding 25 entries tested at Swift Current in 1985 appeared among the highest 25 entries at one or more of the three test sites (Bouider, Breda, and Tel Hadya) in Syria in the 1985-86 growing season. Four of the 19 entries (IC 7021, IC 7157, IC 7183, and IC 9415) appeared among the top 25 at all three sites in Syria. As noted above, IC 7021 has been used in our crossing programme; of the other genotypes used in crosses, IC 9390 was among the top 25 entries at both Breda and Bouider, while IC 8695 was among the top 25 at Bouider. Further testing should be carried out to confirm that screening results are applicable across diverse environments.

## CONCLUSION

Characterizing the entries in wheat germplasm collections will prove useful to breeders, especially for simply inherited traits such as disease resistance and some quality traits. This enlarged information base will encourage better use of germplasm collections. We need more information before we will know whether the general screening of collections for drought tolerance will prove useful. Certainly, Srivastava et al. (1988) suggest that there are indications of a general association between region of origin and stress tolerance in durum wheat, and our own results indicate that some apparently drought-tolerant genotypes perform well under very dry conditions in both Canada and in Syria.

## *Discussion*

J. G. WAINES: What is the morphological basis of drought resistance that you have identified in durum wheat?

J. M. CLARKE: We are not sure what the morphological basis is for the differences in leaf water loss. Some of our preliminary work indicates that epicuticular waxes seem to play a small part in the differences.

J. G. WAINES: Why are you not looking at roots, as Hurd did?

J. M. CLARKE: Quantification of root system size is difficult. We are trying to develop and use techniques that can be used more easily for a large number of accessions.

J. G. WAINES: But could you look at the roots of the 18 you have isolated?

J. M. CLARKE: Yes.

T. T. CHANG: If I may interject here, one of the latest gadgets is the 'Rhizotron' which may be useful for this work, but is quite expensive.

L. PECETTI: If leaf water loss rate is the most useful trait for screening accessions for drought tolerance, can it be easily recorded by any scientist in a national programme?

J. M. CLARKE: Leaf water loss rate can be measured easily and rapidly without expensive equipment. In fact, we have had some success with visual ranking of lines. This could be useful in screening of segregating lines in a breeding programme.

M. N. SANKARY: Can you elaborate on your method for measuring leaf water loss?

J. M. CLARKE: Leaves are removed from field-grown plants, taken to the laboratory, weighed, wilted under controlled environmental conditions for 1-2 hours, and weighed again. Dry weights are then determined, and the rate of water loss is expressed on a leaf dry weight basis.

L. TANASCH: I would like to comment on the question Professor Sankary raised about the method and effectiveness of the excised leaf water loss rate in screening germplasm. We are using the same method in Austria. This method could be used to screen a large number of accessions and reduce the numbers to be further tested in the field or glasshouse for drought tolerance.

M. OBANNI: Do you know why you did not have a correlation between earliness and drought tolerance? In the Mediterranean climate there can be an early cycle drought and a late drought, and early-maturing cultivars will yield higher under these conditions.

J. M. CLARKE: The correlation of yield earliness was for one year, during which there were late-season rains. Our normal pattern would be to have drought stress after anthesis, which would tend to correlate earliness with improved yields.

D. R. KNOTT: You presented some interesting results comparing the yield rankings of 100 accessions in Syria and Canada. Have you calculated correlations between yields in the two countries?

J. M. CLARKE: No, we have not, as the correlations tend to be rather meaningless because of problems with the vernalization requirements of some genotypes which may be well adapted in one environment but not in another.

Wheat Genetic Resources: Meeting Diverse Needs
Edited by J. P. Srivastava and A. B. Damania

Published by John Wiley & Sons

# 25

# *Manipulating the Wheat Pairing Control System for Alien Gene Transfer*

C. CEOLONI

Exploitation of the rich genepool of wild wheat relatives, particularly those in which the chromosomal sets are similar (homoeologous) to chromosomal sets of cultivated forms, is hindered by the complex system of wheat genes, particularly *Ph1*, which suppresses homoeologous chromosome pairing, both within wheat itself and in hybrids between wheat and alien Triticeae. To surmount this obstacle, cytogenetic methods have been developed which make use of chromosomal or genetic mutations involving the *Ph1* locus, allowing the transfer of alien chromatin segments carrying desired genes into the cultivated wheat background.

The efficiency of such methods depends on the choice of the most appropriate transfer scheme (with respect to donor and recipient parents, as well as to the crossing sequence selected) to fit specific objectives. To further improve the potential of these methods, one can choose the most suitable pairing condition, opting for a *ph1* high-pairing or a *ph2* intermediate-pairing environment, depending on the relative cytogenetic affinity of the chromosomes being induced to pair. A better understanding of the effects of 'major' and 'minor' pairing genes would help in making the best choice.

A wider range of new possibilities might be opened up by combining the effects of more than one gene. A possible super-high pairing situation (combination of a *ph1* and a *ph2* mutation) and an intermediate-pairing one (combination of a *ph2* mutation with an extra dose of chromosome 2D) have been tested for their effect in hybrids of common wheat with tetraploid *Aegilops* species. In both cases, an additive effect of the separate chromosomal mutations has been detected. This indicates a more frequent participation in pairing of all the homoeologues present in the hybrid. Several hybrid combinations are being tested under these newly developed pairing conditions. The aim is to verify the mode and extent of the chromosomal association of various related genomes with those of wheat, including more distant relatives such as *Secale*, *Agropyron* and *Haynaldia*.

## CHROMOSOME PAIRING IN NATURAL ALLOPOLYPLOIDS

Meiotic chromosome pairing and recombination in eukaryotic organisms are multi-stage processes, each stage having several substages. The processes include the premeiotic association of the pairing partners, whose closer spatial relationship is necessary to achieve the correct sequence of events leading to crossing-over and chiasma formation. The genetic control of such processes is rather complex and only partially understood. It is especially complex in polyploid species, such as the cultivated wheats, in which an organizational level between the different chromosomal sets is also operative. In fact, in spite of the genetic homoeology of their component genomes, in natural allopolyploids chromosome association is almost invariably restricted to homologous bivalents at metaphase I, with the result that meiosis is mechanically and genetically efficient. Divergence between homoeologous chromosomes is usually not sufficient on its own to achieve complete diploidization. This is normally the result of one or more pairing control genes superimposed on genome differentiation (Evans, 1988).

In the case of polyploid wheats, the fact that the homoeologues have so little tendency to pair remained paradoxical until the discovery by Sears and Okamoto (1958) and Riley and Chapman (1958) that homoeologous pairing is suppressed by a gene, later called *Ph1*, located on the long arm of chromosome 5B. As a result of *Ph1* action, pairing is confined to homologous chromosomes, but it does involve homoeologues when *Ph1* is absent or inactive. *Ph1* thus works as a homoeologous pairing suppressor gene.

Subsequent studies on common wheat (*Triticum aestivum*, 2n = 6x = 42) in the 1960s and 1970s (Sears, 1976) showed clearly that additional suppressors (less potent than *Ph1*), as well as genes exerting an opposite effect (referred to as 'pairing promoters') are present and active in the wheat genome. All these genes are part of a complex, well-coordinated and balanced gene system that controls the various processes involved in chromosome pairing and recombination, both in wheat itself and in its hybrids with related Triticeae.

## HOW *Ph* GENES HINDER EXTENSIVE USE OF ALIEN GENES

Cytogenetic affinity and phylogenetic relatedness within the Triticeae ensure a high probability of obtaining viable hybrids. Since homoeologous relationships exist between wheat chromosomes and those of related species and genera, the almost invariable lack of pairing that interspecific and intergeneric hybrids exhibit is mainly because of the *Ph1* gene's strong suppressive effect. This is counteracted by genes of the alien parent only in certain circumstances, such as in crosses with 'high-pairing' types of *Ae. speltoides* and *Ae. mutica*.

The enormous gene reservoir present in wheat relatives, particularly in the largely unexploited wild forms, deserves more extensive and efficient utilization. This would increase the chances of success in gene transfer.

## CHOOSING APPROPRIATE METHODS FOR ALIEN GENE TRANSFER

To make efficient use of agronomically beneficial alien genes, the most suitable procedure must be chosen for their incorporation into the wheat genome (Feldman, 1988). If one looks only at those species which share some cytogenetic affinity with cultivated wheats, a variety of chromosome manipulations of both the alien donor and the wheat recipient lines can be performed (Ceoloni, 1987), depending on the final goal — for instance, on whether the goal is to create completely new chromosome(s) and gene assortments or to introduce only a given trait, with the least possible modification of the original wheat background. In the former case, one would proceed through direct hybridization between the candidate parents, possibly followed by amphidiploidization; in the latter case, addition or substitution lines for the selected alien chromosome are preferable as donor lines, making the procedure somewhat more sophisticated but much more straightforward (Ceoloni et al., 1988). In one case, the exchange of genetic material is potentially extended to all the corresponding chromosomes of the parental genomes; in the other, a single alien chromosome is exposed to recombination with its wheat counterparts.

Whatever number of alien genes one wishes to insert into a wheat background, in most instances two main prerequisites have to be met:

- transfer must not occur at random, because the closer the new 'genetic environment' of the alien segment to its original one, the higher are the chances of its expression and the lower are the chances of problems caused by the absence of the replaced genes;
- unless one has prior knowledge about particular positive gene associations, the size of the alien chromatin segment(s) should as far as possible be restricted to the desired gene alone, with the undesired linked ones left out.

The only way these two needs can be met, and thus provide useful transfer products in most cases, is through genetic recombination.

## CHOOSING APPROPRIATE PAIRING CONDITIONS

### Intermediate pairing

Obviously, chromosome pairing and subsequent recombination encounter few or no genetic limitations in the case of homologous chromosomal sets (as in *T. dicoccoides* vs. *T. durum*). In most cases, however, this is not the situation, as the most frequent type of relationship between wheat genomes and those of alien Triticeae species (both wild and cultivated) is the only partially homologous — that is, homoeologous (Feldman, 1979).

Because *Ph1* is the most potent and also the best known single factor controlling homoeologous pairing and recombination, nullisomy for chromosome 5B or deleted condition at the *Ph1* locus (Sears, 1977; Giorgi, 1978; Giorgi and Ceoloni, 1985) have been the methods chosen in most attempted introgressions of alien genes (Ceoloni, 1987, 1988). However, numerous other genes affecting intraspecific and interspecific pairing and recombination are known besides *Ph1* (Sears, 1976; Ceoloni et al., 1986).

In certain cases, mutations of such genes, used individually or in combination, would fit the specific 'pairing needs' better than a *ph1* mutation. For instance, mutants at the second major *Ph* locus, *Ph2*, on chromosome arm 3DS (Sears, 1982, 1984) are able to induce about half the amount of homoeologous pairing caused by the *ph1b* mutation in interspecific hybrids (Sears, 1982) (*see* Table 25.1). Such intermediate-pairing mutations may be more appropriate than the high-pairing *ph1b* for promoting pairing between closely related homoeologues. To favour such associations while producing minimal unwanted background rearrangement, it may be advantageous to use an extra dose of chromosomes from group 2, particularly 2D, which contains quite strong promoters (Miller and Reader, 1985; Ceoloni et al., 1986). A further advantage of choosing less potent genes than *Ph1* is that, because the fertility of wide hybrids is associated mainly with the formation of restitution nuclei, which are less frequent as the level of pairing increases, a reasonably good compromise between the promotion of pairing and the recovery of hybrid, recombined products can be reached.

For the above reasons, it was decided to test the effect of combining a *ph2* mutation with an extra dose of chromosome 2D in intergeneric hybrids. Of added interest was the opportunity to observe the result of manipulating genes that possibly control different stages or events in the pairing-recombination process. *Ph2* is likely to be involved (as *Ph1* is) in premeiotic chromosomal arrangement and alignment (Ceoloni and Feldman, 1987), whereas the most prominent effect of the 2D gene(s) is to promote chiasma formation (Ceoloni et al., 1986) (*see* Table 25.1).

A chromosome 2D tetrasomic line (T2D) of *T. aestivum* cv. Chinese Spring was crossed to the intermediate-pairing mutant (IPm) carrying the *ph2a* mutation (Sears, 1982). $F_1$ plants with 2n = 43, with an extra 2D chromosome and heterozygous at the *Ph2* locus, were then pollinated with a number of different related species of wheat (*see* Figure 25.1). In each case, four types of progeny were expected. From the point of view of pairing, apart from the low pairing class (only one 2D, *Ph2*), both the extra dose for 2D and the *ph2a* mutation were expected to correspond to an intermediate type, with the *Ph2a* condition being perhaps somewhat more effective (Sears, 1982; Ceoloni et al., 1986). The fourth type (two doses of 2D, *ph2a*) remained to be investigated.

The preliminary results, presented in Table 25.1, refer to the hybrid combination involving *Ae. kotschyi* (2n = 28, genomes $US^v$). The two IP classes, though not equally represented (one plant only was of the 2n = 36 type), showed about the same amount

**Table 25.1** Mean ± S.E.M. and range (in parentheses) of chromosome pairing at metaphase I in hybrids (ABDUS$^{V)}$ between *Aegilops kotschyi* and common wheat genotypes carrying different *Ph2* alleles and/or different doses of chromosome 2D

| Genotype | No. plants | No. cells | Uni-valents | Bivalents Rod | Bivalents Ring | Bivalents Total | Multivalents III | Multivalents IV | Multivalents V | Chias-mata |
|---|---|---|---|---|---|---|---|---|---|---|
| 2D"/*Ph2* | 1 | 100 | 22.12 | 5.08 | 1.13 | 6.21 | 0.42 | 0.05 | — | 8.47 |
| (2n = 36) | | | ±0.38 | ±0.17 | ±0.05 | ±0.18 | ±0.06 | ±0.02 | | ±0.22 |
| | | | (13-32) | (1-9) | (0-2) | (1-10) | (0-2) | (0-1) | (3-12) | |
| 2D"/*ph2a* | 2 | 200 | 21.07 | 5.25 | 0.18 | 5.43 | 1.01 | 0.20 | 0.05 | 8.45 |
| (2n = 36) | | | ±0.25 | ±0.12 | ±0.03 | ±0.13 | ±0.05 | ±0.03 | ±0.02 | ±0.15 |
| | | | (9-29) | (0-12) | (0-1) | (0-12) | (0-4) | (0-1) | (0-1) | (4-14) |
| *ph2a* | 4 | 200 | 23.38 | 4.88 | 0.13 | 5.01 | 0.46 | 0.05 | 0.005 | 6.24 |
| (2n = 35) | | | ±0.27 | ±0.12 | ±0.03 | ±0.12 | ±0.05 | ±0.02 | ±0.005 | ±0.16 |
| | | | (12-31) | (1-9) | (0-2) | (1-9) | (0-4) | (0-1) | (0-1) | (3-13) |
| *Ph2* | 3 | 250 | 32.62 | 1.15 | 0.01 | 1.16 | 0.02 | — | — | 1.20 |
| (2n = 35) | | | ±0.14 | ±0.07 | ±0.01 | ±0.07 | ±0.01 | | | ±0.07 |
| | | | (23-35) | (0-6) | (0-1) | (0-6) | (0-1) | | | (0-6) |

**Figure 25.1** Combining an extra dose for chromosome 2D with the *ph2a* mutation in hybrids with *Aegilops kotschyi*

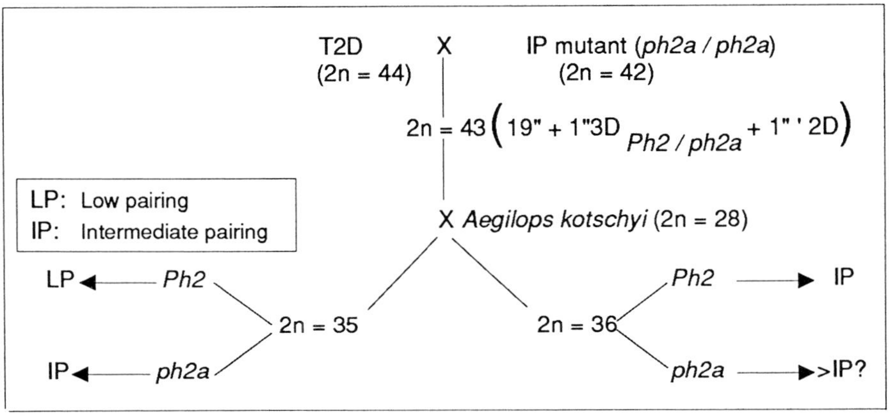


of pairing increase. A small difference was observed between the two genotypes in the relative incidence of multivalent associations. However, the mean number of chiasmata per cell was increased by the presence of the extra 2D (8.47 vs. 6.24), although

2.03 of the 8.47 chiasmata of the 2n = 36 *Ph2* hybrid were formed by the 2D pair itself. Chromosome-specific chiasma formation, which is of less practical interest, was not involved, as the patterns of pairing and chiasmata of the other 2n = 36 pairing class (most probably carrying a *ph2a* mutation) were quite different. In this genotype, the 2D homologous ring bivalent was not formed, but the high number of chiasmata is maintained in spite of this. The 2D ring bivalent has not been converted into a multivalent, in which its own chiasmata might have been retained, because the majority of trivalents and all the higher rank multivalents are of the open type. Therefore, the 8.45 chiasmata were 'redistributed', compared to the *Ph2* 2n = 36 situation. Thus, although the total number of paired chromosomes is augmented only in the 2n = 36 *ph2a* type, the pattern has evidently changed; in particular, multivalent associations (and hence their relative weight in the overall pairing figure) have increased.

In attempting to interpret these observations, I suggest that the extra dose of the promoter gene(s) on chromosome 2D reinforces the *ph2* effect, in terms of making truly chiasmatic a higher number of potential associations. Such an interpretation obviously remains tentative, because of the preliminary nature of the results. Further data are being accumulated, using additional plants and different hybrid combinations. It may be possible to maintain an intermediate level of pairing overall, and yet induce a high degree of crossing-over through an increase in the number of sites which will eventually form chiasmata out of a given potential population (Jones, 1984).

## Super-high pairing

There are, of course, situations in which it would be desirable to obtain recombination frequencies higher than those achievable by the use of *ph1* mutations. Such frequencies, reported to be around 7 per cent for an *Agropyron elongatum* chromosome (Sears, 1972), an *Ae. longissima* telosome (Ceoloni et al., 1988) and their wheat homoeologues, dropped to about 1 per cent when a *Secale cereale* arm was induced to pair (Koebner and Shepherd, 1985). It seems that for distantly related alien chromosomes, such as those of rye, the efficiency of *Ph1* in inducing homoeologous pairing is reduced. In order to investigate whether the *ph1* ceiling could be raised by the simultaneous presence of a *ph2* mutation, a 'double mutant' was developed in common wheat (*T. aestivum* cv. Chinese Spring) (*see* Figure 25.2). From a direct cross between the corresponding single mutants, a true double mutant euploid line could have been obtained. However, because of the difficulties of selection in cases where there was no additive effect, a line has been produced with a *ph1b* mutated to 5B, but lacking the 3DS arm (DT3DL), instead of bearing a *ph2* mutation. The pairing background would not have been significantly altered and the presence or absence of the 3DL telocentric would have served as a cytological marker. This has been particularly useful in testing the effects of *ph1b* vs. *ph1b*/*ph2*- in hybrids, because the two alternatives, derived from the same 2n = 40 + t mother plant (*see* Figure 25. 2), could be immediately identified.

**Figure 25.2** Development of a double mutant line of common wheat, carrying the *ph1b* mutation and lacking the 3DS arm with the *Ph2* suppressor (2n = 40 + 2t) and its hybridization with alien species

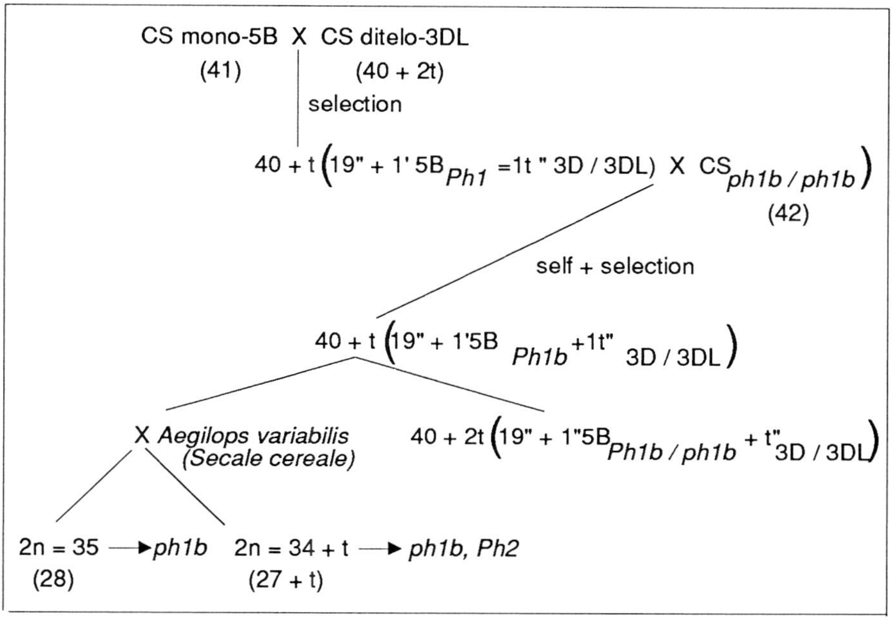


Data are presented in Table 25.2 (*overleaf*) on a hybrid combination involving *Ae. variabilis* (2n = 28, genomes $US^v$). A comparison between the two types of hybrids reveals an 8 per cent increment in the total number of chromosomes paired, which is entirely because of an increase in multivalent configurations. These represent more than 45 per cent of the total chromosome associations in the *ph1b* genotype which also lacks the *Ph2* suppressor. The figure would probably be even higher if both the genotypes were euploid (2n = 35), as the presence of only one arm of chromosome 3D has been observed to reduce its involvement in multivalent associations. The combined effect of the two *ph* mutations is thus most evident when the pattern of chromosomal association is observed. In this respect, the threefold increase in the number of pentavalents per cell in the 2n = 34 + t genotype is noteworthy, as it indicates more efficient participation in pairing by all the five homoeologues of the different genomes present (A, B, D ,$S^v$ and U).

Mello-Sampayo and Canas (1973) found no difference in pairing, based on chiasmata/cell values, between two wheat/*Ae. sharonensis* combinations; the two combinations differed in the presence or absence of the 3DS arm but were both deficient in

**Table 25.2** Mean ± SEM and range (in parentheses) of chromosome pairing at metaphase I in hybrids between *Aegilops variabilis* and common wheat genotypes carrying different *ph1b* mutation alone or in combination with 3DS *(Ph2)* absence (genome formula ABDUS$^{V}$)

| Genotype | No. plants | No. cells | Uni-valent | Bivalents Rod | Ring | Total | Multivalents III | IV | V | Others | Chias-mata |
|---|---|---|---|---|---|---|---|---|---|---|---|
| *ph1b/Ph2* (2n = 35) | 7 | 316 | 11.06 ±0.18 (4-23) | 6.08 ±0.10 (2-11) | 1.48 ±0.08 (0-5) | 7.56 ±0.11 (2-14) | 2.22 ±0.07 (0-6) | 0.46 ±0.03 (0-2) | 0.06 ±0.01 (0-1) | 0.003 ±0.003 (0-1) | 15.12 ±0.14 (7-23) |
| *ph1b/Ph2*$^{-}$ (2n = 34+t) | 6 | 267 | 8.33 ±0.17 (1-18) | 5.62 ±0.11 (2-12) | 1.63 ±0.07 (0-5) | 7.31 ±0.12 (2-13) | 2.58 ±0.08 (0-6) | 0.79 ±0.05 (0-3) | 0.21 ±0.03 (0-2) | 0.02 ±0.01 (0-1) | 17.43 ±0.13 (11-23) |

chromosome 5B. A major difference between their study and the one reported here is the absence in the former of the complete 5B with its strong pairing promoter(s) on the short arm. It is also possible that the detection of increased pairing, apart from that caused by *Ph1* deficiency alone, depends on the genomic and genotypic characteristics of the alien species hybridized. Interactions of pairing genes of the alien parent with those of wheat may change the picture in either a ‘positive’ or a ‘negative’ direction. There may also be species differences unrelated to pairing genes but which have an impact on pairing. This could be the case for wheat/rye chromosome pairing, which is possibly reduced by both taxonomic distances and different amounts and distribution of telomeric heterochromatin (Schlegel, 1979). Indeed, in contrast to the behaviour of the *Ae. variabilis* hybrids described here, no significant difference has emerged in the amount and pattern of pairing between the 2n = 28 and 2n = 27 + t (= -3DS) *ph1b* wheat/rye hybrid plants obtained from the same crossing scheme illustrated in Figure 25.2. To measure the relative effects of the *ph* conditions (single or double mutation) in a variety of situations, the spectrum of hybrid combinations is currently being widened.

## CONCLUSION

Modern cytogenetic methods allow the chromosome engineering of cultivated wheat, with desired introductions of alien chromatin segments. The high degree of precision that can be achieved in selecting and identifying donor and recipient chromosomes, as well as the possibility of gradually reducing the size of the alien chromatin segment (Sears, 1983), make this approach a practical one for exploiting the rich genepool represented by the wild relatives of wheat. For species related to wheat, this is undoubtedly the appropriate method to choose. In this method, the key mechanism that must be genetically altered to permit the desired exchanges is the one that normally prevents homoeologous pairing and recombination within the wheat genome as well as between wheat and other Triticeae. Because *Ph1* is the gene mainly responsible for the control of this phenomenon in wheat, chromosomal or genetic *ph1* mutations have so far almost invariably been used to promote interspecific and intergeneric homoeologous recombination. However, while in some instances higher recombination frequencies than those achievable in the absence of the *Ph1* gene would be advantageous, in others lower values would be more convenient.

If alien genepools are to be used more efficiently, it is important to improve our understanding of the complex gene systems controlling pairing and recombination, present both in the cultivated polyploid wheat species and in their wild or cultivated relatives. Interestingly, genes promoting or suppressing homoeologous pairing appear to be ubiquitous even among diploid species (Chen and Dvorak, 1984; Dvorak, 1987). Indeed, super-high pairing is known to be induced by the interaction of wheat and alien pairing genes (Feldman and Mello-Sampayo, 1967; Dover and Riley, 1972). However, more information is needed on the genetics and species distribution of 'wild' pairing genes. At the same time, a better knowledge of the effects of major and minor wheat genes on chromosome pairing would not only advance our understanding of the entire phenomenon but could also widen the spectrum of potential pairing situations suitable for a variety of cases. This spectrum can be further enlarged if a combination of mutations for pairing genes is developed. This would have an obvious beneficial impact on the overall efficiency of gene transfer.

## *Discussion*

J. G. WAINES: Did you observe any indication of transferable elements in any of your studies on chromosome pairing in the wheat group?

C. CEOLONI: I cannot give an answer based on the wild material I used because the number of species and accessions was very limited. However, I am aware of the existence of a 'genome restructuring gene', rather than a true transposable element, in an *Ae. longissima* accession. Elements that cause genome alterations must exist but, without careful observation, their detection is difficult.

J. G. WAINES: Are there in Italy, or in other parts of the world, local varieties or landraces which are unstable which may be going through this genome recombination in nature?

C. CEOLINI: My survey was in the wild species and commercial varieties only. In Argentina more than one landrace was found to be unstable, but instability may be found more frequently in the wild types.

Wheat Genetic Resources: Meeting Diverse Needs
Edited by J. P. Srivastava and A. B. Damania

Published by John Wiley & Sons

# 26

# *Utilization of Genetic Resources in the Improvement of Hexaploid Wheat*

B. SKOVMAND and S. RAJARAM

Darwin's theory of 'survival of the fittest' assumes no interference from *Homo sapiens*. However, Darwin could hardly have foreseen modern agricultural practices, and the effect of these and other human activities on the natural forces guiding the evolution and balance of the world's flora and fauna.

The Green Revolution, which began in the late 1960s, has had highly beneficial effects on human food supplies, but adverse effects on the habitats of many crop species, especially in their centres of origin (Frankel, 1989). The displacement of wild relatives and landraces of wheat in West Asia and the Mediterranean region by modern semi-dwarf varieties was essential to forestall widespread malnutrition and starvation (Dalrymple, 1986), but this must be offset by the timely conservation of genetic resources.

Modern agriculture is characterized by extreme homogeneity in cultural practices and crop varieties. This results in the cultivation of millions of hectares to a few varieties and the subsequent extinction of natural variation. It is undeniable that on-farm genetic variability worldwide has declined seriously, and genetic vulnerability remains high on the agenda in breeding programmes. In wheat, although genetic variability has declined, serious disease epidemics have been avoided through the use of genetic resistance.

Genetic resources in the various crop species supply the variability needed to provide adequate food in the future. In some instances, these resources have been lost because of ignorance or the realization, too late, of their importance. However, in the past 30 years efforts have been made to preserve natural variability. Since its establishment in 1974, the International Board for Plant Genetic Resources (IBPGR) has played an active role in this area.

## USE OF GENETIC RESOURCES IN WHEAT IMPROVEMENT

Much has been written about the limited use made of the genetic resources contained in genebanks, but few have attempted to estimate these collections' contributions to crop improvement, including wheat. Chapman (1986) found it difficult to assess the role of genetic resources (defined as wild relatives and landrace material) in wheat breeding. By looking at the occurrence of genetic resources in the pedigree of recently released varieties and by counting the number of references to genetic resources in the *Annual Wheat Newsletter*, he concluded that genetic resources are used in about 10 per cent of crosses. Genetic resource utilization can also be assessed by surveying the genes originating from wild material and landraces currently being used in wheat improvement programmes. Some examples are given here.

One of the prime examples is the use of *Rht* dwarfing genes in wheat improvement, especially the genes *Rht1* and *Rht2*. These two important genes became available through the Japanese wheat Norin 10, which derived its dwarfing genes from one of its parents, Shiro Daruma, a Japanese landrace variety (Kihara, 1983). The incorporation of dwarfing genes into valuable genotypes was a formidable challenge ( Krull and Borlaug, 1970; Borlaug, 1988). Additional desirable characteristics (for example, improved fertility and tillering capacity) other than those sought (such as strong straw) were obtained from the development of this germplasm (Krull and Borlaug, 1970). It is now obvious that the dwarfing genes *Rht1* and *Rht2* have a direct effect on yield over and above the benefits obtained from reduced lodging (Gale and Youssefian, 1986).

Roelfs (1988a) showed that, out of 41 known genes for stem rust resistance, 12 originated in species other than *Triticum aestivum* and *T. turgidum*; and out of 35 known genes for leaf rust resistance, 10 originated in species other than *T. aestivum* and *T. turgidum* (*see* Table 26.1). McIntosh (1983) showed that, out of the genes

**Table 26.1** Sources of known genes for stem and leaf rust resistance in hexaploid wheat

| Source | Number of genes | |
|---|---|---|
| | **Stem rust** | **Leaf rust** |
| *Triticum aestivum* | 21 | 23 |
| *T. turgidum* | 8 | 2 |
| *T. monococcum* | 3 | 0 |
| *T. timopheevi* | 2 | 0 |
| *T. taushii* | 2 | 4 |
| *T. umbellatum* | 0 | 1 |
| *Aegilops elongatum* | 3 | 1 |
| *A. intermedium* | 0 | 2 |
| *Secale cereale* | 2 | 2 |

Source: Modified from Roelfs, 1988a

originating in *T. aestivum* and *T. turgidum* for resistance to either rust, several are from landrace varieties.

One of these genes for stem rust resistance, *Sr2*, originally transferred to hexaploid wheat from Yaroslav emmer by McFadden in 1923 (Stakman and Harrar, 1957), has provided durable resistance. Varieties with *Sr2* in combination with other genes have been grown on many millions of hectares over the past 30 years in North America without stem rust losses (Roelfs, 1988b). The gains in wheat production associated with the Green Revolution in India and Pakistan would probably not have been realized without the protection against stem rust provided by *Sr2* in combination with other genes.

Pl 178383, from Turkey, provides an outstanding example of the need to conserve genetic resources, regardless of their immediate value and usefulness. It is now known to have resistance to four races of stripe rust, 35 races of common bunt and 10 races of dwarf bunt. It also has useful tolerance to flag smut and snow mold (Creech and Reitz, 1971). Pl 178383 is a parent of many of the varieties currently grown in the Pacific Northwest region of the USA (Kronstad, 1986).

Hybrid wheat development has depended heavily on the use of genetic resources. Both male sterility and restoration factors have been provided by *T. timopheevi* (Lucken, 1987).

The usefulness of genetic resources can also be gauged by examining the pedigrees of current varieties, as Chapman (1986) did. However, the genealogy of most varieties needs to be traced fairly far back in order to identify the genetic contributions, as most of these are not of an agronomic type leading to immediate varietal release from a simple cross.

At the Centro Internacional de Mejoramiento de Maíz y Trigo (CIMMYT), a computerized pedigree management system (PMS) has been developed. The PMS produces a genealogical tree, and the theoretical genetic contribution of any line, variety or landrace in that tree is calculated.

A partial tree for the Veery wheats, which have been released in 22 countries, is provided in Figure 26.1 (*overleaf*). This tree goes back only five generations and, at this level, it identifies six landraces which have a combined theoretical genetic contribution of little more than 23 per cent. The complete genealogy of the Veery wheats indicates 46 landraces contributing to this wheat. The theoretical genetic contribution of these 46 landraces, according to their country of origin, is given in Table 26.2 (*overleaf*).

These are just a few examples of the contribution of genetic resources to wheat improvement. Marshall (1989) stated that there appears to be a consensus that germplasm collections are poorly used, but that critics who assert that current levels of use are inadequate seldom specify what level *would* be adequate. He proposes that the use made of collections be measured by the number of requests per 1000 accessions held.

**Figure 26.1** Partial genealogy of the Veery wheats, generated with the Wheat Program's new pedigree management system (PMS)

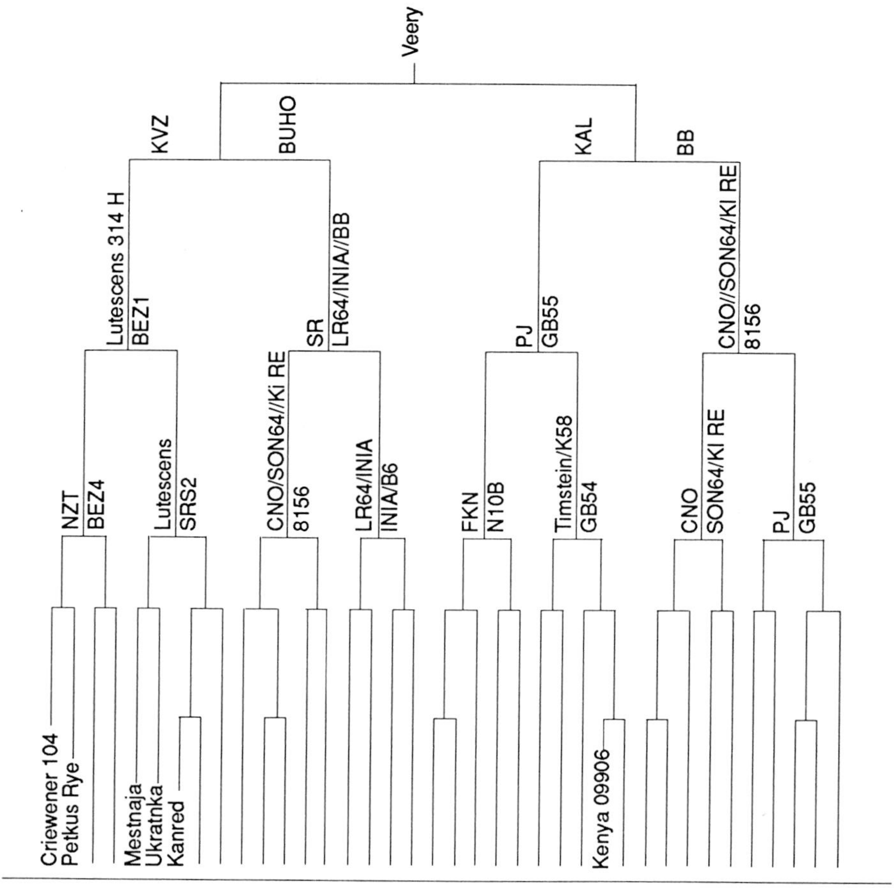


## DEFINITION OF GENETIC RESOURCES

CIMMYT follows a broad categorization of genetic resources, consistent with Frankel (1977) but with the addition of a fifth group:

- cultivated varieties, high yielding and in current use (advanced varieties) and those superseded by advanced varieties (obsolete varieties);
- primitive varieties or landraces of traditional, pre-scientific agriculture;
- wild and weedy relatives of crop species; 27 species are included in this group (Kimber and Feldman, 1987);

**Table 26.2** The theoretical genetic contribution of parental landraces varieties to the Veery wheats by geographic origin

| Species | Origin | No. of varieties | Theoretical genetic contribution |
|---|---|---|---|
| *Triticum aestivum* | Russia | 7 | 23.36 |
| | Poland | 4 | 6.76 |
| | Germany | 1 | 6.18 |
| | Netherlands | 1 | 0.52 |
| | France | 1 | 0.03 |
| | Italy | 2 | 2.52 |
| | USA [a] | 6 | 7.48 |
| | Morocco | 1 | 3.34 |
| | Egypt | 1 | 3.45 |
| | South Africa | 1 | 4.33 |
| | India | 3 | 3.85 |
| | China | 1 | 1.34 |
| | Japan | 2 | 6.02 |
| | Australia | 1 | 5.18 |
| | Argentina | 3 | 4.33 |
| | Brazil | 2 | 2.81 |
| | Uruguay | 2 | 2.99 |
| | Peru | 1 | 0.59 |
| *T. turgidum* | USA [a] | 2 | 2.59 |
| *Secale cereale* | Germany | 1 | 0.07 |

a Introductions into the USA from the Mediterranean region

- special genetic and cytogenetic stocks, including induced mutants;
- elite and advanced breeding lines.

The accessible genepool of wheat is remarkably wide. Autogamy does not stimulate the development of genetic isolating mechanisms, and allopolyploidy offers a network of intergeneric relationships (Mackey and Qualset, 1986). The polyploid wheats have a fairly high phenotypic buffering that confers tolerance to diverse genetic deletions and additions, and the available genepool is being enlarged with hybrids with other genera such as *Haynaldia*, *Secale*, *Agropyron*, *Elymus*, *Elytrigia* and *Hordeum* (Gale and Miller, 1987; Mujeeb-Kazi, 1988)

## EVALUATION OF GENETIC RESOURCES

The effective use of genetic resources depends on these resources being properly characterized and evaluated. If the purpose of evaluation is to increase the usefulness of genetic resources, then it must be carefully planned in consultation with breeders so that it is consistent with current breeding goals. Some evaluation carried out at present may not be applicable even in the near future; also, if breeders already have sufficient genetic variation for a trait in their active programme and working collections, further evaluation for that trait is unnecessary for the time being.

The case of resistance to stem rust illustrates this point. We currently have available genes or combinations of genes for resistance that have been effective for 30 years. There is no need to carry out extensive evaluation to identify other genes for resistance until the present satisfactory complex of genes is rendered impotent by a new race. Furthermore, it is very unlikely that we would find new, better genes for stem rust resistance in landrace varieties; any new sources of resistance will be identified in alien species, and may or may not be expressed in a tetraploid or hexaploid background.

Characters of concern to plant breeders have been divided into two groups that are genetically distinct: qualitative characters (resistance and tolerance to biotic and abiotic stresses) and quantitative characters (adaptation and productivity) (Frankel and Brown, 1984). Both groups contribute to productivity, but the former can be viewed as helping to protect the organism against adversity, while the latter affects productivity per se. Frankel and Brown propose the formation of 'core collections' to rationalize the numbers of accessions to be evaluated, but we would argue that although a core collection might be useful for evaluating certain quantitative characters, it would be of less use for qualitative characters.

## Qualitative characters

Evaluation for qualitative characters can be rationalized by using passport and characterization data, the distribution of a particular stress condition, and any other criteria proposed by breeders to form a subset of the collection for each character under evaluation. In a sense, this forms a specific 'core' for each trait. This might be more useful than a core collection as proposed by Frankel and Brown because it would take into account the other traits needed by breeders. For example, CIMMYT's breeding programme needed resistance to Septoria, which was required in combination with early maturity. A subset of 2500 accessions (about 7 per cent of the total collection) possessing earliness was then evaluated for Septoria reactions. About 250 accessions were identified that merited further evaluation.

The frequency of some characters is so low that the chances are slim that accessions having the trait would be included in a core. The frequency of resistance to the whitebacked planthopper in the *Oryza sativa* collection of more than 48 000 accessions held by the International Rice Research Institute (IRRI) was 0.08 per cent (Chang, 1989).

Evaluation for certain qualitative characters, such as tolerance to some abiotic stresses, where future changes in the environment may have little effect on the character, can be done when there is opportunity or need.

Thus we see the evaluation of observable characters more as a net that can be expanded to larger and larger subsets of a collection, adding characters when needed.

### Quantitative characters

Evaluation for yield-enhancing characters is more difficult and costly than evaluation for observable characters, and information to guide such evaluation is scarce.

It is in the evaluation for these characters that the formation of a core collection such as that proposed by Frankel and Brown (1984) would be of real value. We agree with these authors that, to identify yield-enhancing characters, a systematic programme of test crosses will have to be carried out to recognize combining ability and heterosis in landrace varieties and other genetic resources that have not been widely used in wheat improvement. This will have to be done on a core collection or on a subset thereof. The improved parent to be used must be chosen carefully, as grass dwarfness (*D* genes) and hybrid necrosis (*Ne* genes) can be expected to cause problems.

## FUTURE PROSPECTS FOR THE UTILIZATION OF GENETIC RESOURCES

Genetic resources have played a significant role in wheat improvement. We believe this role will expand, with the emphasis on providing the variability needed for future wheat improvement. Variability will be needed to further increase yield potentials, provide new sources of resistance to diseases and pests, provide adaptation to marginal environments and ensure improved quality.

### Increasing yield potentials

The yield potentials reached in wheat breeding have been achieved through empirical selection for yield. However, this method cannot be used to evaluate landraces and wild relatives because of their inherently low yield. One way of identifying yield-enhancing characters may be to make test crosses, as mentioned above, and measure the heterosis and combining ability of landrace varieties and other genetic resources that can be crossed with hexaploid wheat.

One system for choosing the landrace varieties to include in a programme of test crosses may be the PMS described earlier, which can identify the parental hierarchical relationships of modern varieties. It may be worth investigating more closely landraces that have so far made no or only a limited contribution to modern varieties. Also, more attention could be given to landraces originating in regions that have not contributed to wheat improvement.

Another way would be to evaluate for the physiological characters involved in yield (Ewans, 1981; Day and Lupton, 1987), or to define an ideotype (Donald, 1981) and then evaluate for such characters. However, physiological characters are numerous and a careful analysis of their relative merits would have to be carried out before embarking on such an endeavour.

## Sources of resistance to biotic stress

Genetic resources have played a major role in providing sources of resistance to pathogens and insects, and will probably play a greater role in this area in years to come. Most wheat breeding programmes spend a major proportion of their efforts on protecting the yield potentials already achieved by incorporating new and better genes or combinations of genes for resistance.

Collections of adapted and unadapted wheats have been rich sources of resistance to various diseases, and their underlying value is as a reserve of hitherto undetected genes for resistance (Williams, 1989). For most diseases, there is a need to identify more genes for resistance of the hypersensitive type to achieve combinations of genes that produce the type of resistance we now have for stem rust, which appears to be very effective. The wild relatives of wheat will make major contributions to this type of resistance.

There is a need to identify quantitative types of resistance (partial resistance), characterized by a reduced rate of epidemic build-up and durability (Parlevliet, 1988). This kind of resistance may be very important for diseases such as stripe rust, where race-specific resistance has not lasted. Landraces and obsolete varieties that have been grown extensively over many years in the endemic areas of particular diseases are the likely sources of this kind of resistance.

## Tolerance to abiotic stresses

The expansion of wheat cropping into marginal areas will present many stress challenges. Mineral ion deficiencies and toxicities, drought, wind, salinity and temperature extremes are some of the factors that will limit wheat production in these environments. Many of the primitive wheats and wild relatives originate in environments characterized by such stresses and can be expected to provide sources of tolerance to abiotic stresses.

## Improving quality

There is now evidence that good quality depends on high-molecular-weight (HMW) subunits of glutenin, grain protein content (Payne, 1987) and, to a lesser degree, other proteins in the endosperm (Law and Krattiger, 1987). Genetic resources, especially the landrace varieties, should be evaluated for quality characteristics. Farmers continue to grow landrace varieties in spite of their lower yields, claiming that these varieties have better quality (Skovmand, pers. obs.).

## CONCLUSION

Many authors have claimed that genetic resources are not being used sufficiently, but have not defined what they mean by adequate utilization. In any event, genetic resources have been used in wheat improvement when there has been a need to introduce new characters or genes to provide resistance to a new disease or a new race of a disease, to furnish tolerance to stresses, and to provide adaptation to new conditions.

However, problems of non-adaptation and poor agronomic types complicate the use of genetic resources for breeding new varieties. The breeder's job is to produce superior varieties for the benefit of farmers, and consequently a 'good' genetic resource is one that will lead to a variety within a reasonable time frame. The reason why genetic resources are not used as much or as well as some authors would like is that there are real difficulties involved, and not because of 'a lack of interest' (Peeters and Williams, 1984.

In the past 50 years of wheat breeding, a particular genetic stock, such as Norin 10, has always been the catalyst in breaking through the yield plateau or removing the biotic or abiotic constraints. During the remainder of this century and beyond, the genes that make up these stocks will be needed more than ever, especially as we approach new frontiers in wheat breeding. New methods are likely to permit the transfer of genes from alien genera, with the result that wild relatives will become still more important than they are today (Frankel, 1989). Gene transfer would allow us to exploit the slow accumulation of mutations that have taken place over long periods of time. The use of wild forms could broaden the genetic bases of our crops more rapidly and more extensively than any other form (Harlan, 1984). Whether the character expressions will be measurable or not remains to be seen. The most important point in having genetic resources collections is not that we use them now, but that we have these assets available when the need arises.

## *Discussion*

A. BIESANTZ: To what extent will genetic engineering replace conventional breeding at CIMMYT in the future?

B. SKOVMAND: We are establishing a biotechnology laboratory but I do not think genetic engineering will replace conventional breeding. It will merely aid conventional methods in identifying and mapping or transferring genes.

J. G. WAINES: To what extent does CIMMYT utilize landrace material in its breeding programmes?

B. SKOVMAND: We do not have an organized programme for assessing the value of landraces. We are still thinking about it. There are some problems with using landraces, such as poor economic value. You sometimes transfer undesirable traits together with desirable ones in crosses. But we have not totally ignored landraces. We have utilized landraces from Brazil, for example, to obtain tolerance to Al++ and we have screened many accessions for resistance to Karnal bunt.

Wheat Genetic Resources: Meeting Diverse Needs
Edited by J. P. Srivastava and A. B. Damania

Published by John Wiley & Sons

# 27

# *Utilization of Triticeae for Improving Salt Tolerance in Wheat*

J. GORHAM and R. G. WYN JONES

Salt tolerance is a complex phenomenon, not only because different plants respond to saline conditions in fundamentally different ways, but also because of the great variation in the stress itself. In different locations the total conductivity, chemical composition and pH of the saline soil may vary in three dimensions and also with time, depending on the sources of the salts (natural deposit or irrigation water), the management practices employed, and the local rainfall and evaporation conditions. Salinity is often combined with other stresses such as drought, waterlogging, alkalinity and nutrient deficiency. Plants which grow naturally in saline or sodic soils possess a range of morphological, physiological and biochemical mechanisms, some of which ensure survival in extreme conditions rather than high growth rates, and would not be appropriate in a crop plant.

This paper discusses the potentially useful mechanisms which may be found in both wild and cultivated relatives of wheat, and which may be transferred to wheat by conventional genetic techniques.

## EFFECTS OF SALTS ON PLANTS

Before discussing the mechanisms of salt tolerance which we have been investigating, we will consider briefly the damaging effects of salts on plant growth.

There are two major causes of salt damage in plants. The first is the 'physiological drought' caused by the decrease in soil water potential. This is usually important only when a stress is imposed suddenly, since plants take salt up into their tissues, or manufacture organic solutes, to decrease the water potential in their leaves and maintain a gradient of water potential along the transpirational pathway from the soil, through the plant and into the atmosphere. This accumulation of solutes within the plant is called 'osmotic adjustment'.

The second cause of salt damage is the toxicity of the salts entering the plant. This results mainly from high concentrations of sodium or chloride, but a secondary effect is the suppression of uptake of essential nutrients such as potassium. The toxicity of salts within leaves also depends on the sensitivity of cellular processes to high salt concentrations and on the distribution of salts between and within leaves and between different cellular compartments. In general, it would appear that the biochemistry of cells is not very different in halophytes and glycophytes, but that compartmentation of solutes within cells is very important. Details of the theory of compartmentation may be found in Wyn Jones (1981), Flowers and Läuchli (1983) and Leigh and Wyn Jones (1986).

It is not clear which ion is most toxic, $Na^+$ or Cl-. Since the uptake of anions and cations by plant roots is electrochemically linked, it is also not easy to distinguish between mechanisms for the uptake (or exclusion) of these two ions. Experiments using different combinations of cations and anions suggest that $Na^+$ is more toxic in wheat (Kingsbury and Epstein, 1986), but other work implicates Cl- in barley (Greenway, 1962; Boursier et al., 1987). It is against this background of complexity and uncertainty about the fundamental mechanisms of salt tolerance that we are seeking a logical approach to the problem of improving the salt tolerance of wheat.

## MECHANISMS OF SALT TOLERANCE

Two contrasting types of salt tolerance can be recognized (Munns et al., 1983). These are:

- substantial salt uptake, accompanied by efficient compartmentation of salts into large vacuoles in plants with a succulent morphology, as exhibited by extremely halophytic Chenopodiaceae;
- salt exclusion from the shoots, with an accumulation of sugars to reduce leaf water potentials, as seen in many, but not all, monocotyledonous halophytes (*sensu lato*).

Detailed descriptions and discussions on taxonomically and ecologically related physiotypes may be found in Gorham et al. (1980) and Kinzel (1982), while more general reviews of the physiology of salt tolerance can be found in Greenway and Munns (1980) and Flowers et al. (1986).

In cereal crops, most attention has focused on salt exclusion mechanisms, partly because of empirically determined relationships between exclusion and growth or yield in saline conditions, and partly because of technical difficulties in measuring compartmentation and cellular salt tolerance. Although this approach has been partially successful, it represents an over-simplified view of the problem, as we will show later in this paper.

## PYRAMIDING: THE THEORETICAL FRAMEWORK

The simplest way of screening for salt tolerance would appear to be to grow plants in a saline field and select the best lines or individuals. However, there are two main problems with this approach.

The first problem is the heterogeneity of salt in soils, both spatially and in relation to plant development. Not only may different plants be more or less susceptible at different stages in their growth cycles, but the intensity of the stress may also change with time. The patchy nature of salt in soils is further complicated by the plant's ability to extract water from the least saline areas explored by its roots. The problem is such that two plants next to each other may experience significantly different levels of stress. It has been suggested that breeding for high yield may be more important than breeding for extreme stress resistance, as the yield from less saline patches would compensate for the losses in more salty areas (Richards, 1983; Richards et al., 1987). It has also been suggested that there is a strict relationship between yield and the severity of the stress (yield thresholds and slopes), but we believe that this may also be an over-simplification, both of the complexity of the yield-environment curves and of the genetic potential of the plants.

The second argument against relying solely on field screening is that salt tolerance is such a complex (multigenic) phenomenon that it is very unlikely that currently available materials contain the most efficient and well-integrated versions of the various mechanisms involved. Growth and yield will be determined by whichever factor is limiting. Recognition of this problem has led to the pyramiding approach to breeding for salt tolerance (Yeo and Flowers, 1986).

The pyramiding approach involves identification of individual physiological/biochemical mechanisms, design of suitable screening procedures for each mechanism, selection of elite lines (or even species) for each mechanism, and incorporation into the target crop. At this stage it may be necessary to re-evaluate the effect of the mechanism on the yield of the target crop, although this would need to be done in an appropriate genetic background because of the 'limiting factor' problem. The ultimate aim of the pyramiding approach is to produce both theoretical ideotypes (designed for specific conditions and integrating these mechanisms most effectively) and plant material which can be used as building blocks to construct such 'ideal' salt-tolerant crops.

Two major impediments to implementing pyramiding are our incomplete understanding of the physiology/biochemistry of salt tolerance, and the need for cooperation between a wide range of disciplines. The need for a multidirectional transfer of information, ideas and plant material between physiologists, biochemists, cytogeneticists, plant breeders and agronomists has been discussed previously (Epstein, 1985; Wyn Jones and Gorham, 1989; Gorham and Wyn Jones, in press), and is illustrated in Figure 27.1 (*overleaf*). Our philosophical approach to the problem of breeding for salt tolerance in wheat is not unlike that adopted at the International Center for Agricultural Research for the Dry Areas (ICARDA) for breeding for drought tolerance (*sensu lato*) in barley (Acevedo and Ceccarelli, in press), except that at Bangor we are concerned primarily with the analytical aspects of the process.

**Figure 27.1** Interrelationships between different disciplines in the pyramiding approach to breeding for salt tolerance

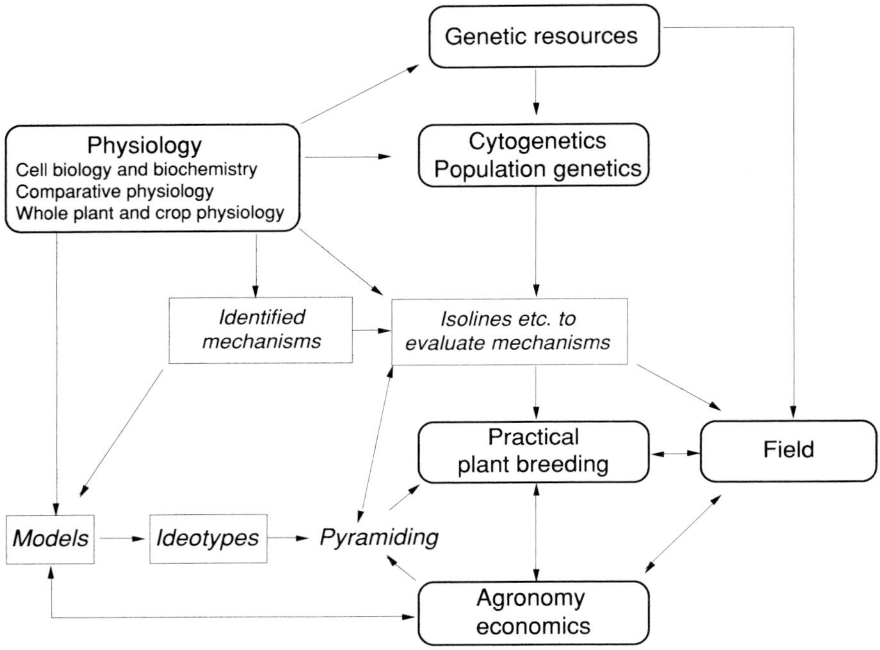


## SOURCES OF VARIATION

For effective research on salt tolerance, a wide range of plant material is needed, including wild relatives, primitive landraces and cultivars, products of wide hybridization and genetic manipulation, and the commercial crop varieties themselves. These are needed for exploratory experiments in comparative physiology, for testing hypotheses about possible mechanisms and their effects on growth and other factors, for direct incorporation into breeding programmes and for constructing the intermediate building blocks needed for the pyramiding of different characters. Variation for salt tolerance (*sensu lato*) and for various physiological parameters has been described within varieties and hybrid populations (Yeo et al., 1988), between varieties (Qureshi et al., 1980; Ashraf and McNeilly, 1988), between species (McGuire and Dvorak,

1981) and between related genera (Boursier et al., 1987; Richards et al., 1987). Our work has concentrated on two main areas: wild perennial wheat grasses; and annual species more closely related to wheat and wheat ancestors. The taxonomy and genome nomenclature of these two groups is somewhat confused. For the perennial wheat grasses we have adopted the genomic classification of Dewey (1984), while for the annual species we are using the nomenclature employed by Miller (1987).

Table 27.1 lists some genera of perennial Triticeae, shows their potential as sources of stress resistance, and indicates their ease of hybridization with wheat. We have concentrated on the section *Thinopyrum* of the genus *Thinopyrum* (Dewey, 1984), and in particular on *Th. bessarabicum*, a diploid, caespitose, glaucous grass from coastal habitats bordering the Black Sea. The potential of most other perennial Triticeae as sources of stress resistance has not been investigated frequently or systematically.

**Table 27.1** Genera of perennial Triticeae and their potential contribution to stress tolerance in wheat

| Genus | Genome | No. of species | Hybrids with wheat | Stress resistance |
|---|---|---|---|---|
| *Agropyron* | P | 10 | - | Drought |
| *Pseudoroegneria* | S | 15 | + | Drought |
| *Psathyrostachys* | N | 10 | - | Drought, alkalinity |
| *Critesion* | H | 30 | +++ | Salinity, drought, alkalinity |
| *Elytrigia* | SX | 5 | + | Salinity |
| *Elymus* | SHY | 150 | ++ | ? |
| *Leymus* | JN | 30 | +++ | Salinity, alkalinity, drought |
| *Pascopyrum* | SHJN | 1 | - | Salinity, alkalinity |
| *Thinopyrum* | J(E) | 20 | +++ | Salinity, alkalinity, drought |

Source: Classification according to Dewey, 1984

Our ignorance of the salt tolerance of the annual Triticeae is almost as great. Much work has been done on the genetics and cytogenetics of the ancestors of wheat but until recently little had been done on the salt tolerance of *Aegilops squarrosa*, a direct ancestor of wheat, although more is known about *Triticum dicoccoides*. There are many other annual species which we are only just beginning to explore in this context.

## SPECIFIC SALT TOLERANCE MECHANISMS

### Tolerance in *Thinopyrum* species

Several *Thinopyrum* species have been shown to be quite salt tolerant, or at least able to survive prolonged exposure to salt (McGuire and Dvorak, 1981; Gorham et al.,

1985). Full or partial hybrids between these species and hexaploid wheat are morphologically similar to wheat but show enhanced salt tolerance (Storey et al., 1985; Dvorak and Ross, 1986; Gorham et al., 1986; Forster et al., 1987). When various addition lines of *Th. elongatum* chromosomes in wheat were examined it was found that several chromosomes contributed to the salt tolerance of the amphidiploid (Dvorak et al., 1988). Of the addition lines of *Th. bessarabicum* in hexaploid wheat, the group 2 chromosomes were detrimental to salt tolerance, whereas the group 5 addition line showed greater tolerance than the parent wheat (Forster et al., 1988). Our experiments only partially confirmed the latter observation, but revealed that the 5J ($5E^j$) chromosome carries a gene or genes for enhanced total salt exclusion at high salinities, a character also displayed by the full amphidiploid hybrid. We detected no difference in salt uptake between the parent wheat and the addition line below 200 mol m-3 NaCl. This character is totally different from the K/Na discrimination character in wheat.

## K/Na discrimination

A series of experiments has demonstrated that *Aegilops squarrosa*, the D genome ancestor of wheat, possesses a gene(s) which controls the ratio of K to Na accumulated in the shoots and is located on the long arm of chromosome 4D (Gorham et al., 1987; Shah et al., 1987; Wyn Jones and Gorham, 1989). This chromosome and the associated trait are present in conventional and synthetic hexaploid wheats, but not in tetraploid, AABB-genome wheats. The discrimination character operates over a wide range of external salinities and affects the transport of Na to the shoots to a much greater extent than fluxes into or out of the roots themselves. Salt accumulation in roots is not affected by the character, nor are fluxes and concentrations of anions in shoots or roots, at least at moderate salinity. We would suggest that xylem loading is the site of discrimination (Gorham et al., in press). The main differences between this character and the one located on chromosome 5 of *Th. bessarabicum* are that the K/Na discrimination character operates at very low salinities and, at least at moderate salinities, does not affect the total salt load in the leaves.

## Tolerance of a high leaf salt load

When tetraploid wheat, which lacks the K/Na discrimination character, was compared with hexaploid wheat and barley, it was found that sodium accumulated in the leaves of tetraploid to a much greater extent than in the leaves of hexaploid wheat. In barley leaves, sodium concentrations were almost as high as in tetraploid wheat leaves, but barley was much more salt tolerant than either of the two wheat species. Thus barley possesses some mechanism which allows it to tolerate high leaf salt loads, possibly by efficient compartmentation of salts into vacuoles at the cellular level and into older leaves at the tissue level. In view of this finding, the relationship between Cl-exclusion and yield in saline conditions (Greenway, 1962) is being re-investigated. We are also examining the presence of this character, and of K/Na discrimination, in wild barleys.

## EXPLOITATION OF WILD TRITICEAE

Investigation of the barley character is still at a very early stage, and it is not yet clear how it can be transferred to wheat. Theoretical considerations suggest, however, that it offers exciting possibilities for improving growth in saline environments.

The aneuploid lines with deletions, additions or substitutions involving chromosome 4D which are currently availabl are all much less vigorous and fertile than the parent wheats. It has thus been impossible to prove unequivocally the beneficial effect of the K/Na discrimination character on salt tolerance, although there are signs of enhanced tolerance (Wyn Jones and Gorham, 1989; Gorham et al., in press). Work in India indicates that the K/Na discrimination character may be of even more value for tolerance to sodicity. To help improve the tolerance of tetraploid wheat to salinity or sodicity, this trait must be transferred to modern durum wheats without the other genes present on chromosome 4D. This is being attempted by colleagues in Italy and the USA. At the same time we are examining the variation for this trait in *Ae. squarrosa* and in both tetraploid and hexaploid wheats, within as well as between varieties. That useful variation exists in this quantitative character in species which lack the 4D chromosome, as well as those in which it is present, can be seen from Figure 27.2.

**Figure 27.2** Frequency of accessions of *Aegilops squarrosa* and *Triticum dicoccoides* in different classes of sodium uptake relative to *Triticum turgidum* cv. Langdon as determined by accumulation of $^{22}$Na in shoots of 'low-salt' seedlings exposed to mol m$^{-3}$NaCl + 0.1 mol m$^{-3}$$KNO_3$ + 0.5 mol m$^{-3}$$CaSO_4$

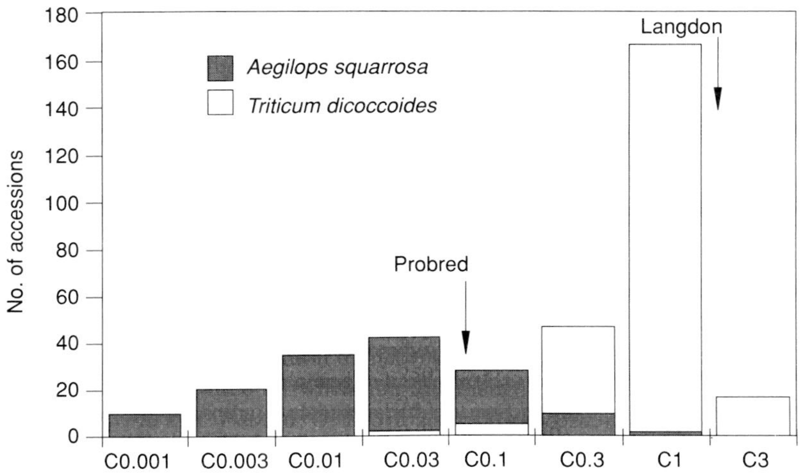

Furthermore, the enhanced K/Na discrimination character is present in a wide range of annual Triticeae (*see* Table 27.2), some of which, including diploid wheat, may be suitable sources for its transfer to tetraploid wheat.

**Table 27.2** Distribution of the enhanced K/Na discrimination character in the annual Triticeae

| Representative species | Genome | Presence of trait |
|---|---|---|
| *Triticum monococcum* | A | + |
| *T. turgidum* | AB | - |
| *T. timopheevi* | AG | + |
| *T. aestivum* | ABD | + |
| *Aegilops squarrosa* | D | + |
| *Ae. speltoides* | S | - |
| *Ae. comosa* | M | + |
| *Ae. umbellulata* | U | ± |
| *Ae. caudata* | C | - |
| *Hordeum vulgare* | I | - |
| *Secale cereale* | R | + |
| *Tritico secale* | ABDR | + |
| | ABR | + |
| *Critesion jubatum* | H | ? |
| *Dasypyrum villosum* | V | ? |

The enhanced salt tolerance of *Thinopyrum* species is being exploited in two ways. Various backcross progeny have been developed by our colleagues in Cambridge and at the Centro Internacional de Mejoramiento de Maíz y Trigo (CIMMYT), with the ultimate aim of introgressing useful characters into commercial wheats. The other possibility is to use the amphidiploid hybrids directly as new crops for stressful environments. This requires broadening the genetic inputs on both the wheat and the alien wheat grass sides, and improvement of the agronomic qualities of the 'tritipyrum' hybrids for fodder or grain production.

## CONCLUSION

We have been able to describe only a few of the many mechanisms which contribute to salt tolerance, and to discuss only some of their implications. The three mechanisms described above have been identified through careful interpretation of the data from experiments in comparative physiology. Such experiments should continue to provide a deeper understanding of the processes which contribute to salt tolerance, provided that the right questions are asked and the right plant material is available. In this context, the more distant relatives of wheat may be as important as the landraces and cultivars of wheat itself. The collection of more accessions of wild species, together with data on climate and soil chemistry (and preferably soil samples), and the maintenance of stocks of aneuploid lines of wheat involving alien chromosomes are worthwhile activities.

## *Discussion*

S. JANA: Is it possible that you did not come across the Na/K discrimination character in wild emmer wheat because your accessions originated from a relatively narrow spectrum of genetic variation?

J. GORHAM: We would like to have access to a collection complete with data on soil salinity and other characteristics. It would be useful if collectors of wild species also took soil samples.

J. G. WAINES: The Na reaction appeared to show wide variation in *T. turgidum* and another species. Have you studied the genetics of high and low levels of this character?

J. GORHAM: Not yet. We need to establish that the $^{22}$Na screen is specific for the Na/K discrimination character before we investigate the genetics of this character within species.

J. G. WAINES: Roughly how many accessions were you screening for each species?

J. GORHAM: That depended on the availability of material. In some cases there was only one accession, so the work needs to be repeated, obviously. In other cases there were 20 or 30 accessions. I have not presented data for all the experiments but we do run at least three experiments for each class.

A. MADDUR: Did you observe any variation to salt tolerance at different stages of plant growth? Are different genes of groups of genes operating?

J. GORHAM: It is well known that there are differences in plant response to salinity at various stages of growth. As to whether there is any significant variation for this among species, we have very little information. We do know that barley and wheat have a different response to salinity, at least in the seedling stage. But within wheats I do not know.

A. MADDUR: Do you have any standard or systematic screening technique for large numbers of plants during their entire life cycle?

J. Gorham: Response to salinity in the field can be tested during the life cycle of the plant. But, as I have indicated, there are difficulties in doing this in the field. There are other techniques which apply reasonably uniform stress using hydroponics or gravel culture, providing certain precautions are taken. If you are looking for short duration experiments in the laboratory which can be extended to the entire life cycle, the answer is no.

Wheat Genetic Resources: Meeting Diverse Needs
Edited by J. P. Srivastava and A. B. Damania

Published by John Wiley & Sons

# 28

# *Evaluation and Utilization of* Dasypyrum villosum *as a Genetic Resource for Wheat Improvement*

C. De Pace, R. Paolini, G. T. Scarascia Mugnozza,
C. O. Qualset and V. Delre

*Dasypyrum villosum* (L.) Cand. (syn. *Triticum villosum*, *Haynaldia villosa*) is an outcrossed, diploid, wild relative of wheat. It is found throughout the Mediterranean Basin, particularly in Italy and the Yugoslavian islands (De Pace, 1987). Evaluation studies have revealed a large variability in several biochemical and agronomical traits (Qualset et al., 1981; De Pace, 1987). Compared to wheat, *D. villosum* has a high seed and plant protein content (Della Gatta et al., 1984; De Pace et al., 1988), early seed maturity and resistance to certain soil-borne fungal diseases, such as take-all (Scott, 1981) and some strains of powdery mildew (De Pace et al., 1988). It also appears to have considerable above-ground biomass development and good recovery after grazing. In general, these traits appear to be species-specific and not confined to particular genotypes (De Pace, 1987).

In view of this versatility, several workers carried out wheat x *Dasypyrum* crosses. Strampelli obtained a fertile $F_1$ after crossing Frumento Nero with *Triticum villosum* (Strampelli, 1932), apparently to confer lodging resistance to non-dwarf wheat varieties. Subsequently, other workers (Tschermak, 1930; Sando, 1935; Kostoff, 1937; Halloran, 1966; Nakajima, 1966; Blanco et al., 1983; Jan et al., 1986) obtained wheat x *Dasypyrum* amphiploids and backcross-derived lines with the disomic addition of *D. villosum* chromosomes (Hyde, 1953; Sears, 1953; Blanco et al., 1987).

At the University of Tuscia in Italy, a research programme is in progress to:

- establish a technique for the rapid detection of *D. villosum* DNA in lines derived from backcrossing (wheat x *Dasypyrum*) x wheat;

- assess the potential of *D. villosum* and of a wheat x *Dasypyrum* hexaploid amphiploid as a forage crop;
- evaluate the response of the amphiploid and its parents to important agronomic factors such as plant density, N fertilization and sowing depth;
- to identify the traits transferred from *D. villosum* which make the amphiploid superior to the cultivated wheat parent.

The initial results of this research programme are reported here. The following abbreviations are used in this report: CS (Chinese Spring); Dv (*Dasypyrum villosum*); M (*Triticum turgidum* var. *durum* cv. Modoc); and B (*Hordeum vulgare* cv. Barberousse).

## MATERIALS AND METHODS

In 1981, two amphiploids were obtained (Jan et al., 1986):

- the hexaploid amphiploid M x Dv (*Triticum turgidum* var. *durum* cv. Modoc x *D.villosum*, 2n = 42, AABBVV);
- the octoploid amphiploid CS x Dv (*T. aestivum* cv. Chinese Spring x *D.villosum*, 2n = 56, AABBDDVV).

The two amphiploids showed a wheat-like plant type (Jan et al., 1986; De Pace et al., 1988). However, M x Dv had a stable chromosome number, whereas CS x Dv did not. Thus, M x Dv was used in experiments to determine the potential practical use of the amphiploids, while CS x Dv was used in backcrossing to CS in order to produce wheat x *Dasypyrum* derived lines with a varying amount of Dv alien genetic material in a supposedly unaltered CS wheat genomic background. M and a Dv population collected near Viterbo, Italy, were also used, and B was used as a reference crop in the evaluation of the biomass and N content of M x Dv. Several accessions of wheat (2x, 4x and 6x), *Aegilops*, *Secale*, *Dasypyrum*, wheat x *Dasypyrum* amphiploids (M x Dv and CS x Dv) and the Sears' disomic addition lines CS+1V, CS+2V, CS+3V, CS+4V, CS+5V, CS+6V and CS+7V were used to extract DNA and detect the amount of Dv genomic DNA by using DNA dot blot hybridization tests. The probe used was a Dv species-specific DNA (pAS2). Other experiments carried out on this material included the biotinilated DNA-probe preparation and responses to soil N availability, seedling emergence, and survival under low soil water conditions in a greenhouse and sowing depth in a growth chamber. The reported differential response of species were tested for significance using the F statistics. Details of the procedures for these experiments can be obtained from the senior author.

## RESULTS

Three observations from the experiments using dot-blot hybridization are worth mentioning here (*see* Figure 28.1). The first is that the hybridization signals of DNA from Dv, the wheat x Dv amphiploids and the Sears' disomic addition lines with the biotinilated probe pAS2 were almost identical. This indicates that the presence of only two V chromosomes in the wheat genetic background will be easily detected, although the visual examination of hybridized DNA dots does not show differences. It is probable that the use of a $^{32}P$ labelled probe, followed by scintillation counting, will better discriminate among derived lines containing 1 to 14 V chromosomes.

**Figure 28.1** Dot blot hybridization of 1μg DNA from different Triticinae and *Triticum* x *Dasypyrum* amphiploids and derived lines using the plasmid pAS2 containing a Dv repeated sequence as biotinilated probe

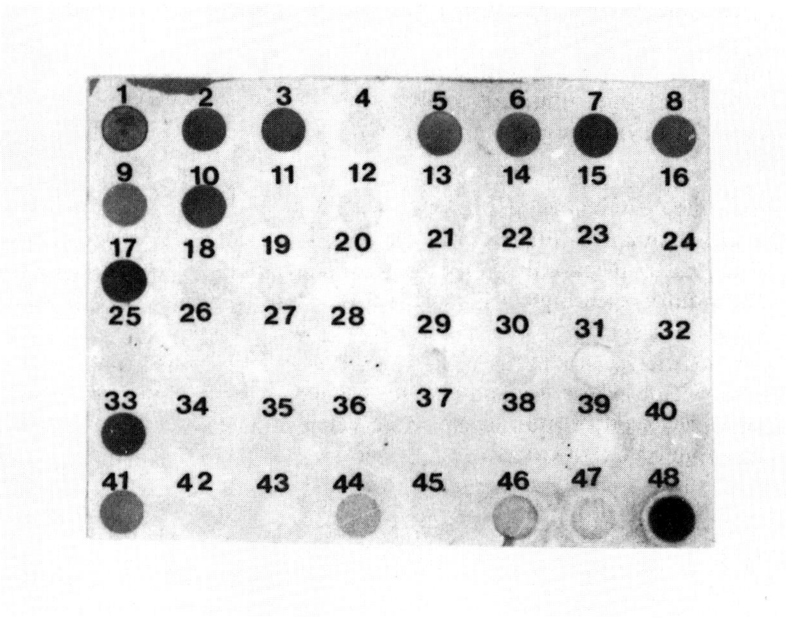


| | | | |
|---|---|---|---|
| 1 | Dv line 16B24 - 11/Line20 | 12 | *Triticum aestivum* cv. CS |
| 2 | CS x Dv Amphiploid | 13-16 | *Triticum* species (2n = 42) |
| 3 | M x Dv | 17 | Dv Line 1207 - 16/Line9 |
| 4 | *Triticum durum* cv. Modoc | 18 | *Secale cereale* |
| 5 | CS x Dv Amphiploid | 19-24 | *Triticum* species (2n = 28) |
| 6 | CS + 1Dv Addition Line | 25-32,34,38 | *Aegilops* species |
| 7 | CS + 2Dv Addition Line | 33 | Dv Line 39 - 9/Line12 |
| 8 | CS + 4Dv Addition Line | 35-38 | *Triticum* species (2n = 28) |
| 9 | CS + 5Dv Addition Line | 39-47 | *Triticum* x Dv derived lines |
| 10 | CS + 6Dv Addition Line | | |
| 11 | CS + 7Dv Addition Line | 48 | AS2 Plasmid (control) |

The second observation is that the wheat x Dv derived lines can be easily divided into two groups: those which contain Dv DNA and those which do not. In addition, by comparing the hybridization signals of the latter group with those from Sear's disomic addition lines, it can be seen that the amount of Dv DNA in the CS x Dv derived lines is less than two chromosomes. Therefore, lines containing 42 or 56 wheat chromosomes with only a small fragment of V chromosome derived from recombination will be detected.

Thirdly, the preparation of dot-blot filters (once the DNAs have been extracted) with DNA from 100 different genotypes takes about two hours, and the subsequent hybridization and its signal takes 17 hours. This means that, on average, the analysis for each sample takes about 11 minutes. This time can be further reduced if two filters are hybridized simultaneously. The use of these methods is therefore faster, easier and more powerful than cytogenetical observation.

However, as indicated above, the Sears' disomic addition line CS+7V did not show any hybridization signal. There are two possible explanations for this: either the chromosome 7V does not contain the repeated sequences detected in other V chromosomes; or the CS+7V disomic addition line has lost the 7V chromosome. So far, we have been unable to distinguish between those two situations, although the second seems unlikely on the basis of morphology and the presence of hairiness on the ligule of the CS+7V leaves. If this is the case, it means that the available probe is able to detect Dv DNA in 6 out of 7 pairs of V chromosomes of the diploid genome complement of Dv, thus decreasing the effectiveness of the developed screening strategy of wheat x Dv derived lines.

## Evaluation of biomass yielding ability and N content of M x Dv

Considerable differences for dry matter yield, leaf/plant dry matter ratio and plant N content were observed between species and at various growth stages (*see* Table 28.1). There were also differences in N yield at various growth stages. The species x growth stage interaction variance was significant only for dry matter yield and leaf/plant dry matter ratio. The tested plant density levels did not significantly influence dry matter yield and N content, and interactions involving plant density were not significant.

The highest accumulated rates of dry matter seemed to occur between 20 and 23 weeks after sowing (WAS), which on average corresponded to the interval between the end of tillering and the heading phase (*see* Figure 28.2 *overleaf*). During this interval, Dv showed the lowest dry matter accumulation. At the 23rd WAS, the dry matter yield of M and M x Dv did not differ significantly from B, and Dv produced 2.6 t/ha less than B. However, after the 23rd WAS, the dry matter yield of Dv increased by 1.53 t/ha. Although the data for the maturity stage are not available, it is likely that the relative dry matter yield observed at the 26th WAS will remain unchanged up to maturity. This is because no more than 25-30 per cent of the final dry matter is expected to be accumulated in the plant after heading, as experiments involving barley (Cervato et al.,

**Table 28.1** Mean squares and F-test in a factorial experiment involving four species (Dv, Mxv, M and B), three plant densities (170, 340 and 680 plants per m$^2$) and five growth stages (10, 16, 20, 23, and 26 weeks after sowing). Plant N content (%) and N yield were determined only at 16, 23 and 26 weeks after sowing

| Sources of variation | DF[a] | Dry matter (t/ha) (x10$^5$) | Leaf/plant DM[b] ratio (%) | DF | Plant N content (%) | N-yield (g/m$^2$) |
|---|---|---|---|---|---|---|
| Species (S) | 3 | 127.23 [d] | 638.531 [d] | 3 | 1.261 [d] | 4.969 ns |
| Density (D) | 2 | 15.233 ns | 61.742 [c] | 2 | 0.822 [c] | 15.502 ns |
| Growth stage (G) | 4 | 2810.10 [d] | 8637.176 [d] | 2 | 28.323 [d] | 979.37 [d] |
| S x D | 6 | 9.60 ns | 16.313 ns | 6 | 0.117 ns | 9.754 ns |
| S x G | 12 | 31.66 [d] | 149.216 [d] | 6 | 0.379 ns | 3.090 ns |
| D x G | 8 | 7.13 ns | 15.998 ns | 4 | 0.124 ns | 11.246 ns |
| S x D x G | 24 | 7.23 ns | 8.522 ns | 2 | 0.205 ns | 8.226 ns |
| Error | 59 | 6.10 | 15.651 | | 0.178 | 9.158 |

a DF = Degrees of Freedom; b DM = Dry Matter; c P = 0.05; d 0.01, respectively;
ns = not significant

1985) and durum wheat (Watson et al., 1963; Wittmer et al., 1982; Massantini et al., 1985) have shown.

From an agronomic point of view, the lack of significant differences among species for N content in the 16-26 WAS interval indicates that M x Dv and Dv have some potential as new crop species for earlier green fodder production in monoculture herbage. This potential will have practical value if M x Dv shows a rapid recovery ability. In addition, M x Dv performance for dry matter and N yield did not differ from that of barley at 26 WAS, which indicates that M x Dv dry matter production is comparable to that of a high-yielding, non-regenerating forage crop.

## Response to soil N availability

For the leaf/plant dry matter ratio (*see* Figure 28.3 *overleaf*), M x Dv gave the maximum value (22.1) at the 25 ppm N level. At this N level, M showed a similar value (14.1) to that expressed at 0 N level application (15.7). At 50 and 100 ppm, the average value remained at 16 for both genotypes. Thus, it seems that soil N levels above 25 ppm have no significant effect on increasing leaf/plant dry matter ratio in either genotype. However, the maximum increase observed at 25 ppm in M x Dv for total dry matter and plant N content indicated that a higher proportion of dry matter originated in the leaf tissues. This suggests that M x Dv has the potential to produce a greater amount of digestible green fodder than M under low soil N availability.

Although the amount of dry matter and N content of M x Dv is similar to M, the former has a high leaf/plant dry matter ratio and thus a higher nutritional value than M.

**Figure 28.2** Dry matter accumulation trend of Dv, M x Dv, M and B in field experiment

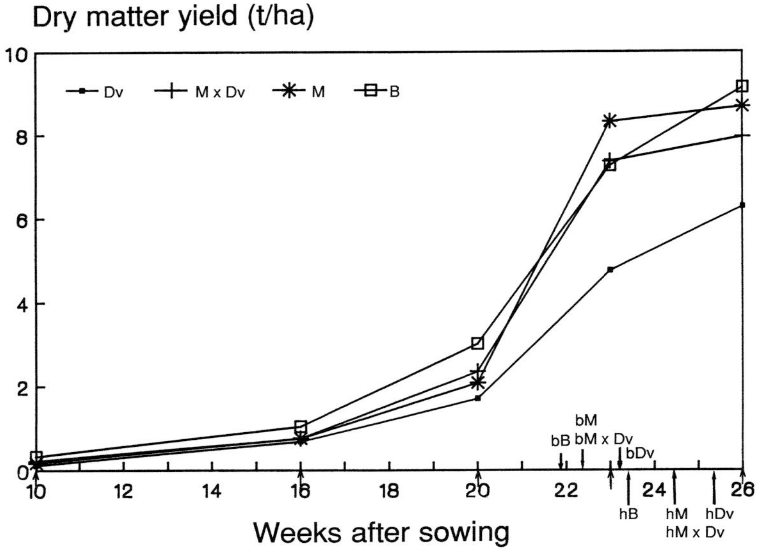


| | |
|---|---|
| bB | number of weeks to the booting stage in barley |
| bM | number of weeks to the booting stage in Modoc |
| bM x Dv | number of weeks to the booting stage in M x Dv |
| bDv | number of weeks to the heading stage in Dv |
| hB | number of weeks to the heading stage in barley |
| hM | number of weeks to the heading stage in Modoc |
| hM x Dv | number of weeks to the heading stage in M x Dv |
| hDv | number of weeks to the heading stage in Dv |

This higher leaf/plant dry matter ratio may be caused by the action of genes on Dv chromosomes, because Dv showed the highest leaf/plant dry matter ratio at heading time (*see* Figure 28.4 *overleaf*).

## Response of seedling emergence and survival to low soil water availability

In this experiment, for the first four days the soil moisture was kept above 30 per cent of field capacity. This was enough to ensure over 90 per cent seedling emergence in both Dv and M x Dv (*see* Figure 28.5 *overleaf*). Over the following 25 days, the soil moisture was maintained at below 30 per cent of field capacity. The plants suffered serious drought stress; 83 and 72 per cent of the Dv and M x Dv plants, respectively, died. Only 6 per cent and 13 per cent of the emerged Dv and M x Dv plants recovered.

**Figure 28.3** Plant N content (% of dry matter) of M x Dv and M grown in pots fertilized with different N rates in the glasshouse experiment

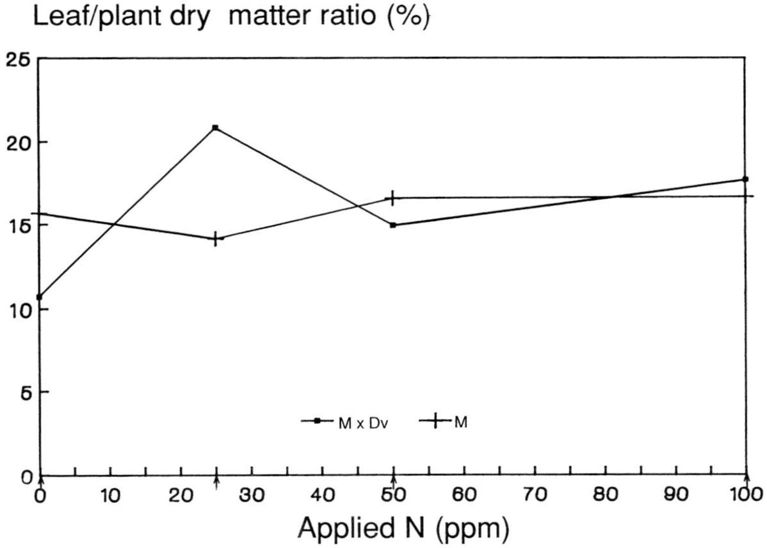


The percentage of drought-stressed plants remained almost unchanged when a 4-day or 30-day watering level above 30 per cent of field capacity followed the 25-day drought stress period. In the water regime C, the survival rate was 4 per cent for Dv and 4.5 per cent for M x Dv; and in the water regime D it was 4 per cent for Dv and 0 per cent for M x Dv.

## Response of seedling emergence to sowing depth

The five sowing depths tested had a significant effect on the seedling emergence and coleoptile length of Dv, M x Dv and M (*see* Table 28.2 *overleaf*). The highest percentage of seedling emergence was observed at 1.05 cm sowing depth, and was 68.8 per cent for Dv, 80.3 per cent for M x Dv and 93.7 per cent for M (*see* Figure 28.6 *page 288*). The seedling emergence of Dv halved at 1.9 cm sowing depth and was almost nil at 3.4 cm sowing depth. The seedling emergence of M and Dv decreased abruptly to about 25 per cent at 6 cm sowing depth, and was reduced to about 6 per cent at 10.3 cm sowing depth. It is interesting to note that M x Dv had a seedling emergence value intermediate between the low value of Dv and the high value of M when the depth was between 1.05 and 3.4 cm. Therefore, the gene complex controlling the seedling emergence in Dv and M acts additively in M x Dv, but this depends on the sowing

**Table 28.2** Mean squares and F-test for two characters in a factorial experiment involving three species (Dv, Mxv and M) and five sowing depths (1.05, 1.95, 3.40, 6.00 and 10.30 cm)

| Sources of variation | DF [a] | Seedling emergence (%) | Coleoptile length (cm) |
|---|---|---|---|
| Species (S) | 2 | 4.964[c] | 37.91[c] |
| Sowing depth (D) | 4 | 9.853[c] | 2.546[c] |
| S x D | 8 | 0.674[b] | 5.883[c] |
| Error | 28 | 0.248 | 0.552 |

a DF = Degrees of Freedom; b P = 0.05; c P = 0.01

**Figure 28.4** Leaf/plant dry matter ratio (%) in Dv, M x Dv, M and B measured at different week intervals from sowing in field experiment

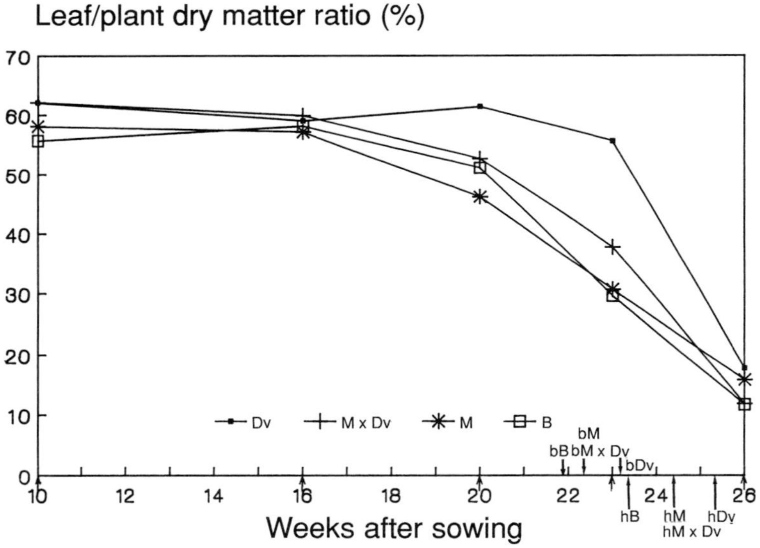


| | |
|---|---|
| bB | number of weeks to the booting stage in barley |
| bM | number of weeks to the booting stage in Modoc |
| bM x Dv | number of weeks to the booting stage in M x Dv |
| bDv | number of weeks to the heading stage in Dv |
| hB | number of weeks to the heading stage in barley |
| hM | number of weeks to the heading stage in Modoc |
| hM x Dv | number of weeks to the heading stage in M x Dv |
| hDv | number of weeks to the heading stage in Dv |

**Figure 28.5** Effect of different water regimes on Dv and M x Dv plant emergence and survival

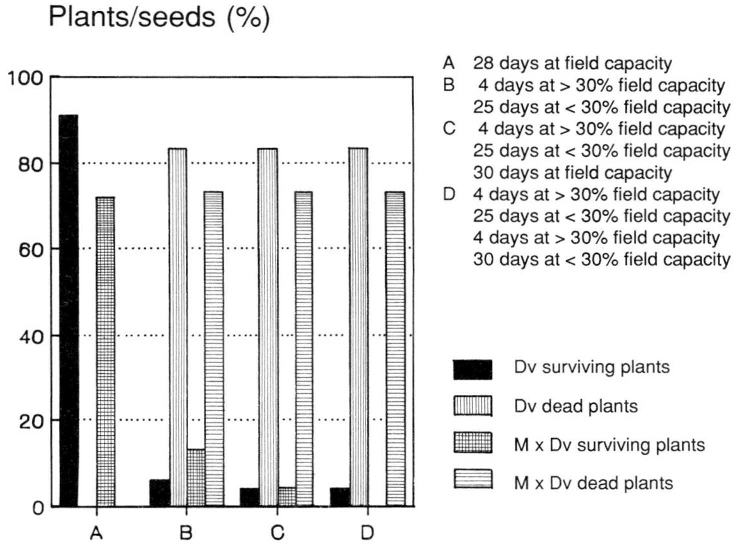


depth. There were also indications of strong interactions (that is, of M dominance) between the Dv and M genes that controlled coleopltile length in M x Dv at all sowing depths. This resulted in M x Dv having similar coleoptile length to M (± 3.5 cm) and longer than Dv coleoptile (± 2.5 cm) up to a sowing depth of 3.4 cm.

## CONCLUSION

In terms of the objectives of the study, the following conclusions can be drawn.

- Compared to the best performing parent (wheat) and to a reference species (barley), M x Dv showed a good biomass yield, mainly at early stages (heading), in terms of both quality (N yield) and quantity (biomass). Therefore, it seems that M x Dv has good potential as a forage crop.
- Neither increase of soil N availability above 25 ppm nor the increase of plant density up to 680 plants/m$^2$ has an effect on dry matter yield. However, in sandy soils, sowing depth is critical for the Dv and M x Dv seedling emergence, and a depth greater than 2 cm should be avoided.
- Both field and glasshouse tests showed that M x Dv has higher plant N content and leaf/plant dry matter ratio than M, indicating that the genes on the V genome of the amphiploid contributed positively to these traits. The leaf/plant dry matter ratio is more favourable under low soil N availability conditions.

**Figure 28.6** Seedling emergence (% of sown seeds) of Dv, M x Dv and M planted at five soil depths in pots kept in growth chamber

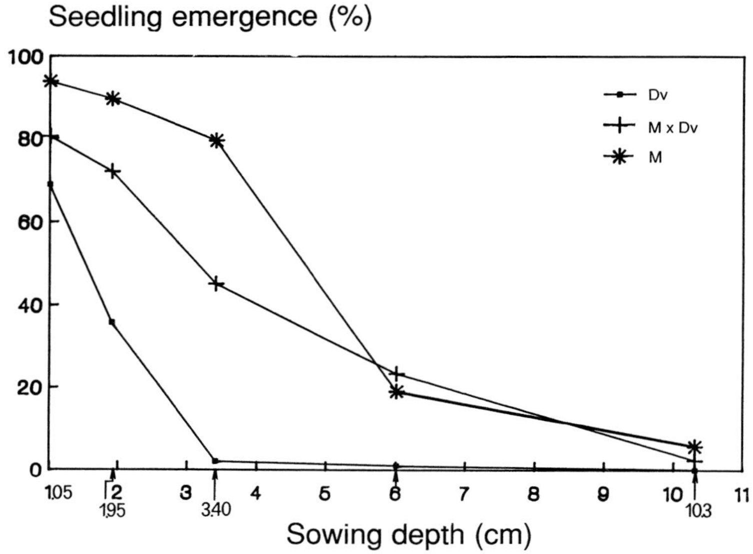


Our experiments also showed the following results.

- In the amphiploids, the expression of alleles for low emergence ability at high sowing depths and shortness of coleoptile present on Dv chromosomes are masked or counterbalanced by the action of the alleles on M chromosomes, and M x Dv has better seedling emergence ability than Dv.
- Drought stress tolerance is not a genetic feature of all the Dv genotypes, and it will be difficult to improve wheat for this character using conventional gene transfer from Dv, unless the Dv genotypes are properly selected.

## *Discussion*

D. R. Knott: Both your amphiploids were immune to powdery mildew, although in each case both parents were susceptible. How do you explain these results?

C. De Pace: There are seven different genes controlling powdery mildew resistance in wheat. And because of the close gene compatibility between *D. villosum* chromosomes and wheat chromosomes, it may imply that the same loci are present in *D. villosum*. However, the allele at these loci do not confer resistance to *Erysiphe graminis* f. sp. *haynaldiae* but to *E. graminis* f. sp. *tritici*. In the wheat chromosomes, it is the other way around. When the whole genome complement of wheat and *D. villosum* is present in the amphiploids, they will have resistance to both types of *E. graminis*.

M. OBANNI: Did you test your parent plants against individual powdery mildew races?

C. DE PACE: Yes, we did, using a very simple test. We took spores from postules on *D. villosum* plants and infected isolated plants of *D. villosum* and Modoc. We did the same for spores taken from the latter.

Wheat Genetic Resources: Meeting Diverse Needs
Edited by J. P. Srivastava and A. B. Damania

Published by John Wiley & Sons

# 29

# *Achievements and Constraints in the Utilization of Genetic Resources in the USSR*

V. SHEVELUKHA and B. MALINOVSKY

Plant breeding and seed production have played a pivotal role in the grain industry and other branches of agroindustry in the USSR. Using modern technologies to develop new varieties, we have been able to increase yield by 40 to 50 per cent. Well-known indigenous and breeding varieties of winter and spring wheat, as well as other crops, have been widely used by breeders in Western Europe and the USA. Two major breeding varieties, Mirinovskaya-808 and Besostaya-1, were predominant in most European socialist countries and have been used as a genetic resource for the development of the second and third generation varieties.

The history of plant breeding in the USSR has gone through several stages. In the early decades of this century, plant breeders developed a number of important plant varieties using the traditional breeding methods (such as individual and mass selection, and combining these methods with sexual hybridization). For example, V. S. Pustovoit was largely responsible for the development of a new sunflower variety with a high (50-53 per cent) oil content; and P. P. Lukyanenko, V. N. Remeslo and V. N. Mamontova developed varieties which had a high protein content and doubled yields.

This stage was followed by the period during which the eminent Russian geneticist, N. I. Vavilov, assembled a significant genetic resources collection in Leningrad's Experimental Botanic Bureau, based in the All-Union Research Institute of Plant Industry. All the plant materials were acquired from their centre of origin.

This genebank now has some 380 000 samples, and the collection continues to expand. It needs to be studied in detail, evaluated and used internationally for plant breeding. At the same time, through the international exchange of the plant samples, the collection needs to supplemented with new materials which would help solve some of the critical problems facing plant breeding and production in the USSR.

We are interested in obtaining plant forms which combine drought- and disease-resistance with high productivity. The current emphasis in our plant breeding institutes is on breeding varieties which are resistant to what are considered the most important problems — drought and pathogens. Many new cereal varieties with high resistance to some pathogens have been developed; these include varieties of spring wheat such as Jigulevskaya, Rodina, Novosibirskaya 67, Besenchukskaya 139, Svetlana and Gordeiforme 53. Among the successful varieties developed for drought resistance are Albidum 28, Saratovskaya 55, Celinnaya 26, Celinnaya 60, Celinaya Ubileinnya, Svetlana and Orenburgskaya 10.

However, we still face a number of problems in breeding for resistance to drought and pathogens. With the intensification of cultivation, the increase in the use of mineral fertilizers and the spread of cultivation to fields irrigated with sewage water, crops are more vulnerable to attack by pests and diseases. Simultaneously, the world agro-meteorological centres are predicting a significant increase in drought occurrence, which makes the search for more drought-resistant material all the more urgent. The use of genetic resources to develop varieties with the necessary traits is the most effective means of tackling this problem, but experience has shown that this is a slow process and that there are difficulties in obtaining varieties which combine high productivity with good resistance to drought and disease. Attempts at hybridization with wild forms containing the resistant genes have met with little success, and many more years, if not decades, of work are necessary before there is a breakthrough in this field.

## BIOTECHNOLOGY IN PLANT BREEDING

In view of these problems, the first thing to turn our attention is biotechnology. Using biotechnological methods it is possible, in a relatively short period of time, to transfer monogene traits from one plant to another. The use of these methods has resulted in the development of salt-resistant lines and varieties, and plants which are resistant to herbicides, pests and diseases. Unfortunately, the origin of most important resistance traits is poligenic, which makes the intensification, cloning and transfer of the genes very difficult.

At present, all biotechnological research in the USSR is based on cell and tissue engineering techniques. We have technological processing lines, using these techniques, for microclone multiplication of the valuable genotypes of sugar beet, potato, cereals and other crops. Using haploid methods, we have obtained valuable lines of several crops, including barley, spring and winter wheat, rice, clover and alfalfa. The new varieties of barley and rice, which combine good quality with disease resistance and high productivity, took only five years to develop and release. The rice variety known as Bioriza is notable for high disease resistance; the barley variety, Odessky 155, is drought resistant, and is grown in 13 areas of Russia and the Ukraine. The technology of the production of somaclone hybrids of wheat, potato, rice, tomato and other crops has been developed in order to obtain the distant interspecific hybrids resistant to biotic and abiotic stresses.

The development of biotechnological research is being undertaken at the research institutes and at the 15 interdepartmental biotechnological centres in the different zones of the USSR. The theoretical investigations are carried out at the leading institutes of the USSR Academy of Sciences. Collaboration between scientists of the Academy of Sciences and Vashknil is very important. Mutual cooperative programmes with socialist countries allows us to obtain new specialized plant forms. We also collaborate with the scientists of other countries on agricultural biotechnology and, in this regard, wish to develop our relationship with the International Center for Agricultural Research in the Dry Areas (ICARDA).

## RECENT DEVELOPMENTS IN PLANT BREEDING

Despite the advances made in biotechnology, traditional breeding methods are still important, and will continue to be so for some time to come. The scale of the breeding work carried out at Vashknil is reflected in the following figures. In the 1971-75 period, 460 new varieties and hybrids were released. This figure rose to 578 between 1976 and 1980, and reached 821 in the1981-85 period. In the 1986-89 period, 727 were released. Every year, 250-300 new varieties and hybrids are sent for strain testing; of these, between 130 and 170 varieties and hybrids are released.

The main developments in plant breeding in recent years can be summarized into 11 types of activities. The first is the creation of the short-stem varieties of wheat, rye, oats, barley and other crops with one, two or three genes of short-stem plants; the most widespread of these are the two-gene varieties. Short-stem varieties are very suitable for intensive agriculture, particularly because they are not prone to lodging.

Second is the development of pea varieties with resistance to husking, using a resistant donor indigenous to the USSR. The breeding work on almost all varieties of peas now involves the use of this donor. These high-quality varieties have been released in all the republics of the USSR.

Third is the development of winter varieties of durum wheat. Produced and released by the All-Union Plant Breeding and Genetics Institute in Odessa, the Zernograd Plant Breeding Centre in the Rostov district, and the Krasnodar Research Agricultural Institute, these varieties have high-quality grain suitable for the macaroni industry and, compared to spring durum wheat varieties, have high productivity.

Fourth is the development of varieties and hybrids of barley and maize with a high amino acid (lysine) content (4.1-4.5 per cent) and high protein content. We use the well-known lines Opeik-2 and Flowri-2 as the donors of high lysine content. However, there are a number of problems associated with using these donors: they do not give high inheritance of this trait; and the genes are linked with the other undesirable genes, resulting in poor agricultural performance. We are currently searching for more reliable donors and origins of high lysine content.

Fifth is the development of crop varieties and hybrids which have a good response to fertilizers. Extensive research in this area has produced genotypes of winter and spring wheat which, after the application of 1 kg of fertilizer, increase crop yields from 6-10 to 15-20 kg.

Sixth is the research into ecologically sound ways of applying mineral fertilizers and pesticides. This requires investigation into the problems associated with developing disease- and pest-resistant hybrids; into developing genotypes with high nitrogen fixation; and into producing new pesticides and fertilizers. Research is being conducted in all these areas. New plant varieties and micro-organisms with high nitrogen fixation have been developed and are being used in agroindustry. Four effective pesticides have been developed, but the scale of their application is still limited.

Seventh is increasing the use of heterosis breeding. A major focus of research in this area is the development of sunflower hybrids. We have developed and released sunflower hybrids with a high achene yield (4 t/ha) and a 50 per cent increase in oil content. Nationally and internationally developed sunflower hybrids now occupy about 1 million ha in the USSR. Nationally developed maize and grain hybrids occupy about 18.5 million ha and 4.5 million ha, respectively. Research is also being carried out on developing hybrids of vegetables, rice and other crops; recently, valuable heterosis hybrids of sugar beet have been produced. In general, heterosis breeding is conducted on a relatively small scale, but it is likely to grow considerably in the near future.

Eighth is the development of varieties of winter wheat and rye for Siberia, the Kazakh Republic and the Far East. This has been a long-standing and important problem. Initially, we tried to solve it by using Triticeae alien genes (cross-breeding *Agropyrum repens* with common wheat); this resulted in fairly resistant perennial forms of wheat, but they lacked the required winter hardiness. Biotechnological methods are now being used in attempts to solve this problem.

Ninth is the development of the varieties and hybrids with valuable individual traits. Among the varieties which have been produced in this area are: two-knot barley forms, which are resistant to drought or low temperatures; self-pollinating forms of alfalfa and clover; clover forms with two ovaries in one flower; drought resistant forms of plants with a well-developed root system; and pea forms without leaves.

Tenth is the development of halophytic plants which are suitable for desert conditions and saline soils. The emphasis in breeding work here is on pasture crops. These crops are being used on the large-scale, state sheep farms. The large halophyte collection is being studied and evaluated with reference to sea water irrigation. We are carrying out research to produce oil and forage halophytic varieties.

Finally, a considerable amount of work is being carried out on developing varieties and hybrids which will thrive on reclaimed land. Some 40 million ha of sewage lands have been turned over to agriculture, and this area is likely to increase significantly in the future. High-yielding varieties of winter wheat, maize, alfalfa, soybean, vegetables and other crops have been produced for cultivation on fields irrigated with sewage water.

## CONCLUSION

All these developments in plant breeding in the USSR have increased the effectiveness of our research efforts. In the next decade, the main objectives in our research institutions will be to develop forms with increased pesticide resistance and to use technologies which reduce yield loss and increase domestic food production. With Perestroika, these objectives stand a far greater chance of being realized. The benefits of Perestroika are not only that individual initiative among scientists and among workers in collectives is being encouraged but also that there is likely to be an improvement in the quality of technical equipment and research facilities.

In addition, we must take advantage of the increased opportunity for scientific cooperation with researchers from other countries. It is very important for the USSR to establish strong collaboration with scientific institutions in arid countries, such as Syria, so that we can exchange knowledge and resources on plant forms which have good drought resistance and other important traits. We look forward to establishing fruitful international cooperation in plant breeding, which will allow us to realize our economic potential and increase food production.

## *Discussion*

S. JANA: How are wheat accessions screened for drought tolerance in the USSR?

V. SHEVELUKHA: In the laboratory, physiological methods are mainly used, such as concentration of cell sap, presence of proline and other amino acids, and intensity of water exchanges. Fitotrons are also used with regulation of light, temperature and humidity. In the field, the methods used include sowing in drought-prone regions, and the application of auxanagraphics (measuring of the plant's growth functions). If any scientist or institution wants to collaborate with us in this study they are most welcome.

A. DAALOUL: Is tissue culture and biotechnology used for wheat improvement in the USSR?

V. SHEVELUKHA: We have found that wheat is a difficult crop to work with using biotechnological techniques such as tissue culture. Plants are regenerated from cells but this takes a long time and there is little success. However, we are using suspension of cells for callus production on substrates with a high salt concentration and other growth limiting factors. In this way, we were able to obtain resistance to soil salinity.

Wheat Genetic Resources: Meeting Diverse Needs
Edited by J. P. Srivastava and A. B. Damania

Published by John Wiley & Sons

# 30

# *Evaluation and Utilization of Exotic Wheat Germplasm in China*

TONG DAXIANG

Wheat production occupies an important place in food crop production in China. In terms of sown area and total output, it ranks second to rice. It is grown almost throughout the country, from Heilong Jiang province in the north to Guangdong province in the South, from the east coast to the Pamier plateau in the west, and from the Tibetan plateau, with an elevation of over 4000 m, to Tulufan Basin lying at less than 150 m below sea level.

Since the 1950s, wheat production in China has grown rapidly. In 1949 the acreage under wheat was about 21.5 million ha, with a yield of 645 kg/ha and a total output of 13.81 million tonnes; by 1985 the figures were 29.218 million ha, 2937 kg/ha and 85.805 million tonnes. This represents a 30 per cent increase in acreage, a fourfold increase in yield per hectare and a fivefold increase in total output. By 1985, the area under wheat in China was only slightly less than that in the USSR, but in terms of output China ranked highest in the world.

For the purposes of wheat research, the production area is divided into 10 regions, each containing several sub-regions, according to ecological and climatic conditions.

1. *Winter wheat region in northern China*. There are three sub-regions: the north-eastern plain of Hebei province, and the loess plateau and coastal area in eastern parts of Shandong and Liaoning provinces. The average annual temperature is 7-12.5°C; annual rainfall is 300-900 mm, with 100-250 mm during the wheat-growing season. The sown area in the whole region is about 3 million ha, representing 11 per cent of the total area under winter wheat.
2. *Huanghuai winter wheat region*. The three sub-regions are the western hilly and plain land, the northern plain of China and Huaibei plain. The average annual temperature is 9-15°C; annual rainfall is 500-700 mm, with 120-250 mm during

the wheat-growing season. This is the largest wheat production area in China; the sown area, 11.4 million ha, represents about 40 per cent of the total area under winter wheat.

3. *Winter wheat region of the middle and lower reaches of the Yangzi river.* The average annual temperature is 15.2-17.7°C; annual rainfall is 770-1480 mm, with 430-750 mm during the wheat-growing season. The sown area is about 4.8 million ha, representing 17 per cent of the total area under winter wheat.
4. *Winter wheat region in south-western China.* This includes the Sichuan basin, Guizhou plateau and Yunnan plateau sub-regions. In this region there is wide geographical and climatic diversity, with significant differences in elevation, temperatures and rainfall between or within the sub-regions.
5. *Winter wheat region in southern China.* There are two sub-regions: the coastal plain and interior mountain range. The average annual temperature is 16-24°C; annual rainfall is 1000-2400 mm, with 250-265 mm during the wheat-growing season. The sown area is about 0.8 million ha.
6. *Spring wheat region in north-eastern China.* The western sub-region has a dry and hot climate, the northern sub-region has a cooler climate and the eastern sub-region is humid. The average annual temperature is -3 to 7°C; annual rainfall is 350-900 mm, with 300 mm during the wheat-growing season. This is the largest spring wheat production area in China; the sown area, about 2.54 million ha, represents about 50 per cent of the total area under spring wheat.
7. *Spring wheat region in northern China.* This includes plain and river land, and hilly cold land. The average annual temperature is 1.4-13°C; annual rainfall is 100-600 mm. The sown area is about 1.2 million ha, representing 25 per cent of the total area under spring wheat.
8. *Spring wheat region in north-western China.* The sub-regions are the river basins, the mountainous area, the cool and wet area, and the Hexi corridor. The average annual temperature and rainfall in the cool and wet area are 6°C and 500-600 mm; in the other sub-regions the average annual temperature is 5-10°C and the annual rainfall is 50-400 mm. The sown area is 1.1 million ha, representing almost 25 per cent of the total area under spring wheat.
9. *Spring and winter wheat region in north-western China.* The sub-regions are the Qinghai and Tibetan plateaux. This region is characterized by high elevation and low rainfall, and the wheat produced here has a very high thousand-kernel-weight (TKW). The Qinghai plateau is 2600-3200 m above sea level; the average annual temperature is 2.6-5.1°C, and the annual rainfall is less than 100 mm. The average elevation of the Tibetan plateau is over 4000 m, but wheat is grown mainly in the valleys, which are 2600-4100 m above sea level and have an average annual temperature of 4.7-8.6°C. The sown area of the region is 0.15-0.16 million ha.
10. *Winter and spring wheat region in XinJiang province.* The region is divided into the northern and southern sub-regions. Spring wheat is grown mainly in the northern sub-region, which has severe winters and an annual rainfall of 200 mm. The warmer southern region, where most of the sown area is under winter wheat,

has an average annual temperature of about 10°C and an annual rainfall of less than 50 mm. The sown area in the whole region is 1.3 million ha.

## INTRODUCTION AND EVALUATION OF EXOTIC WHEAT GERMPLASM

Wheat germplasm resources were first introduced into China in the 1920s. By the 1940s, nearly 4000 wheat samples had been introduced. After the foundation of the People's Republic of China, and particularly since the 1970s, wheat introductions have increased significantly. The germplasm has been acquired from about 80 countries. The number of viable and valuable exotic germplasm samples now exceeds 10 000, which represents about one third of the total wheat germplasm collection.

The introduced wheat germplasm is grown in isolated fields, to check for dangerous diseases and pests. Healthy materials are then sent to relevant institutes for investigation and characterization. This takes 2-3 years, after which the materials are ready for further evaluation for pest- and disease-resistance, tolerance to cold, drought and saline soils, grain quality, yield potential and stability, and adaptation. The evaluations are carried out in different ecological conditions. For example, late varieties from high latitude areas are evaluated in XinJiang. Disease resistance evaluation focuses mainly on rusts; it is now also concerned with powdery mildew, which has extended over much of northern China in recent years. The purpose of evaluation for tolerance to cold, drought and saline soils is to select varieties or breeding materials which could survive under unfavourable ecological and farming conditions.

All research institutes, at both national and provincial level, carry out evaluations according to their speciality and geographical position. The Institute of Crop Germplasm Resources (ICGR), which is part of the Chinese Academy of Agricultural Sciences (CAAS), is responsible for the overall coordination of exotic germplasm evaluation. Located in Beijing (39°48'N), it also carries out evaluations for winter hardiness. The Institute of Plant Protection is responsible for evaluating resistance to diseases and pests. Some institutes situated near the lower reaches of the Yangzi river are responsible for scab resistance evaluation. Agronomic and morphological characters are investigated mainly by ICGR and other related institutes belonging to provincial academies of agricultural sciences.

The evaluation results of exotic wheat germplasm over the past 50 years have shown that varieties originating in Italy (such as Mentana, Abbondanza, Funo and Tevere) are suited mainly to regions with mild winters. They are now grown widely in China, particularly in the area around the middle and lower reaches of the Yangzi river. They have performed well and possess many important traits, such as big spikes with many grains, vigorous growth, high yield and good resistance to some diseases. They have also been used as parental stock in breeding programmes.

The varieties from the USA have been adapted to suit Chinese conditions and are resistant to various diseases, particularly rusts. They have more tillers and more spikes per plant than the European varieties, but the spike size is slightly smaller. They are usually taller than European varieties, although in recent years shorter USA varieties have been introduced.

In general, the varieties from the USSR are late-maturing, with a limited adaptability in China; they are grown mainly in XinJiang and Gansu provinces. Some of the USSR varieties with strong resistance to lodging possess large spikes and grains, and have been used as parental material in the breeding programme for the north winter wheat region.

The varieties from Mexico and India have adapted well in Yannan province. They are characterized by early maturity, shorter height and grainy spikes. In some varieties, the height is less than 70 cm, the grain number per spike is almost 100 and the TKW exceeds 40 g. These materials are therefore important sources of elite wheat germplasm.

Among the varieties from other countries which have adapted well to relevant wheat regions in China are Orofen from Chile, Quality from Australia and Lovrin 10.13 from Romania. Some of these varieties have become the main cultivars in certain wheat regions. Others have been crossed with Chinese wheat varieties or have been used as a genetic stocks for disease resistance.

## UTILIZATION OF EXOTIC WHEAT GERMPLASM

The history of agriculture has shown that the international exchange of agricultural technology and germplasm plays a crucial role in agricultural development. In the case of wheat production in China, the utilization of exotic germplasm has been very important (JAAS, 1983; Jing Shanbao, 1983; Zeng Xueqi et al., 1987; Zheng Diansheng, 1988). Exotic wheat germplasm has been utilized in China in the following ways: for direct utilization; as parental material in wheat breeding programmes; in crossing with rye to produce primary octoploid Triticale; and in genetic research.

### Direct utilization

The degree of direct utilization of exotic wheat germplasm depends on the adaptation and productive potential of the variety. Over 80 exotic wheat varieties are cultivated directly in China, some of them over a wide area. The variety Mentana was introduced into China in 1932 and reselected by Chinese agronomists. It was one of the main varieties in China during the 1950s, with a sown area of over 4.6 million ha (20 per cent of the total area under wheat, making it the second most widely grown variety).

The area sown to Italian varieties, such as Abbondanza, Funo, St1472/502, Ardito, Villa Glori 25 H 10, Tevere and Giuliari, is about 100 000 ha. The area sown to the variety CI 12203 from the USA is 0.6 million ha; this variety played an important part in controlling stem rust in the north-eastern spring wheat region during the 1950s. Varieties from Australia, Chile, the USSR and Mexico have also made an important contribution to the increase in wheat production.

# Summary and Recommendations

# *Summary and Recommendations*

The Symposium reviewed the present status of genetic resources evaluation, documentation and utilization in wheat improvement, both in national programmes and in international institutions. The limitations of current efforts in evaluation and utilization were highlighted, and recommendations were made on how to improve these efforts.

1. Nearly two decades of intensive genetic resources collecting activities has resulted in the large number of accessions now being conserved in the major genebanks. Systematic evaluation of these collections for desirable traits is necessary to facilitate their use in breeding programmes.

2. It is necessary to distinguish between 'characterization' and 'evaluation'. The former refers to the recording of information on those traits which are highly heritable, can identify an accession and are phenotypically expressed in all environments. Evaluation refers to the recording of desirable characters for potential use in crop improvement programmes.

3. Provenance and characterization data are important for the selection of germplasm for specific trait evaluation.

4. Large-scale evaluation of collections over several seasons at different sites is useful. Constraints to large-scale multi-site field evaluations can be minimized by using appropriate experimental designs (such as modified augmented design and nearest neighbour analysis), consistent checks, well-trained evaluators and appropriate documentation and statistical packages for data analysis.

5. Where resources are limited for a large-scale evaluation study, prioritization of characters, collaboration with other scientists in relevant disciplines and the use of visual observations such as 'agronomic score' is recommended.

6. Morphological and physiological markers can be used to select accessions from a collection for further study. For example, the presence of awn and pubescent

glumes is associated with sprouting resistance, and low rates of excised-leaf water loss are associated with high yield in dry environments. Further research should be undertaken to identify other such correlations between traits.

7. Data from characterization and evaluation research should be entered into computerized databases which are accessible to potential users of the information. This task could be performed, at national level, by germplasm banks and, internationally, by the international agricultural research centres for their mandate crops or by the International Board for Plant Genetic Resources (IBPGR).

8. Scientists handling crop germplasm (including curators, evaluators, geneticists, pathologists and breeders) should provide detailed identification of the samples and information on evaluation sites. They could also provide information on the population structure (for example, bulk or pure line), and the number of plants involved. Mentioning the accession numbers of the materials will result in greater use of the information and germplasm.

9. To promote the efficient transfer of information between genebanks, internationally agreed descriptor states (not more than 10) should accompany accession data. Additional information could be supplied upon request.

10. Increased communication between national and international genebanks will promote sharing of germplasm and information, thereby enhancing utilization.

11. Requests for seed and information should be made on a prescribed form. This will ensure that requests are unambiguous and emanate from bonafide sources. It will also prevent multiple requests coming from the same person/institution for identical material.

12. Any evaluation at an international level should be directed towards specific constraints to production. These constraints may differ from country to country and can be determined on a regional basis.

13. Elite germplasm of lines selected for specific traits could be distributed to the national agricultural research systems, which would grow the material in their own environment and share information with interested national programmes and the relevant international agricultural research centre.

14. A centrally located institution should be able to receive and collate data from national genebanks and make it easily available to all interested institutions. A basic format for data exchange needs to be determined.

15. Genebanks need not be places where genetic diversity remains in a static state. Evolutionary and other dynamic forms of preserved germplasm, as well as samples which augment adaptive genetic diversity, should receive serious attention.

16. *In situ* conservation of evolutionary germplasm should be given greater consideration. The practicality of this measure needs further discussion.

17. Genetic introgression between elite germplasm, improved genetic stocks produced by breeders and the largely unexploited genepools of indigenous genetic resources should be promoted. This will not only facilitate the transfer of desirable traits but will also enhance genetic diversity.

18. For reasons of safety, it was agreed that duplicate seed collections of major crops (such as wheat, rice, maize and barley) and their databases will be established at different institutions.

19. The user community of germplasm resources consists of both plant breeders and scientists engaged in basic research. The requirements of this community should be kept in mind by scientists engaged in comprehensive evaluation and characterization activities.

20. Free accessibility of germplasm for basic research as well as for crop improvement should be further encouraged. The international agricultural research centres are committed to the distribution of germplasm from their collections to bonafide users upon request.

21. To achieve further progress in breeding and to address new demands being made of wheat, it has become increasingly necessary to broaden the crop's genetic base with the introduction of genes from non-conventional germplasm.

22. There is increasing interest in the evaluation, documentation and utilization of wild progenitors and relatives of wheat, as well as other members of the Triticeae, particularly for utilization to overcome biotic and abiotic stresses.

23. The exploitation of wild types in wheat breeding for tolerance to stresses in unfavourable environments has been inadequate, for three reasons. Firstly, collections of wild species have been fragmentary and scanty, and the material available is not representative of the spectrum of variability of the species. Secondly, work on wild forms has concentrated mainly on evolutionary and taxonomic studies. Thirdly, variability within populations of wild species has not been examined in adequate detail and utilization has been limited.

24. As the availability of genes for disease resistance in cultivated forms decreases, scientists should increase their efforts to find such genes in the wild species.

25. In several cases, identical accessions of wild forms representing a very narrow base have been distributed upon request to different institutions. It is necessary to explore diverse ecological zones and collect representative samples for evaluation. This will result in the distribution of more authentic information.

26. It is necessary to evaluate wild species in the field for at least two seasons and at more than one location to be certain of the true characteristics of the germplasm.

27. Protein content is important and should be improved in wheat, especially for the West Asia and North Africa (WANA) region, but not at the cost of grain yield. In the WANA region, vitreousness and quality are more important than protein content.

28. There has been a reduction in seed protein content from the wild diploid *Triticum boeoticum* to the tetraploid *dicoccoides*, and from the cultivated *dicoccum* to durum landraces and, subsequently, to modern durum cultivars. For starch content and thousand-kernel-weight (TKW) the reverse is true.

29. If *T. dicoccoides* genes are utilized to increase the protein content of durum wheat, care should be taken to select only those progenies which have inherited good gluten strength from the durum parent in the cross. Good gluten strength in durum wheat is as important as high protein content and TKW.

30. Band 45 in the polyacrylamide gel electrophoresis (PAGE) profiles of gliadins (storage protein of wheat) does not indicate good pasta quality when found in *T. dicoccoides* samples. The two bands have identical mobility and intensity but indicate the presence of two different proteins.

31. The growing knowledge on the inheritance of a character and its genetic control will increase the efficiency of research to transfer genes of that character into cultivated forms.

32. Additional research should be undertaken to study the morphological and physiological basis of the characters of wild species under evaluation.

33. The phenomenon of 'co-adapted complexes' needs further study. Selection for one correlated response could result in incorporation of other desirable traits in the progenies.

34. Evaluators should include both wild and cultivated/domesticated parents in the evaluation of $F_2$ or higher segregants under field or greenhouse conditions. The evaluation of only one parent or the segregants will provide insufficient or incomplete information.

35. There is a need to carry out a systematic survey of wild relatives of wheat presently conserved in genebanks and to fill the gaps in collections by organizing further explorations in collaboration with international agricultural research centres and the national programmes.

36. The IBPGR and International Center for Agricultural Research in the Dry Areas (ICARDA) have initiated a joint project to conduct a world survey of *Aegilops* species in collections. The project will also identify duplicates and misclassifications andwill establish a database of passport and other information at ICARDA.

37. It is relatively easy to enter a large quantity of data into a computer but, to ensure quality, the task must be carried out carefully by experts. Data-entry personnel undertaking this work must be thoroughly briefed and well trained and supervised.

38. ICARDA has assembled a collection of *Aegilops* and *T. dicoccoides* species and primitive forms of wheat which has been evaluated mainly for drought tolerance, protein content and yellow rust tolerance. Pure lines have been identified from these accessions and crosses will be made with cultivars for the transfer of these traits. The material is available upon request to national programmes and other users.

39. Wild relatives and primitive forms identified as possessing desirable traits should be included in a pre-breeding programme to develop trait specific genetic stocks with good background. These stocks could then be distributed to the national programmes and other scientists for greater utilization.

40. In utilizing wild progenitors, it is relatively easy to transfer disease resistance, as this character is simply inherited. However, dealing with more complex characters, such as protein content or yield, is more difficult. Introgression of wild genes into cultivated forms should be encouraged in order to create genetic stocks. This may not necessarily be a breeding activity.

41. A catalogue of genome donor species for specific traits should be assembled. Such information will also encourage the less developed research programmes to utilize wild progenitors in crop improvement.

42. It is not necessary to have a formal germplasm evaluation network. There is, however, a real need for the sharing of information between genebanks on total holdings, characterization information, passport data, and data on evaluation for specific traits. A more active dialogue between genebank curators and plant breeders is desirable. An informal network of scientists interested in the evaluation, documentation and utilization of genetic resources of wheat should be developed.

43. There are several intermediate steps between the evaluation and utilization of germplasm. The international agricultural research centres could be of assistance by providing the $F_1$ and $F_2$ material resulting from specific, interspecific and inter-generic crosses with the parents supplied or suggested by the national programmes.

44. Support from international centres such as ICARDA, the IBPGR and the Centro Internacional de Mejoramiento de Maíz y Trigo (CIMMYT) for crop genetic resources networks would have the following benefits:
    - improved coordination and collaborative efforts for collection, conservation, documentation and utilization would minimize duplication of efforts and lead to a better sharing of responsibility;
    - new avenues would be available for identifying research priorities and managing and utilizing germplasm;
    - all interested parties would be able to provide their inputs and benefit from greater access to the germplasm and related information;
    - specific needs or constraints could be identified and international agricultural research centres could provide assistance in addressing these needs or identifying donor agencies which could help

45. Scientists should identify specific stresses which limit production in their own region or country and subsequently screen germplasm for resistance to these stresses. Cooperation between scientists in this field could reduce the costs of screening germplasm. Some institutions could assume overall responsibility for the evaluation of those traits most commonly required in their crop improvement programmes.

46. ICARDA has identified several trait-specific elite lines from durum wheat landrace accessions. These lines are available on request.

| | |
|---|---|
| Early heading at dry site | 146 |
| Early heading at moderately favourable site | 70 |
| Long grain-filling period at dry site | 128 |
| Early heading and long grain-filling period at dry site | 61 |
| Good overall agronomic performance at dry site | 187 |

| | |
|---|---|
| Good overall agronomic performance at dry site and high yield at moderately favourable site | 60 |
| Resistant to yellow rust | 190 |
| Resistant to common bunt | 193 |
| Resistant to Septoria | 147 |
| Resistant to Septoria and common bunt | 40 |
| Tolerant to cold in vegetative phase | 185 |
| Tolerant to salinity | 126 |
| Early heading, long grain-filling period and high tillering capacity and tall plants at dry site | 148 |
| Early heading, long grain-filling period and high tillering capacity at moderately favourable site | 227 |
| Early heading, high TKW, high tillering capacity, and high number of seeds/spike at moderately favourable site | 79 |

47. It has been recently agreed that CIMMYT will be responsible for the bread wheat base collection and ICARDA will provide the backup for this programme. Concurrently, ICARDA will take responsibility for the durum wheat base collection and for wild relatives of wheat, with CIMMYT providing the backup.

48. The major seed banks, such as those operated by the United States Department of Agriculture (USDA), the Vavilov All-Union Institute of Plant Industry in the USSR and the Germplasm Institute in Bari, Italy, need to interact more actively in germplasm and information exchange.

49. National programmes need to be strengthened and encouraged through training in systematic germplasm evaluation, conservation and utilization.

50. The ultimate user of the germplasm is not the evaluator or the breeder, but the farmer.

# *References*

## Chapter 1

Pecetti, L., Jana, S., Damania, A. B., and Strivastava, J. P. (in prep.) Practical problems in large-scale germplasm evaluation: A case study in durum wheat. Presented at the Symposium on Evaluation and Utilization of Genetic Resources in Wheat Improvement, 1989, Aleppo, Syria.

Jana, S., Srivastava, J. P., Damania, A. B., Clarke, J. M., Yang, R. C., and Pecetti, L. 1990. Phenotypic diversity and associations of some drought-related characters in durum wheat in the Mediterranean region. In *Wheat Genetic Resources: Meeting Diverse Needs*. Chichester, UK: John Wiley.

## Chapter 2

Brady, N. C. 1975. Rice responds to science. In Brown, A. W. A., et al. (eds.), *Crop Productivity Research Imperatives*. East Lansing, Michigan and Yellow Spring, Ohio, USA: Michigan Agricultural Experiment Station and C.F. Kettering Foundation.

Burgess, S. (ed.). 1971. *The National Program for Conservation of Crop Germplasm*. Athens, Georgia, USA: University of Georgia.

Chang, T. T. 1967. Growth characteristics, lodging and grain development. *International Rice Commission Newsletter*. Special Issue 54-60. Bangkok, Thailand: Food and Agriculture Organization (FAO).

Chang, T. T. 1976. *Manual on Genetic Conservation of Rice Germplasm for Evaluation and Utilization*. Los Baños, Philippines: International Rice Research Institute (IRRI).

Chang, T. T. 1980. The rice genetic resources program of the International Rice Research Institute (IRRI) and its impact on rice improvement. In *Rice Improvement in China and Other Asian Countries*. Los Baños, Philippines: IRRI.

Chang, T. T. 1983. Exploration and collection of the threatened genetic resources and the system of documentation of rice genetic resources at the International Rice Research Institute (IRRI). In Jain, S. K., and Mehra, K. L. (eds.), *Conservation of Tropical Plant Resources*. Howrah, India: Botanical Survey of India.

Chang, T. T. 1984a. The role and experience of an international crop-specific genetic resources center. In *Conservation of Crop Germplasm: An International Perspective.* Crop Science Society of America (CSSA) Special Publication No. 8. Madison, Wisconsin, USA: CSSA.

Chang, T. T. 1984b. Evaluation of germplasm: Rice. In Holden, J. H. W., and Williams, J. T. (eds.), *Crop Genetic Resources: Conservation and Evaluation.* London, UK: George Allen and Unwin.

Chang, T. T. 1985a. Collection of crop germplasm. *Iowa State Journal of Research* 59: 349-364.

Chang, T. T. 1985b. Evaluation and documentation of crop germplasm. *Iowa State Journal of Research* 59: 379-398.

Chang, T. T. 1985c. Germplasm enhancement and utilization. *Iowa State Journal of Research* 59: 399-424.

Chang, T. T. 1988. Taxonomic key for identifying the 22 species in the genus *Oryza. International Rice Research Newsletter* 13(5): 4-5.

Chang, T. T. 1989. The case for large collections. In Brown, A. H. D., et al. (eds.), *The Use of Plant Genetic Resources.* Cambridge, UK: Cambridge University Press.

Chang, T. T., Adair, C. R., and Johnston, T. H. 1982a. The conservation and use of rice genetic resources. *Advances in Agronomy* 35: 37-91.

Chang, T. T., and Loresto, G. C. 1986. Screening techniques for drought resistance in rice. In Chopra, V. L., and Paroda, R. S. (eds.), *Approaches for Incorporating Drought and Salinity Resistance in Crop Plants.* New Delhi, India: Oxford and International Book House

Chang, T. T., Dietz, S. M., and Westwood, M. N. 1989. Management and utilization of plant germplasm collections. In Knutson, L., and Stoner, A. K. (eds.), *Biotic Diversity and Germplasm: Global Imperatives.* Dordrecht, The Netherlands: Kluwer Academis Publishers.

Chang, T. T., Loresto, G. C., O'Toole, J. C., and Armenta-Soto, J. L. 1982b. Strategy and methodology of breeding rice for drought-prone areas. In *Drought Resistance in Crops with Emphasis on Rice.* Los Baños, Philippines: International Rice Research Institute (IRRI).

Chang, T. T., Marciano, A. P., and Loresto, G. C. 1977. Morpho-agronomic variousness and economic potentials of *Oryza glaberrima* and wild species of the genus *Oryza*. In *Meeting on African Rice Species.* Paris, France: Institut de recherches agronomiques tropicales et des cultures vivrières (IRAT)/Institut français de recherche scientifique pour le développement en coopération (ORSTOM).

Chang, T. T., Ou, S. H., Pathak, M. D., Ling, K. C., and Kauffman, H. E. 1975a. The search for disease and insect resistance in rice germplasm. In Frankel, O. H., and Hawkes, J. G. (eds.), *Crop Genetic Resources for Today and Tomorrow.* Cambridge, UK: Cambridge University Press.

Chang, T. T., Seshu, D. V., and Khush, G. S. 1988. The seed exchange and evaluation programs of the International Rice Research Institute (IRRI). In *Rice Seed Health.* Los Baños, Philippines: IRRI.

Chang, T. T., Villareal, R. L., Loresto, G. C., and Perez, A. T. 1975b. IRRI's role as a genetic resources center. In Frankel, O. H., and Hawkes, J. G. (eds.), *Crop Genetic Resources for Today and Tomorrow.* Cambridge, UK: Cambridge University Press.

Curtis, B. C. 1988. The potential for expanding wheat production in marginal and tropical environments. In Klatt, A. R. (ed.), *Wheat Production Constraints in Tropical Environments*. Mexico, DF: United Nations Development Programme (UNDP)/Centro Internacional de Mejoramiento de Maíz y Trigo (CIMMYT).

Dalrymple, D. G. 1980. *Development and spread of semidwarf varieties of wheat and rice in the United States: An international perspective*. Agricultural Economic Report 455. Washington DC, USA: United States Department of Agriculture (USDA)/United States Agency for International Development (USAID).

Epstein, E. 1977. Genetic potentials for solving problems of soil mineral stress: Adaptation of crops to salinity. In Wright, M. J. (ed.), *Plant Adaptation to Mineral Stress in Problem Soils*. Cornell University Agricultural Experiment Station Special Publication. Ithaca, New York, USA: Cornell University.

Flavell, R. B. 1985. Molecular and biochemical techniques in plant breeding. In *Biotechnology in International Agricultural Research*. Los Baños, Philippines: International Rice Research Institute (IRRI).

Frankel, O. H. 1989. Principles and strategies of evaluation. In Brown, A. H. D., et al. (eds.), *The Use of Plant Genetic Resources*. Cambridge, UK: Cambridge University Press.

Frankel, O. H., and Soulé, M. E. 1981. *Conservation and Evolution*. Cambridge, UK: Cambridge University Press.

Frey, K. J., Cox, T. S., Rodgers, D. M., and Bramel-Cox, P. 1984. Increasing cereal yields with genes from wild and weedy species. In Chopra, V. L. et al. (eds.), *Genetics: New Frontiers*. (Vol. 4). New Delhi, India: Oxford and International Book House.

Goodman, M. M. 1985. Exotic maize germplasm: Status, prospects and remedies. *Iowa State Journal of Research* 59: 497-528.

Hallauer, A. R., and Miranda, J. B. 1981. *Quantitative Genetics in Maize Breeding*. Ames, Iowa, USA: Iowa State University Press.

Harinarayana, G., and Rai, K. N. 1989. Use of pearl millet germplasm and its impact on crop improvement in India. In *Collaboration on Genetic Resources*. Patancheru, AP, India: International Center for Research in the Semi-Arid Tropics (ICRISAT).

Harlan, J. R. 1984. Evaluation of wild relatives of crop plants. In Holden, J. H. W., and Williams, J. T. (eds.), *Crop Genetic Resources: Conservation and Evaluation*. London, UK: George Allen and Unwin.

Harlan, J. R., and Starks, K. J. 1980. Germplasm resources and needs. In Maxwell, F. G., and Jennings, P. R. (eds.), *Breeding Plants Resistant to Insects*. New York, USA: John Wiley.

Hawkes, J. G. 1977. The importance of wild germplasm in plant breeding. *Euphytica* 26: 615-621.

Heinrichs, E. H., Medrano, F. G., and Rapusas, H. R. 1985. *Genetic Evaluation for Insect Resistance in Rice*. Los Baños, Philippines: International Rice Research Institute (IRRI).

Hermsen, J. G. 1984. Some fundamental considerations on interspecific hybridization. *Iowa State Journal of Research* 58: 461-474.

Hermsen, J. G. 1989. Current use of potato collections. In Brown, A. H. D., et al. (eds.), *The Use of Plant Genetic Resources.* Cambridge, UK: Cambridge University Press.

Hibino, H. (in press) Towards stable resistance to rice viruses. In *Proceedings of the Symposium on Crop Protection in the Tropics: 11th International Congress on Plant Protection, 1987, Manila, Philippines.*

Hibino, H. 1987. Rice tungro virus disease: Current research and prospects. In *Proceedings of the Workshop on Rice Tungro Virus, 1987.* Agency for Agricultural Research and Development (AARD)-Maros Research Institute for Food Crops, Maros, Indonesia.

Huaman, Z. 1984. The evaluation of potato germplasm at the International Potato Center (CIP). In Holden, J. H. W., and Williams, J. T. (eds.), *Crop Genetic Resources: Conservation and Evaluation.* London, UK: George Allen and Unwin.

Ikehashi, H., and Ponnamperuma, F. N. 1978. Varietal tolerance of rice for adverse soils. In *Soils and Rice.* Los Baños, Philippines: International Rice Research Institute (IRRI).

International Board for Plant Genetic Resources (IBPGR). 1983. *Annual Report 1982.* Rome, Italy: IBPGR.

International Rice Research Institute (IRRI). (in press) *Annual Report 1988.* Los Baños, Philippines: IRRI.

International Rice Research Institute (IRRI). 1980. *Five Years of the International Rice Testing Programme (IRTP): A Global Rice Exchange and Testing Network.* Los Baños, Philippines: IRRI.

Khan, Z. R., and Saxena, R. C. 1984. Electronically recorded waveforms associated with the feeding behaviour of *Sogatella furcifera* (Homoptera: Delphacidae) on susceptible and resistant rice varieties. *Journal of Economic Entomology* 77: 1479-1482.

Knott, D. R., and Dvorak, J. 1976. Alien germplasm as a source of resistance to disease. *Annual Review of Phytopathology* 14: 211-235.

Lin, S. C., and Yuan, L. P. 1980. Hybrid rice breeding in China. In *Innovative Approaches to Rice Breeding.* Los Baños, Philippines: International Rice Research Institute (IRRI).

Ling, K. C. 1972. *Rice Virus Diseases.* Los Baños, Philippines: International Rice Research Institute (IRRI).

Loresto, G. C., and Chang, T. T. 1981. Decimal scoring systems for drought reactions and recovery ability in rice screening nurseries. *International Rice Research Newsletter* 6(2): 9-10.

Marshall, D. R. 1989. Limitations to the use of germplasm collections. In Brown, A. H. D., et al. (eds.), *The Use of Plant Genetic Resources.* Cambridge, UK: Cambridge University Press.

Mazumder, M. K., Seshu, D. V., and Shenoy, V. V. (in press) Implications of fatty acids and seed dormancy in a new screening procedure for cold tolerance in rice. *Crop Science.*

O'Toole, J. C. 1982. Adaptation of rice to drought-prone environments. In *Drought Resistance in Crops with Emphasis on Rice.* Los Baños, Philippines: International Rice Research Institute (IRRI).

Paguia, P. M., Pathak, M. D., and Heinrichs, E. A. 1980. Honeydew excretion measurement techniques for determining differential feeding activity of biotypes of *Nilaparvata lugens* on rice varieties. *Journal of Economic Entomology* 73: 35-40.

Palmer, R. G. 1989. Germplasm collections and the experimental biologist. In Brown, A. H. D., et al. (eds.), *The Use of Plant Genetic Resources.* Cambridge, UK: Cambridge University Press.

Panda, N., and Heinrichs, E. A. 1983. Levels of tolerance and antibiosis in rice varieties having moderate resistance to the brown planthopper, *Nilaparvata lugens* (Stal) (Hemiptera: Delphacidae). *Environmental Entomology* 12: 1204-1214.

Pathak, M. D. 1975. Utilization of insect plant interactions in pest control. In Pimental, D. (ed.), *Insects, Science and Society.* New York, USA: Academic Press.

Peeters, J. P., and Williams, J. T. 1984. Towards better use of genebanks with special reference to information. *Plant Genetic Resources Newsletter* 60: 22-32.

Peloquin, S. J. 1984. Utilization of exotic germplasm in potato breeding: Germplasm transfer with haploids and 2n gametes. In *Report of the 1983 Plant Breeding Research Forum.* Des Moines, Iowa, USA: Pioneer Hi-Bred International.

Prasada Rao, K. E., Mengesha, M. H., and Reddy, V. G. 1989. International use of a sorghum germplasm collection. In Brown, A. H. D., et al. (eds.), *The Use of Plant Genetic Resources.* Cambridge, UK: Cambridge University Press.

Rick, C. M. 1979. Potential improvement of tomatoes by controlled introgression of genes from wild species. In Zeven, A. C., and Harten, A. M., (eds.), *Proceedings of the Conference on Broadening the Genetic Base of Crop*s. Wageningen, The Netherlands: Centre for Agricultural Publishing and Documentation (PUDOC).

Rick, C. M. 1984. Conservation and use of exotic tomato germplasm. In *Report of the 1983 Plant Breeding Research Forum.*, Iowa, USA: Pioneer Hi-Bred International.

Riley, R., and Lewis, K. R. (eds.). 1966. *Chromosome Manipulations and Plant Genetics.* New York, USA: Plenum Press.

Satake, T., and Toriyama, K. 1979. Two extremely cool tolerant varieties. *International Rice Research Newsletter* 4(2): 9-10.

Saxena, R. C., and Khan, Z. R. 1984. Comparison between free-choice and no-choice seedling bulk tests for evaluating resistance of rice cultivars to the whitebacked planthopper. *Crop Science* 24: 1204-1206.

Saxena, R. C., and Pathak, M. D. 1977. Factors affecting resistance of rice varieties to the brown planthopper, *Nilaparvata lugens* (Stal). In *Proceedings of the 8th Annual Conference of the Pest Control Council, 1977, Bacolod, Philippines.*

Seshu, D. V. 1985. International rice testing program: A mechanism for international cooperation in rice improvement. *International Rice Testing Program (IRTP) Bulletin 1.* Los Baños, Philippines: International Rice Research Institute (IRRI).

Seshu, D. V., and Kauffman, H.E. 1980. Differential response of rice varieties to the brown planthopper in international screening trials. International Rice Research Institute (IRRI) Research Paper Series 52. Los Baños, Philippines: IRRI.

Seshu, D.V., and Cady, F.B. 1984. Response of rice to solar radiation and temperature estimated from international field trials. *Crop Science* 24: 649-654.

Srivastava, J. P., and Damania, A. B. 1989. Use of collections in cereal improvement in semi-arid areas. In Brown, A. H. D., et al. (eds.), *The Use of Plant Genetic Resources*. Cambridge, UK: Cambridge University Press.

Stalker, H. T. 1980. Utilization of wild species for crop improvement. *Advances in Agronomy* 33: 111-147.

Velusamy, R., Heinrichs, E. A., and Medrano, F. G. 1986. Greenhouse techniques to identify field resistance to the brown planthopper in rice cultivars. *Crop Protection* 5: 328-333.

Vergara, B. S., and Chang, T. T. 1985. The flowering response of the rice plant to photoperiod. *IRRI Technical Bulletin 8*. Los Baños, Philippines: International Rice Research Institute (IRRI).

Vidyabhushanam, R. V., Rana, B. S., and Reddy, B. V. S. 1989. Use of sorghum germplasm and its impact on crop improvement in India. In *Collaboration on Genetic Resources*. Patancheru, AP, India: International Center for Research in the Semi-Arid Tropics (ICRISAT).

Williams, J. T. 1989. Practical considerations relevant to effective evaluation. In Brown, A. H. D., et al. (eds.), *The Use of Plant Genetic Resources*. Cambridge, UK: Cambridge University Press.

## Chapter 3

Frankel, O. H., and Brown, A. D. H. 1984. Plant genetic resources today: A critical appraisal. In *Genetics: New Frontiers: Proceedings of the 15th International Congress of Genetics, 1984, New Delhi*.

Mackay, M. C. 1986. Utilizing wheat genetic resources in Australia. In *Proceedings of the 5th Assembly of the Wheat Breeding Society of Australia, 1986, Perth/Merredin, Australia*.

## Chapter 4

Blum, A. 1988. *Plant Breeding for Stress Environments*. Boca Raton, Florida, USA: CRC Press.

Bowman, K. O., Hutcheson, K., Odum, P., and Shenton, L. R. 1971. Comments on the distribution of indices of diversity. *Statistical Ecology* 3: 315-366.

Clarke, J. M. 1987. Use of physiological and morphological traits in breeding programmes to improve drought resistance of cereals. In Srivastava, J.P., et al. (eds.), *Drought Tolerance in Winter Cereals*. Chichester, UK: John Wiley.

Clarke, J. M., and McCaig, T. N. 1982a. Evaluation of techniques for screening for drought resistance in wheat. *Crop Science* 22: 503-506.

Clarke, J. M., and McCaig, T. N. 1982b. Excised-leaf water capability as an indicator of drought resistance of *Triticum* genotypes. *Canadian Journal of Plant Science* 55: 369-378.

Clarke, J. M., McCaig, T. N., Gautam, P. L., and Jana, S. 1987. Evaluation and characterization of durum wheats for rainfed production. *Rachis* 6: 31-32.

Fienberg, S. E. 1980. *The Analysis of Cross-Classified Categorical Data.* Cambridge, Massachusetts, USA: MIT Press.

Fisher, R. A. 1921. On the 'probable error' of a coefficient of correlation deduced from a small sample. *Metron* 1: 3-22.

Gummuluru, S., Hobbs, S. L. A., and Jana, S. 1989. Genotypic variability in physiological characters and its relationship to drought tolerance in durum wheat. *Canadian Journal of Plant Science* 69: 703-712.

Hutcheson, K. 1970. A test for comparing diversities based on the Shannon formula. *Journal of Theoretical Biology* 29: 151-154.

Jain, S. K., Qualset, C. O., Bhatt, G. M., and Wu, K. K. 1975. Geographic patterns of phenotypic diversity in a world collection of durum wheats. *Crop Science* 15: 700-704.

Kahler, A. L., Allard, R. W., Krazakowa, M., Wehrhahn, C. F., and Nevo, E. 1980. Associations between isozyme phenotypes and environments in the slender wild oat, *Avena barbata*, in Israel. *Theoretical and Applied Genetics* 56: 31-47.

Pecetti, L., Jana, S., Damania, A.B., and Strivastava, J. P. (in prep.) Practical problems in large-scale germplasm evaluation: A case study in durum wheat. Presented at the Symposium on Evaluation and Utilization of Genetic Resources in Wheat Improvement, 1989, Aleppo, Syria.

Porceddu, E., and Srivastava, J. P. 1990. Evaluation, documentation and utilization of durum wheat germplasm at the the International Center for Agricultrual Research in the Dry Areas (ICARDA) and the University of Tuscia, Italy. In *Wheat Genetic Resources: Meeting Diverse Needs.* Chichester, UK: John Wiley.

Smith, E. L. 1987. A review of plant breeding strategies for rainfed agriculture. In Srivastava, J. P., et al. (eds.), *Drought Tolerance in Winter Cereals.* Chichester, UK: John Wiley.

Sokal, R. R., and Rohlf, F. J. 1981. *Biometry.* San Francisco, California, USA: W. H. Freeman.

Vavilov, N. I. 1951. The origin, variation, immunity and breeding of 'cultivated plants'. *Chronica Botanica* 13: 1-366 (trans. from Russian by Starr, C.K.)

Zhang, Q., and Allard, R. W. 1986. Sampling variance of the genetic diversity index. *Journal of Heredity* 77: 54-56.

Zohary, D., and Hopf, M. 1988. *Domestication of Plants in the Old World.* Oxford, UK: Clarendon Press.

## Chapter 5

Biesantz, A., Kyzeridis, N., and Limberg, P. 1985-1987: GTZ - Bericht: Erhaltung wertvollen Genmaterials von *Triticum durum.* Berlin: German Agency for Technical Cooperation (GTZ)/Technical University of Berlin.

Christiansen-Weniger, D. F. 1970. *Ackerbauformen im Mittelmeerraum und Nahen Osten, dargestellt am Beispiel der Türkei.* Frankfurt am Main, Germany FR: Deutsche Landwirtschafts Gesellschaft (DLG).

Damania, A. B., and Somaroo, B. H. 1988. Preliminary screening of Greek durum wheat varieties using PAGE. *Rachis* 7: 54-56.

Frankel, O. H. 1973. *Survey of Crop Genetic Resources in Their Centres of Diversity. First Report.* Rome: Food and Agriculture Organization (FAO)/International Biological Programme (IBP).

Kling, C. H. 1985. Durumanbau: Auch eine Frage der Qualität. *Deutsche Landwirtschafts Gesellschaft (DLG)-Mitteilungen* 4: 189-195.

Schuster, W., and Von Lochow, J. 1979. *Anlage und Auswertung von Feldversuchen.* Frankfurt am Main, Germany FR: Deutsche Landwirtschafts Gesellschaft (DLG).

Zamanis, A. 1982. *Collecting in Greece.* Salonica, Greece: Cereal Institute.

Zamanis, A. 1983. *Collecting of Crop Germplasm in the East Aegan Islands.* Salonica, Greece: Cereal Institute (and pers. comm.)

## Chapter 6

Avivi, L. 1979. Utilization of *Triticum dicoccoides* for the improvement of grain protein quantity and quality in cultivated wheats. *Genetica Agraria* 4: 27-38.

Damania, A. B., and Srivastava, J. P. (in press) Genetic resources for optimal input technology: International Center for Agricultural Research in the Dry Areas (ICARDA)'s perspectives. *Plant and Soil* .

Dhaliwal, H. S., Gill, K. S., Singh, P. J., Multani, D. S., and Singh, B. 1986. Evaluation of germplasm of wild wheats, and *Aegilops, Agropyron* for resistance to various diseases. *Crop Improvement* 13: 145-149.

Lange, W., and Balkema-Boomstra, A. G. 1988. The use of wild species in breeding barley and wheat, with special reference to the progenitors of the cultivated species. In Jorna, M. L., and Slootmaker, L. A. J. (eds.), *Cereal Breeding Related to Integrated Cereal Production.* Wageningen, The Netherlands: Eucarpia.

Lawrence, J. M., Day, K. M., Huey, E., and Lee, B. 1958. Lysine content of wheat varieties, species and related genera. *Cereal Chemistry* 35: 169-179.

Pantanelli, E. 1944. *Cultivated Plants.* Bari, Italy: Facoltà di Agraria.

Sharma, H. C., Waines, J. G., and Foster, K. W. 1981. Variability in primitive and wild wheats for useful genetic characters. *Crop Science* 21: 555-559.

Srivastava, J. P., and Damania, A. B. 1989. Use of collections in cereal improvement in semi-arid areas. In Brown, A. H. D., et al. (eds.), *The Use of Plant Genetic Resources.* Cambridge, UK: Cambridge University Press.

Srivastava, J. P., Damania, A. B., and Pecetti, L. 1988. Landraces, primitive forms and wild progenitors of durum wheat: Their use in dryland agriculture. In Miller, T. E., and Koebner, R. M. D. (eds.), *Proceedings of the 7th International Wheat Genetics Symposium.* Cambridge, UK: Institute of Plant Science Research (IPSR).

Waines, J. G., Ehdaie, B., and Barnhart, D. 1987. Variability in *Triticum* and *Aegilops* species for seed characteristics. *Genome* 29: 41-46.

Zohary, D. 1969. The progenitors of wheat and barley in relation to domestication and agricultural dispersal in the Old World. In Ucko, P.J., and Dimbleby, G.W. (eds.), *The Domestication and Exploitation of Plants and Animals*. London, UK: Duckworth.

Zohary, D., and Hopf, M. 1988. *Domestication of Plants in the Old World*. Oxford, UK: Clarendon Press.

## Chapter 7

Beckman, J. S., and Soller, M. 1983. Restriction fragment polymorphisms in genetic improvement: Methodologies, mapping and costs. *Theoretical and Applied Genetics* 67: 35-43.

Chao, S., Sharp, P. J., and Gale, M. D. 1988. A linkage map of wheat homoeologous group 7 chromosomes using RFLP markers. In Miller, T. E., and Koebner, R. M. D. (eds.), *Proceedings of the 7th International Wheat Genetics Symposium*. Cambridge, UK: Institute of Plant Science Research (IPSR).

Dvorak, J., McGuire, P. E., and Cassidy, B. 1988. Apparent sources of the A genomes of wheats inferred from polymorphism in abundance and restriction fragment length of repeated nucleotide sequences. *Genome* 30: 680-689.

Figdore, S. S., Kennard, W. C., Song, K. M., Slocum, M. K., and Osborn, T. C., 1988. Assessment of the degree of restriction fragment length polymorphism in *Brassica*. *Theoretical and Applied Genetics* 75: 833-840.

Gale, M. D., and Sharp, P. J. 1988. Genetic markers in wheat: Developments and prospects. In Miller, T. E., and Koebner, R. M. D. (eds.), *Proceedings of the 7th International Wheat Genetics Symposium*. Cambridge, UK: Institute of Plant Science Research (IPSR).

Gebeyehu, G., Rao, P. Y., Soochan, P., Simms, D. A., and Klevan, L. 1987. Novel biotinylated nucleotide: Analogs for labelling and colorimetric detection of DNA. *Nucleic Acids Research* 15 (11): 4513-4534.

Landry, B. S., Kesseli, R. V., Farrara, B., and Michelmore, R. W. 1987. A genetic map of lettuce, *Lactuca sativa* L., with restriction fragment length polymorphism, isozyme, disease resistance and morphological markers. *Genetics* 116: 331-337.

Mason, J. P., and Williams, J. G. 1985. Hybridization in the analysis of recombinant DNA. In Hames, B. D., and Higgins, S. J. (eds.), *Nucleic Acid Hybridization: A Practical Approach*. Washington DC, USA: IRL Press.

Neuffer, M. G., and Coe, E. H. 1974. Corn (maize). In King, R. C. (ed.), *Handbook of Genetics 2*. New York, USA: Plenum Press.

Rick, C. M. 1974. The tomato. In King, R.C. (ed.), *Handbook of Genetics 2*. New York, USA: Plenum Press.

Southern, E. 1975. Detection of specific sequences among DNA fragments separated by gel electrophoresis. *Journal of Molecular Biology* 98: 503-509.

Wahl, G. M., Stern, M., and Stark, G. R. 1979. Efficient transfer of large DNA fragments from agarose gels to diazobenzloxymethal-paper and rapid hybridization by using dextran sulphate. *Proceedings of the National Academy of Sciences* 76: 3683-3685.

## Chapter 8

Bietz, J. A. 1983. Separation of cereal proteins by reversed-phase high-performance liquid chromatography. *Journal of Chromatography* 255: 219-238.

Bietz, J. A. 1985. High-performance liquid chromatography: How proteins look in cereals. *Cereal Chemistry* 62: 201-212.

Bietz, J. A. 1987. Genetic and biochemical studies of nonenzymatic endosperm proteins. In Heyne, E. G. (ed.), *Wheat and Wheat Improvement*. (2nd edn.) Madison, Wisconsin, USA: American Society of Agronomy.

Bietz, J. A., and Burnouf, T. 1985. Chromosomal control of wheat gliadin: Analysis by reversed-phase high-performance liquid chromatography. *Theoretical and Applied Genetics* 70: 599-608.

Bietz, J. A., and Wall, J. S. 1972. Wheat gluten subunits: Molecular weights determined by sodium dodecyl sulfate-polyacrylamide gel electrophoresis. *Cereal Chemistry* 49: 416-430.

Bietz, J. A., Burnouf, T., Cobb, L. A., and Wall, J. S. 1984. Wheat varietal identifcation and genetic analysis by reversed-phase high-performance liquid chromatography. *Cereal Chemistry* 61: 129-135.

Bietz, J. A., Shepherd, K. W., and Wall, J. S. 1975. Single-kernel analysis of glutenin: Use in wheat genetics and breeding. *Cereal Chemistry* 49: 416-430.

Brown, R. J., Kemble, R. J., Law, C. N., and Flavell, R. B. 1979. Control of endosperm proteins in *Triticum aestivum* (var. Chinese Spring) and *Aegilops umbellulata* by homoeologous group 1 chromosomes. *Genetics* 93: 189-200.

Burnouf, T., and Bietz, J. A. 1985. Chromosomal control of glutenin subunits in aneuploid lines of wheat: Analysis by reversed-phase high-performance liquid chromatography. *Theoretical and Applied Genetics* 70: 610-619.

Burnouf, T., and Bietz, J. A. 1987. Identification of wheat cultivars and prediction of quality by reversed-phase high-performance liquid chromatographic analysis of endosperm storage proteins. *Seed Science and Technology* 15: 79-99.

Cole, E. W., Fullington, J. G., and Kasarda, D. D. 1981. Grain protein variability among species of *Triticum* and *Aegilops*: Quantitative SDS-PAGE studies. *Theoretical and Applied Genetics* 60: 17-30.

Cox, T. S., and Harrell, L. G. 1988. Geographical variation among 212 accessions of *Aegilops squarrosa* for gliadin PAGE patterns. In McGuire, P. E. (ed.), *Population Genetics and Germplasm Resources in Crop Improvement*. (Supplementary Vol.) Davis, USA: University of California.

Croy, R. R. D., and Gatehouse, J. A. 1985. Genetic engineering of seed proteins: Current and potential applications. In Dodds, J. H. (ed.), *Plant Genetic Engineering*. Cambridge, UK: Cambridge University Press.

Damania, A. B., Porceddu, E., and Jackson, M. T. 1983. A rapid method for the evaluation of variation in germplasm collections of cereals using polyacrylamide gel electrophoresis. *Euphytica* 32: 877-883.

Damidaux, R., Autran, J. C., Grignac, P., and Feillet, P. 1978. Mise en évidence de relations applicable en sélection entre l'électrophorégramme des gliadines et les propriétés viscoélastiques du gluten de *Triticum durum* Desf. *Comptes Rendus de l'Académie des Sciences (Série D)* 287: 701-704.

Damidaux, R., Autran, J. C., Grignac, P., and Feillet, P. 1980. Déterminisme génétique des constituants gliadines de *Triticum durum* Desf. *Comptes Rendus de l'Académie des Sciences (Série D)* 291: 585-588.

Du Cros, D. L., Joppa L. R., and Wrigley, C. W. 1983. Two-dimensional analysis of gliadin proteins associated with quality in durum wheat: Chromosomal location of genes for their synthesis. *Theoretical and Applied Genetics* 66: 297-302.

Dvorak, J., Kasarda, D. D., Dietler, M. D., Lew, E. J., Anderson, O. D., Litts, J. C., and Shewry, P. R. 1986. Chromosomal location of seed storage protein genes in the genome of *Elytrigia elongata*. *Canadian Journal of Genetics and Cytology* 28: 818-830.

Feldman, G., Galili, G., and Levy, A. A. 1986. Genetic and evolutionary aspects of allopolyploidy in wheat. In Barigozzi, C. (ed.), *The Origin and Domestication of Cultivated Plants*. Amsterdam, The Netherlands: Elsevier.

Galili, G., and Feldman, M. 1983. Diploidization of endosperm protein genes in polyploid wheats. In Sakamoto, S. (ed.), *Proceedings of the 6th International Wheat Genetics Symposium*. Kyoto, Japan: Plant Germplasm Institute, University of Kyoto.

Galili, G., Felsenburg, T., Levy, A. A., Altschuler, Y., and Feldman, M. 1988. Inactivity of high-molecular-weight glutenin genes in wild diploid and teratrploid wheats. In Miller, T. E., and Koebner, R. M. D. (eds.), *Proceedings of the 7th International Wheat Genetics Symposium*. Cambridge, UK: Institute of Plant Science Research (IPSR).

Gepts, P. 1988. Phaseolin as an evolutionary marker. In Gepts, P. (ed.), *Genetic Resources of* Phaseolus *Beans*. Amsterdam, The Netherlands: Kluwer Academic.

Gupta, R. B., and Shepherd, K. W. 1988. Low-molecular-weight glutenin subunits in wheat: Their variation, inheritance and association with bread-making quality. In Miller, T. E., and Koebner, R. M. D. (eds.), *Proceedings of the 7th International Wheat Genetics Symposium*. Cambridge, UK: Institute of Plant Science Research (IPSR).

Johnson, B. L. 1972. Seed protein profiles and the origin of the hexaploid wheats. *American Journal of Botany* 59: 952-960.

Johnson, B. L., and Hall, B. 1965. Analysis of phylogenetic affinities in the Triticineae by protein electrophoresis. *American Journal of Botany* 52: 506-513.

Johnson, V. A., Wilhelmi, K. D., Kuhr, S. L., Mattern, P. J., and Schmidt, J. W. 1978. Breeding progress for protein and lysine in wheat. In Ramanujam, S. (ed.), *Proceedings of the 5th International Wheat Genetics Symposium*. New Delhi, India: Indian Society of Genetics and Plant Breeding.

Kasarda, D. D. 1980. Structure and properties of α gliadins. *Annales de Technologie Agricole* 29: 151-173.

Kasarda, D. D., Lafiandra, D., Morris, R., and Shewry, P. R., 1984. Genetic relationships of wheat gliadin proteins. *Kulturpflanze* 32: 33-52.

Kushnir, U., Du Cros, D. L., Lagudah, E., and Halloran, G. M. 1984. Origin and variation of gliadin proteins associated with pasta quality in durum wheat. *Euphytica* 33: 289-294.

Ladizinsky, G., and Johnson, B. L. 1972. Seed protein homologies and the evolution of polyploidy in *Avena*. *Canadian Journal of Genetics and Cytology* 14: 875-888.

Lafiandra, D. 1988. Combined high-performance liquid chromatography and electophoretical analysis of storage proteins as a tool for wheat genetical studies. *Cereal Foods World* 33: 689-690 (Abstract 186).

Lafiandra, D., and Kasarda, D. D. 1985. One- and two-dimensional (two-pH) polyacrylamide gel electrophoresis in a single gel: Separation of wheat proteins. *Cereal Chemistry* 62: 314-319.

Lafiandra, D., Benedettelli, S., and Porceddu, E. 1988. Null forms for storage proteins in bread wheat and durum. In Miller, T. E., and Koebner, R. M. D. (eds.), *Proceedings of the 7th International Wheat Genetics Symposium*. Cambridge, UK: Institute of Plant Science Research (IPSR).

Lafiandra, D., Benedettelli, S., Spagnoletti-Zeuli , P. L., and Porceddu, E. 1983. Genetical aspects of durum wheat gliadins. In Porceddu, E. (ed.), *Breeding Methodologies in Durum Wheat and Triticale*. Viterbo, Italy: University of Tuscia/Institute of Agricultural Biology.

Lafiandra, D., Colaprico, G., Kasarda, D. D., and Porceddu, E. 1987a. Null alleles for gliadin blocks in bread and durum wheat cultivars. *Theoretical and Applied Genetics* 74: 610-616.

Lafiandra, D., Kasarda, D. D., and Morris, R. 1984. Chromosomal assignment of genes coding for the wheat gliadin protein components of the cultivars Cheyenne and Chinese Spring by two-dimensional (two-pH) electrophoresis. *Theoretical and Applied Genetics* 68: 531-539.

Lafiandra, D., Perrino, P., and Margiotta, B. 1987b. Storage proteins variation in wheat germplasm and its possible use for gluten quality improvement. In Borghi, B. (ed.), *Hard Wheat: Agronomic, Technological, Biochemical and Genetic Aspects*. Brussels, Belgium: European Economic Commission.

Lafiandra, D., Splendido, R., Tomassini, C., and Porceddu, E. 1987c. Lack of expression of certain storage proteins in bread wheats: Distribution and genetic analysis of null forms. In Lasztity, R., and Békés, F. (eds.), *Third International Workshop on Gluten Proteins*. Budapest, Hungary: Technical University of Budapest.

Lagudah, E. S., and Halloran, G. M. 1988. Phylogenetic relationships of *Triticum tauschii* the D genome donor to hexaploid wheat. 1. Variation in HMW subunits of glutenin and gliadins. *Theoretical and Applied Genetics* 75: 592-598.

Lagudah, E. S., Flood, R. G., and Halloran, G. M. 1987. Variation in high molecular weight glutenin subunits in landraces of hexaploid wheat from Afghanistan. *Euphytica* 36: 3-9.

Lawrence, G. J., and Shepherd K. W. 1980. Variation in glutenin protein subunits of wheat. *Australian Journal of Biological Sciences* 33: 221-233.

Lawrence, G. J., and Shepherd, K. W. 1981. Chromosomal location of genes controlling seed proteins in species related to wheat. *Theoretical and Applied Genetics* 59: 25-31.

Levy, A. A., and Feldman, M. 1988. Ecogeographical distribution of HMW glutenin alleles in populations of the wild tetraploid wheat, *Triticum turgidum* var. *dicoccoides*. *Theoretical and Applied Genetics* 75: 651-658.

Levy, A. A., Galili, G., and Feldman, M. 1988. Polymorphism and genetic control of high molecular weight glutenin subunits in wild tetraploid wheat, *Triticum turgidum* var. *dicoccoides*. *Heredity* 61: 63-72.

Mansur-Vergara, L., Konzak, C. F., and Gerechter-Amitai, Z. K. 1984. A computer-assisted examination of the storage protein genetic variation in 841 accessions of *Triticum dicoccoides*. *Theoretical and Applied Genetics* 69: 79-86.

Margiotta, B., Colaprico, G., and Lafiandra, D. 1987. Variation for protein components associated with quality in durum wheat lines and varieties. In Lasztity, R. and Békés, F. (eds.), *Third International Workshop on Gluten Proteins*. Budapest, Hungary: Technical University of Budapest.

Margiotta, B., Lafiandra, D., Tomassini, C., Perrino, P., and Porceddu, E. 1988. Variation in high molecular weight glutenin subunits in a hexaploid wheat collection from Nepal. In Miller, T. E., and Koebner, R. M. D. (eds.), *Proceedings of the 7th International Wheat Genetics Symposium*. Cambridge, UK: Institute of Plant Science Research (IPSR).

Mecham, D. K., Kasarda, D. D., and Qualset, C. O. 1978. Genetic analysis of wheat gliadin proteins. *Biochemical Genetics* 16: 831-853.

Metakovsky, E. V., Novoselskaya, A. Y., Kopus, M. M., Sobko, T. A., and Sozinov, A. A. 1984. Blocks of gliadin components in winter wheat detected by one-dimensional polyacrylamide gel electrophoresis. *Theoretical and Applied Genetics* 67: 559-568.

Nevo, E., and Payne, P. I. 1987. Wheat storage proteins: Diversity of HMW glutenin subunits in wild emmer. 1. Geographical patterns and ecological predictability. *Theoretical and Applied Genetics* 74: 827-836.

Nevo, E., Beiles, A., Storch, N., Doll, H., and Andersen, B. 1983. Microgeographic edaphic differentiation in hordein polymorphisms of wild barley. *Theoretical and Applied Genetics* 64: 123-132.

Osborne, T. B. 1907. *The Proteins of the Wheat Kernel*. Washington DC, USA: Carnegie Institute of Washington.

Payne, P. I., and Lawrence, G.J. 1983. Catalogue of alleles for the complex gene loci, *Glu-A1*, *Glu-B1*, and *Glu-D1* which code for high-molecular-weight subunits of glutenin in hexaploid wheat. *Cereal Research Communications* 11: 29-35.

Payne, P. I., Holt, L. M., and Law, C. N. 1981. Structural and genetical studies on the high-molecular-weight subunits of wheat glutenin. 1. Allelic variation in subunits amongst varieties of wheat, *Triticum aestivum*. *Theoretical and Applied Genetics* 60: 229-236.

Payne, P. I., Holt, L. M., and Law, C. N. 1984a. Wheat storage proteins: Their genetics and their potential for manipulation by plant breeding. *Philosophical Transactions of the Royal Society of London (Biological Series)* 304: 359-371.

Payne, P. I., Holt, L. M., and Lister P. G. 1988. *Gli-A3* and *Gli-B3*, two newly designated loci coding for omega-type gliadins and D subunits of glutenin. In Miller, T. E., and Koebner, R. M. D. (eds.), *Proceedings of the 7th International Wheat Genetics Symposium*. Cambridge, UK: Institute of Plant Science Research (IPSR).

Payne, P. I., Holt, L. M., Lawrence G. J., and Law, C. N. 1982. The genetics of gliadin and glutenin, the major storage proteins of the wheat endosperm. *Qualitas Plantarum — Plant Foods for Human Nutrition* 31: 229-241.

Payne, P. I., Jackson, E. A., and Holt, L. M. 1984b. The association between ψ gliadin 45 and gluten strength in durum wheat varieties: A direct causal effect or the result of genetic linkage? *Journal of Cereal Science* 2: 73-81.

Payne, P. I., Jackson, E. A., Holt, L. M., and Law, C. N. 1984c. Genetic linkage between endosperm storage protein genes on each of the short arms of chromosomes 1A and 1B in wheat. *Theoretical and Applied Genetics* 67: 235-243.

Payne, P. I., Law, C. N., and Mudd, E. E. 1980. Control by homoeologous group 1 chromosomes of the high-molecular-weight subunits of glutenin, a major protein of wheat endosperm. *Theoretical and Applied Genetics* 58: 113-120.

Pogna, N., Lafiandra, D., Feillet, P., and Autran, J. C. 1988. Evidence for a direct causal effect of low molecular weight subunits of glutenins on gluten viscoelasticity in durum wheats. *Journal of Cereal Science* 7: 211-214.

Porceddu, E., and Lafiandra, D. 1986. Origin and evolution of wheats. In Barigozzi, C. (ed.), *Origin and Domestication of Cultivated Plants*. Amsterdam, The Netherlands: Elsevier.

Przyblska, J. 1988. Identification and classification of the *Pisum* genetic resources with the use of electrophoretic proteins analysis. *Seed Science and Technology* 14: 529-543.

Sapirstein, H. D., and Bushuk, W. 1986. Computer-aided wheat cultivar identification and analysis of densitometric scanning profiled of gliadin electophoregrams. *Seed Science and Technology* 14: 489-517.

Sergio, L., Spagnoletti-Zeuli, P. L., and Volpe, N. 1988. Genotypical frequency variation in durum wheat seed multiplication. *Genetica Agraria* 42:485.

Shepherd, K. W. 1988. Genetics of wheat endosperm proteins in retrospect and prospect. In Miller, T. E., and Koebner, R. M. D. (eds.), *Proceedings of the 7th International Wheat Genetics Symposium*. Cambridge, UK: Institute of Plant Science Research (IPSR).

Shewry, P. R., and Miflin, B. J. 1985. Seed storage proteins of economically important cereals. In Pomeranz, Y. (ed.), *Advances in Cereal Science and Technology*. (Vol. 3) St Paul, Minnesota, USA: American Association of Cereal Chemists.

Shewry, P. R., Tatham, A. S., Forde, J., Kreis, M., and Miflin, B. J. 1986. The classification and nomenclature of wheat gluten proteins: A reassessment. *Journal of Cereal Science* 4: 97-106.

Singh, N. K., and Shepherd, K. W. 1988. Linkage mapping of genes controlling endosperm storage proteins in wheat. 1. Genes on the short arms of group 1 chromosomes. *Theoretical and Applied Genetics* 75: 628-641.

Sozinov, A. A., and Poperelya, F. A. 1980. Genetic classification of prolamines and its use for plant breeding. *Annales de Technologie Agricole* 29: 229-245.

Thompson, R. D., Bartels, D., and Flavell, R. B. 1983. Characterization of the multigene family coding for HMW glutenin subunits in wheat using cDNA clones. *Theoretical and Applied Genetics* 67: 81-96.

Tomassini, C., Colaprico, G., Benedettelli, S., and Lafiandra, D. 1989. Two dimensional analysis of durum wheat gliadins in cultivar identification. In book of poster abstracts (Poster 28-5), 12th Eucarpia Congress, Science for Plant Breeding. Gottingen, Germany FR: European Association for Research on Plant Breeding.

Vallega, V., and Waines, J. G. 1987. High-molecular-weight glutenin subunit variation in *Triticum turgidum* var. *dicoccum*. *Theoretical and Applied Genetics* 74: 706-710.

Waines, J. G., and Payne, P. I. 1987. Electrophoretic analysis of the high-molecular-weight glutenin subunits of *Triticum monococcum*, *T. urartu*, and the A genome of bread wheat, *T. aestivum*. *Theoretical and Applied Genetics* 74: 71-76.

Woychick, J. H., Boundy, J. A., and Dimler, R. J. 1961. Starch gel electrophoresis of wheat gluten proteins with concentrated urea. *Archives of Biochemistry and Biophysics* 94: 477-482.

Wrigley, C. W., 1980. The genetic and chemical significance of varietal differences in gluten composition. *Annales de Technologie Agricole*29: 213-227.

Wrigley, C. W., and Bietz, J. A. 1988. Proteins and amino acids. In Pomeranz, Y. (ed.), *Wheat Chemistry and Technology*. (Vol .1) St Paul, Minnesota, USA: American Association of Cereal Chemistry.

Wrigley, C. W., and Shepherd, K. W. 1973. Electrofocusing of grain proteins from wheat genotypes. *Annals of the New York Academy of Sciences* 209: 154-162.

## Chapter 9

Barrai, I. 1986. Quaderni di biologia pura e applicata. Introduzione all'analisi multivariata. Bologna, Italy: Edagricole.

Benedettelli, S., Ciaffi, M., Tomassini, C., Lafiandra, D., and Porceddu., E. (in prep.) Assessment of genetic variation of durum wheat populations by means of their gliadin patterns.

Benzecry, J. P. 1973. *L'Analyse de Donées*. Paris, France: Dunod.

Burt, C., and Banks, C. 1974. A factorial analysis of body measurements for British adult males. *Annals of Eugenics* 13: 238-256.

Bushuk, W., and Zillman, R. R. 1978. Wheat cultivar identification by gliadin electrophoregrams. I. Apparatus, method and nomenclature. *Canadian Journal of Plant Science* 58: 505-515.

Feldman, M., and Sears E. R. 1981. The wild gene resources of wheat. *Scientific American* 244: 102-112.
Hill, N. O. 1974. Correspondence analysis: A neglected multivariate method. *Applied Statistics* 23: 340-354.
Lafiandra, D., Benedettelli, S., Spagnoletti-Zeuli, P. L., and Porceddu, E. 1983. Genetical aspects of durum wheat gliadins. In Porceddu, E. (ed.), *Breeding Methodologies in Durum Wheat and Triticale*. Viterbo, Italy: University of Tuscia/Institute of Agricultural Biology.
Lafiandra, D., Colaprico, G., Kasarda, D. D., and Porceddu, E. 1987a. Null alleles for gliadin blocks in bread and durum wheat cultivars. *Theoretical and Applied Genetics* 74: 610-616.
Lafiandra, D., Perrino, P., and Margiotta, B. 1987b. Storage proteins variation in wheat germplasm and its use for gluten quality improvement. In Borghi, B. (ed.), *Hard Wheat: Agronomic, Technological, Biochemical and Genetic Aspects.* Brussels, Belgium: European Economic Commission (EEC).
Lagudah, E.S., Flood, R. G., and Halloran, G. M. 1987. Variation in high molecular weight glutenin subunits in landraces of hexaploid wheat from Afghanistan. *Euphytica* 36: 3-9.
Margiotta, B., Lafiandra, D., Tomassini, C., Perrino. P., and Porceddu, E. 1988. Variation in high molecular weight glutenin subunits in a hexaploid wheat collection from Nepal. In Miller, T. E., and Koebner, R. M. D. (eds.), *Proceedings of the 7th International Wheat Genetics Symposium.* Cambridge, UK: Institute of Plant Science Research (IPSR).
Nevo, E., and Payne, P. I. 1987. Wheat storage proteins: Diversity of HMW glutenin subunits in wild emmer. 1. Geographical patterns and ecological predictability. *Theoretical and Applied Genetics* 74: 827-836.
Porceddu, E., Ceoloni, C., Lafiandra, D., Tanzarella, O. A., and Scarascia Mugnozza, G. T. 1988. Genetic resources and plant breeding: Problems and prospects. In Miller, T. E., and Koebner, R. M. D. (eds.), *Proceedings of the 7th International Wheat Genetics Symposium.* Cambridge, UK: Institute of Plant Science Research.
She, J. X., Autem, M., Kotulas, G., Pasteur, N., and Bonhomme, F. 1987. Multivariate analysis of genetic exchanges between *Solea aegyptiaca* and *S. senegalensis* (Teleosts, Soleidae). *Biological Journal of the Linnean Society* 32: 357-371.

## Chapter 10

Austin, R. B., Morgan, C. L., and Ford, M. A. 1986. Dry matter yields and photosynthetic rates of diploid and hexaploid *Triticum* species. *Annals of Botany* 57: 847-857.
Bowden, M. W. 1959. The taxonomy and nomenclature of wheats, barleys, ryes and their wild relatives. *Canadian Journal of Botany* 37: 657-684.
Bradshaw, A. D. 1965. Evolutionary significance of phenotypic plasticity in plants. *Advances in Genetics* 13: 115-155.

Davis, P. H. 1985. *Aegilops*. In Davis, P. H. (ed.), *The Flora of Turkey and the East Aegean Islands*. (Vol. 9) Edinburgh, UK: Edinburgh University Press.

Dvorak, J. 1977. Transfer of leaf rust resistance from *Aegilops speltoides* to *Triticum aestivum*. *Canadian Journal of Genetics and Cytology* 19: 133-141.

Dvorak, J., McGuire, P. E., and Cassidy, B. 1988. Apparent sources of the A genomes of wheats inferred from polymorphism in abundance and restriction fragment length of repeated nucleotide sequences. *Genome* 30: 680-689.

Ehdaie, B., and Waines, J. G. 1989. Variation in growth, flowering and seed set under high temperatures in *Aegilops* species. *Wheat Information Service* (Kyoto) 68: 9-12.

Fletcher, H. R. et al. *International Code of Nomenclature for Cultivated Plants. Regnum Vegetabile*. (Vol. 10) Utrecht, The Netherlands: Bohn, Schelkma and Holkema.

Gill, B. S., and Raupp, W. J. 1987. Direct genetic transfers from *Aegilops squarrosa* L. to hexaploid wheat. *Crop Science* 27: 445-450.

Gill, B. S., and Waines, J. G. 1978. Paternal regulation of seed development in wheat hybrids. *Theoretical and Applied Genetics* 51: 265-270.

Guzy, M. R., Ehdaie, B., and Waines, J. G. 1989. Yield and its components in diploid, tetraploid and hexaploid wheats in diverse environments. *Annals of Botany* 64: 635-642.

Hammer, K. 1980. Zur Taxonomie und Nomenclatur der Gattung *Aegilops* L. *Feddes Repertorium* 91: 225-258.

Johnson, B. L., and Dhaliwal, H. S. 1976. Reproductive isolation of *Triticum boeoticum* and *T. urartu* and the origin of the tetraploid wheats. *American Journal of Botany* 63: 1088-1094.

Kaloshian, I. 1988. Genetics of resistance to root knot nematodes, *Meloidogyne* spp., in wheat. PhD dissertation. University of California, Riverside, California, USA.

Kaloshian, I., Roberts, P. A., and Thomason, I. J. 1989a. Resistance to *Meloidogyne* spp. in allohexaploid wheat derived from *Triticum turgidum* and *Aegilops squarrosa*. *Journal of Nematology* 21: 42-47.

Kaloshian, I., Roberts, P. A., and Thomason, I. J. 1989b. Resistance in *Triticum turgidum* and *Aegilops squarrosa* to *Meloidogyne chitwoodi*. Supplement to *Journal of Nematology* 21(4S): 632-634.

Kimber, G., and Feldman, M. 1987. *Wild Wheat: An Introduction*. Special Report No. 353. Columbia, Missouri, USA: College of Agriculture, University of Missouri.

Kimber, G., and Sears, E. R. 1987. Evolution in the genus *Triticum* and the origin of cultivated wheat. In Heyne, E. G. (ed.), *Wheat and Wheat Improvement*. (2nd edn) Madison, Wisconsin, USA: American Society of Agronomy.

Kushnir, U., and Halloran, G. M. 1981. Evidence for *Aegilops sharonensis* Eig as the donor of the B genome of wheat. *Genetics* 99: 495-512.

Lanjouw, J. et al. 1956. *International Code of Nomenclature for Cultivated Plants. Regnum Vegetabile*. (Vol. 8) Utrecht, The Netherlands: Bohn, Schelkma and Holkema.

Lucas, H., and Jahier, J. 1987. Differences in meiotic pairing induction between the A genomes of *Triticum boeoticum* Boiss, and *T. urartu* Tum. using diploid species of the subtribe Triticineae as analysers. *Genome* 29: 891-893.

Meletti, P., Onnis, A., and Stefani, A. 1977. *X Haynaltotriticum sardoum* Meletti et Onnis e lua origine. *Giornale Botanico Italiana* III (6): 376. (Abstract).

Merkle, O. G., and Starks, K. J. 1985. Resistance of wheat to the yellow sugarcane aphid. *Journal of Economic Entomology* 78: 127-128.

Morris, R., and Sears, E. R. 1967. The cytogenetics of wheat and its relatives. In Quisenberry, K. S., and Reitz, L. P. (eds.), *Wheat and Wheat Improvement.* (1st edn) Madison, Wisconsin, USA: American Society of Agronomy.

Roberts, P. A., Van Gundy, S. D., and Waines, J. G. 1982. Reaction of wild and domesticated *Triticum* and *Aegilops* species to root knot nematodes (*Meloidogyne*). *Nematologia* 28: 182-191.

Sharma, H. C., and Waines, J. G. 1981. The relationships between male and female fertility and among taxa in diploid wheats. *American Journal of Botany* 68: 449-451.

Smith-Huerta, N. L., Huerta, A. J., Barnhart, D., and Waines, J. G. 1989. Genetic diversity in wild diploid wheats *Triticum monococcum* var. *boeoticum* and *T. urartu* (Poaceae). *Theoretical and Applied Genetics* 78: 260-264.

Suemoto, H. 1968. Cytoplasmic relationships in the Triticineae. In Finlay, K. W., and Shepherd, K. W. (eds.), *Proceedings of the 3rd International Wheat Genetics Symposium.* Canberra, Australia: Australian Academy of Science.

Suemoto, H. 1973. The origin of the cytoplasm of tetraploid wheats. In Sears, E. R., and Sears, L. M. S. (eds.), *Proceedings of the 4th International Wheat Genetics Symposium.* Columbia, Missouri, USA: Agricultural Experiment Station, University of Missouri.

The, T. T., and Baker, E. P. 1975. Basic studies relating to the transference of genetic characters from *Triticum monococcum* to hexaploid wheat. *Australian Journal of Biological Sciences* 28: 189-199.

Thomas, J. B., Kaltsikes, P. J., and Anderson, R. G. 1981. Relation between wheat-rye crossability and seed set of common wheat after pollination with other species in the Hordeae. *Euphytica* 30: 121-127.

Waines, J. G., and Johnson, B. L. 1972. Genetic differences between *Aegilops longissima*, *Ae. sharonensis* and *Ae. bicornis*. *Canadian Journal of Genetics and Cytology* 14: 411-416.

Waines, J. G., and Payne, P. I. 1987. Electrophoretic analysis of the high-molecular-weight glutenin subunits of *Triticum monococcum*, *T. urartu* and the A genome of bread wheat, *T. aestivum*. *Theoretical and Applied Genetics* 74: 71-76.

Waines, J. G., Ehdaie, B., and Barnhart, D. 1987. Variability in *Triticum* and *Aegilops* species for seed characteristics. *Genome* 29: 41-46.

Waines, J. G., Hilu, K. W., and Sharma, H. C. 1982. Species formation in *Aegilops* and *Triticum*. In Estes, J. R., et al. (eds.), *Grasses and Grasslands: Systematics and Ecology.* Norman, Oklahoma: University of Oklahoma Press.

Yaghoobi-Saray, J. 1979. An electrophoretic analysis of genetic variation within and between populations of five species in the *Triticum-Aegilops* complex. PhD dissertation. University of California, Davis, California, USA.

## Chapter 11

No references

## Chapter 12

Ascherson, P., and Graebner, P. 1898-1902. *Synopsis der Mitteleuropaeischen Flora. Zweiter Band erste Abteilung.* Leipzig, Germany FR: Engelmann.

Baum, B. R. 1977. Taxonomy of the tribe Triticeae (Poaceae) using various numerical techniques. I. Historical perspectives, data accumulation, and character analysis. *Canadian Journal of Botany* 55: 1712-1740.

Baum, B. R. 1978a. Taxonomy of the tribe Triticeae (Poaceae) using various numerical techniques. II. Classification. *Canadian Journal of Botany* 56: 27-56.

Baum, B. R. 1978b. Taxonomy of the tribe Triticeae (Poaceae) using various numerical techniques. III. Synoptic key to genera and synopses. *Canadian Journal of Botany* 56: 374-385.

Bor, N. L. 1968. Gramineae. In Townsend, C. C., et al. (eds.), *Flora of Iraq.* (Vol. 9). Baghdad, Iraq: Ministry of Agriculture.

Bowden, W. M. 1959. The taxonomy and nomenclature of the wheats, barleys and ryes and their wild relatives. *Canadian Journal of Botany* 37: 657-684.

Davis, P. H. 1985. *Aegilops* L. In Davis, P. H. (ed.), *Flora of Turkey and the East Aegean Islands.* (Vol. 9) Edinburgh, UK: Edinburgh University Press.

Dewey, D. R. 1984. The genomic system of classification as guide to intergeneric hybridization with the perennial Triticeae. In Gustafson, J. P. (ed.), *Gene Manipulation and Plant Improvement.* New York, USA: Plenum Press.

Dorofeev, V. F., and Korovina, O. N. 1979. Wheat. In Brezhnev, D. D. (ed.), *Flora of Cultivated Plants.* Leningrad, USSR: Kolos.

Eig, A. 1929. Monographical-kritische Uebersicht der Gattung *Aegilops.* Repertorium Specierum Novarum Regni Vegetabilis, Beihafta 55: 1-228.

Feldman, M. 1976. Wheats. In Simmonds, N. W. (ed.), *Evolution of Crop Plants.* Harlow, UK: Longman.

Fraser, J. 1907. *Triticum peregrinum* Hackel: A new species found as an alien near Edinburgh. *Annals of Scottish Natural History Quarterly Magazine* 101-103.

Grenier, M., and Godron, M. 1856. Triticeae. In *Flore de France* (Vol. 3) Paris, France.

Greuter, W. et al. 1988. *International Code of Botanical Nomenclature.* Koenigstein, Germany FR: Koeltz Scientific Books.

Gupta, P. K., and Baum, B. R. 1986. Nomenclature and related taxonomic issues in wheats, triticales and some of their wild relatives. *Taxon* 35: 144-149.

Hackel, E. 1887. *Triticum* L. In Engler, A., and Prantl, K. (eds.) *Die Natuerliche.* Leipzig: Pfanzenfamlelien.

Hammer, K. 1980. Zur Taxonomie und Nomenclatur der Gattung *Aegilops. Feddes Repertorium* 91: 225-258.

Hawkes, J. G. 1978. The taxonomist's role in the conservation of genetic diversity. In Street, H. E. (ed.), *Essays in Plant Taxonomy*. London, UK: Academic Press.

Hawkes, J. G. 1986. Germplasm evaluation with special reference to the role of taxonomy in genebanks. In *The Conservation and Utilization of Ethiopian Germplasm*. Addis Ababa, Ethiopia: Plant Genetic Resources Centre.

Kimber, G., and Feldman, M. 1987. *Wild Wheat: An Introduction*. Special Report 353. Colombia, Missouri, USA: College of Agriculture, University of Missouri.

Kimber, G., and Sears, E. R. 1983. Assignment of genome symbols in Triticeae. In Sakamoto, S. (ed.), *Proceedings of the 6th International Wheat Genetics Symposium*. Kyoto, Japan: Plant Germplasm Institute, University of Kyoto.

Löve, A. 1982. Generic evolution of the wheatgrasses. *Biologisches Zentralblatt* 101: 199-212.

Löve, A. 1984. Conspectus of the Triticeae. *Feddes Repertorium* 95 (7-8): 425-521.

MacKey, J. 1966. Species relationship in *Triticum*. In *Proceedings of the 2nd International Wheat Genetics Symposium*. (Supplementary Vol 2) Lund, Sweden: Hereditas.

Maire, R., and Weiller, M. 1955. *Flore de l'Afrique du Nord*. (Vol. 3: Monocotyledonae, Glumiflorea) Paris, France: Paul Lechevalier.

Morris, R., and Sears, E. R. 1967. The cytogenetics of wheat and its relatives. In Quisenberry, K. S., and Reitz, L. P. (eds.), *Wheat and Wheat Improvement*. Wisconsin, USA: American Society of Agronomy.

Mouterde, P. 1966. *Nouvelle Flore du Laban et de la Syrie*. (Vol. 1) Beirut, Lebanon: L'Imprimerie Catholique.

Rossman, A. Y., Miller, D. R., and Kirkbide, J. H. 1988. Systematics, diversity and germplasm. In *Proceedings of the Beltsville Symposium 12*. Beltsville, Maryland, USA: United States Department of Agriculture (USDA).

Täckholm, V. 1974. *Students' Flora of Egypt*. (2nd edn) Beirut, Lebanon: Cooperative Printing Company.

West J. G., McIntyre, C. L., and Appels, R. 1988. Evolution and systematic relationships in the Triticeae (Poaceae). *Plant Systematics and Evolutio*n 160: 1-28.

## Chapter 13

Chapman, C. G. D. 1984. Wheat collecting: An inventory 1981-1983. *Plant Genetics Resources Newsletter* 58: 1-4.

Chapman, C. G. D. 1985. *A Survey and Strategy for Collection, Genetic Resources of Wheat*. Rome, Italy: International Board for Plant Genetic Resources (IBPGR).

Chapman, C. G. D. 1986. The role of genetic resources in wheat breeding. *Plant Genetics Resources Newsletter* 65: 2-5.

Croston, R. P., and Williams, J. T. 1981. *A World Survey of Wheat Genetic Resources*. Rome, Italy: International Board for Plant Genetic Resources (IBPGR).

Marshall, D. R. 1989. Crop genetic resources: current and emerging issues. In Brown, A. H. D., Clegg, M. T., Kahler, A. L., and Weir, B. S. (eds.) *Plant Population Genetics, Breeding and Genetic Resources*. Sunderland, Massachussetts, USA: Sinauer Associates.

Van Sloten, D. H. 1989. Presentation to the 3rd Meeting of the Food and Agriculture Organization (FAO) Commission on Plant Genetic Resources. Rome, Italy: International Board for Plant Genetic Resources (IBPGR).

## Chapter 14

Bor, P. E. 1968. *Graminae, Flora of Iraq*. (Vol. 9) Baghdad, Iraq: Ministry of Agriculture and Agrarian Reform.

Chapman, C. G. D. 1985. *The Genetic Resources of Wheat: A Survey and Strategies for Collecting*. Rome, Italy: International Board for Plant Genetic Resources (IBPGR).

Kihara, H. 1982. *Wheat Studies, Retrospect and Prospects*. Amsterdam, The Netherlands: Elsevier.

Kimber, G., and Feldman, M. 1987. *Wild Wheat: An Introduction*. Special Report 353. Missouri, Columbia, USA: College of Agriculture, University of Missouri.

Migahid, A. M. 1978. *Flora of Saudi Arabia*. (2nd edn) Riyadh, Saudi Arabia: Riyadh University Publications.

Mouterde, P. 1966. *Nouvelle Flore du Laban et de la Syrie*. (Vol. 1) Beirut, Lebanon: L'Imprimerie Catholique.

Post, G. E., and Dinsmore, J. E. 1932. *Flora of Syria, Palestine and Sinai*. (Vol. 1) Beirut, Lebanon: American University of Beirut Press.

Post, G. E., and Dinsmore, J. E. 1933. *Flora of Syria, Palestine and Sinai*. (Vol. 2) Beirut, Lebanon: American University of Beirut Press.

Sankary, M. N. 1977. *Ecology, Flora and Range Management of Arid and Very Arid Areas of Syria, Conservation and Development*. (In Arabic) Aleppo, Syria: Aleppo University Publications/Directorate of Books.

Sankary, M. N. 1982. *The Development of the Syrian Badia and its Rangelands*. Arab Center for the Studies of Arid Zones and Drylands (ACSAD) Special Paper, delivered at Syrian Badia Development Seminar, 1982, Damascus, Syria.

Sankary, M. N. 1988. Plant association map for the Syrian arid and very arid areas with a plant indicator's legend. Majalat al-Ziraa Wa al-Meiah . (In Arabic) *Journal of Agriculture and Water* 8: 73-84.

Sankary, M. N., and Mouchantat, A. H. 1986. *Field Crops Ecology*. (In Arabic) Aleppo, Syria: Aleppo University Publications.

Witcombe, J. R. 1983. *A Guide to the Species of* Aegilops *L: Their Taxonomy, Morphology and Distribution*. Rome, Italy: International Board for Plant Genetic Resources (IBPGR).

## Chapter 15

Al Alazzeh, A., Hammer, K., Lehmann, C., and Perrino, P. 1982. Report on a travel to the Socialist People's Libyan Arab Jamahiriya 1981 for the collection of indigenous taxa of cultivated plants. *Kulturpflanze* 30: 191-202.

De Cillis, E. 1927. *I Grani d'Italia. Tipografia della Camera dei Deputati, Roma.* Rome, Italy: Ministero dell'Economia Nazionale.

De Cillis, U. 1942. *I Frumenti Siciliani.* Stazione Sperimentale di Granicoltura per la Sicilia. Publ. 9. Catania, Sicily, Italy: Stazione Sperimentale di Granicoltura per la Sicilia.

Hammer, K., Cifarelli, S., and Perrino, P. 1986. Collection of landraces of cultivated plants in south Italy, 1985. *Kulturpflanze* 34: 261-273.

Jedel, P., Perrino, P., Dyck, P. L., and Martens, J. W. (1988) Resistance to *Puccinia recondita* and *Puccinia gramis* in *Triticum* accessions from the National Research Council Germplasm Institute, Bari, Italy. In *Proceedings of the 7th European and Mediterranean Cereal Rust Conference, 1988.*

Lafiandra, D., Margiotta, B., Lioi, L., Polignano, G. B., Laghetti, G., and Perrino, P. 1988. Seed storage proteins: Usefulness of their evaluation in genebank collections. In *Abstracts of an International Symposium on Population Genetics on Germplasm Resources in Crop Improvement, University of California, 1988, Davis, California, USA.*

Lafiandra, D., Perrino, P., Colaprico, G., and Porceddu, E. 1985. Applicazione delle tecniche elettroforetiche per una migliore utilizzazione delle risorse genetiche di frumento duro. *Genetica Agraria, Monografia* 7: 317-325.

Papadakis, J. S. 1929. Formes greques des blés. Publ. 1. Thessaloniki, Greece: Institute of Plant Improvement.

Pasquini, M., Corazza, L., and Perrino, P. (in press) Evaluation of *Triticum monococcum* L. and *Triticum dicoccum* Schubler strains for resistance to rusts and powdery mildew. *Phytopathologia Mediterranea.*

Peeters, J. P. 1988. The emergence of new centres of diversity: Evidence from barley. *Theoretical and Applied Genetics* 76: 17-24.

Perrino, P., and Hammer, K. 1982. *Triticum monococcum* L. and *T. dicoccum* Schubler (syn. of *T. dicoccon* Schrank) are still cultivated in Italy. *Genetica Agraria* 36: 343-352.

Perrino, P., and Hammer, K. 1983. Sicilian wheat varieties. *Kulturpflanze* 31: 227-279.

Perrino, P., and Martignano, F. 1973. The presence and distribution of old wheat varieties in Sicily. *Ecologia Agraria* 9(1): 1-11.

Perrino, P., Hammer, K., and Hanelt, P. 1981. Report of travels to south Italy 1980 for the collection of indigenous material of cultivated plants. *Kulturpflanze* 29: 433-442.

Perrino, P., Hammer, K., and Hanelt, P. 1984a. Collection of landraces of cultivated plants in south Italy. *Kulturpflanze* 32: 207-216.

Perrino, P., Hammer, K., and Lehmann, C. 1982. Collection of landraces of cultivated plants in south Italy 1981. *Kulturpflanze* 30: 181-190.

# Summary and Recommendations

Perrino, P., Polignano, G. B., and Porceddu, E. 1976a. Frumenti del Nord-Africa. II. Risultati di una missione nel Nord-Ovest Algerino e nelle oasi. *Annali della Facolta di Agraria dell'Università di Bari* 28: 13-39.

Perrino, P., Polignano, G. B., Hammer, K,. and Lehmann, C. 1984c. Wheat and barley collected in Libya. *Plant Genetic Resources Newsletter* 58: 39-41.

Perrino, P., Polignano, G. B., Hammer, K., and Lehmann, C. 1984b. Report on a travel to the Socialist People's Libyan Arab Jamahiriya 1983 for the collection of indigenous taxa of cultivated plants. *Kulturpflanze* 32: 197-206.

Perrino, P., Polignano, G. B., Porceddu, E., Olita, G., and Volpe, N. 1976b. Frumenti del Nord-Africa. III. Risultati di una missione nell'Atlante Algerino Tunisino. *Annali della Facoltà di Agraria del'Università di Bari* 28: 401-414.

Perrino, P., Polignano, G. B., Sui-Kwong, Y., and Khouya-Ali, M. 1986. Collecting germplasm in southern Morocco. *Plant Genetic Resources Newsletter* 65: 26-28.

Porceddu, E. 1979a. Genetic variability in durum wheat germplasm of different origins. *Genectia Agraria, Monografia* 4: 19-70.

Porceddu, E. 1979b. Wheat collecting in the Mediterranean region. In *Proceedings of the Conference on Broadening Genetic Base Crops, 1978, Wageningen.*

Porceddu, E., and Olita, G., 1974. Frumenti del Nord-Africa. Risultati di una missione Nord-Est Algerino. *Annali della Facoltà di Agraria dell'Università di Bari* 27: 671-686.

Porceddu, E., and Perrino, P. 1973. Wheat in Ethiopia: Preliminary report of a collecting mission. *Plant Genetic Resources Newsletter* 30: 33-36.

Porceddu, E., Perrino, P., and Olita, G. 1973. Preliminary information on an Ethiopian wheat germplasm collection mission. In *Proceedings of the Symposium on Genetics and Breeding of Durum Wheat, 1973, Bari, Italy.*

Spagnoletti-Zeuli, P. L., and Qualset, C. O. 1987. Geographical diversity for quantitative spike characters in a world collection of durum wheat. *Crop Science* 27: 235-241.

## Chapter 16

Central Statistical Authority (CSA), Ethiopia. 1987. Time series data on area, production and yield of major crops. CSA 56: 1-2.

Ciferri, R., and Giglioli, G. R. 1939. *I Cereali dell'Africa Italiana. I. I Frumenti dell'Africa Orientale Italiana Studiati su Material Originali.* Florence, Italy: Regio Istituto Agronomico per l'Africa Italiana.

Demissie, A. 1986. A decade of germplasm exploration and collection activities by Plant Genetic Resources Center/Ethiopia (PGRC/E). In Engels, J. M. M. (ed.), *Proceedings of International Symposium on the Conservation and Utilization of Ethiopian Germplasm.* Addis Ababa, Ethiopia: PGRC/E.

Harlan, J. R. 1969. Ethiopia: A centre of diversity. *Economic Botany* 23: 309-314.

Harlan, J. R. 1974. *Crops and Man.* Madison, Wisconsin, USA: Crop Science Society of America.

Hatemariam, G., and Mekbib, H. 1988. Characterization and preliminary evaluation of Ethiopian *Triticum polonicum* germplasm accessions. *Plant Genetic Resources Center/ Ethiopia (PGRC/E)/International Livestock Centre for Africa (ILCA) Germplasm Newsletter* 17: 2-7.

Nastasi, V. 1964. Wheat production in Ethiopia. *Information Bulletin on the Near East Wheat and Barley Improvement and Production Project.* 1(13): 13-23.

Percival, J. 1927. *The Wheat Plant.* A monograph. London, UK: Duckworth.

Porceddu, E., Perrino, P., and Olita, G. 1973. Preliminary information on an Ethiopian wheat germplasm collection mission. In *Proceedings of the Symposium on Genetics and Breeding of Durum Wheat, 1973, Bari, Italy.*

Sakamoto, S., and Fukui, S. 1972. Collection and preliminary observation of cultivated cereals and legumes in Ethiopia. *Kyoto University African Studies* 7: 181-189.

Tessema, T. 1987. Durum wheat breeding in Ethiopia. In *Proceedings of the 5th Regional Wheat Workshop for Eastern, Central and Southern Africa and the Indian Ocean, 1987, Madagascar.*

Tessema, T., 1986. Improvement of indigenous wheat landraces in Ethiopia. In Engels, J. M. M. (ed.), *Proceedings of International Symposium on the Conservation and Utilization of Ethiopian Germplasm.* Addis Ababa, Ethiopia: Plant Genetic Resources Center/ Ethiopia (PGRC/E).

Vavilov, N. I. 1929. Wheat of Ethiopia. *Bulletin of Applied Botany, Genetics and Plant Breeding* 20: 224-356.

Vavilov, N. I. 1951. Origin, variation, immunity and breeding of cultivated plants. *Chronica Botanica* 13: 2-29.

Worede, M. 1974. Genetic improvement of quality and agronomic characteristics of durum wheat in Ethiopia. PhD dissertation. University of Nebraska, Nebraska, USA.

Worede, M. 1983. Crop genetic resources in Ethiopia. In Holmes, J. C., and Yahir, W. M. (eds.), *More Food from Better Technology.* Rome, Italy: Food and Agriculture Organization (FAO).

## Chapter 17

No references

## Chapter 18

Beke, B., and Matuz, J. 1982. Öszi és tavaszi durum buza (Winter and spring durum wheat). *Magyar Mezögazdaság* 37/48: 6-7.

Dalloul, A. 1981. Öszi-tavaszi buza hibridek szelekciója tavaszi buzatörzsek elöállitása céljából (Selection of winter-spring wheat crossings for breeding of spring wheat lines). PhD dissertation. Hungarian Scientific Academy, Budapest, Hungary.

Karácsonyi, L. 1970. *Gabona-, listz-, sütö- éŝ tésztaipari vizsgálati módszerek* (Methods of testing grain, flour, bread and pasta making quality of wheat). Budapest, Hungary: Mezögazdasági Kiadó.

Lelley, J. 1973 Einzelpflanzenauslese und mechanisierun in der weizenzüchtung. *Pflanzenzühtung* 69: 129-134.

Matuz J., Bóna L., and Medovarszky, Z. 1986. Egy gyors eljárás a buza lisztminöségre való szelekcióra: A módositott SDS-teszt (A quick method for testing wheat bread-making quality: The modified SDS test). *Növénytermlés* 35: 1, 9-15.

## Chapter 19

Agrawal, R. K. 1986. Development of improved varieties. In Tandon, J. P., and Sethi, A. P. (eds.), *Twenty Five Years of Co-ordinated Wheat Research 1961-86.* New Delhi, India: Wheat Project Directorate, Indian Agricultural Institute.

Austin, A., Hanslas, V. K., and Singh, D. 1982. *Evaluation of Wheat Germplasm Collection for Quality.* New Delhi, India: Wheat Project Directorate, Indian Agricultural Institute.

Gupta, A. K., Saini, R. G., Sharma, S. C., Goel, R. K., and Seth, D. 1983. Screening for disease incidence. In *Exploration, Conservation, Evaluation, Documentation of Indigenous Wheat Genetic Resources.* New Delhi, India: National Bureau of Plant Genetic Resources (NBPRG).

Howard, G. L. C. 1916. Mem. Debt. Agril. Ind. Bot. Series 8: 88.

Kumar, D., and Yadav, J. S. P. 1983. Salt tolerance of mutants from wheat variety HD 1553. *Indian Journal of Agricultural Sciences.* 53(12): 1009-1015.

Kumar, D., Singh, B., and Singh, R. K. 1983. Salt tolerance in wheat varieties. *SABRAO Journal* 15(1): 71-76. Society for the Advancement of Breeding Research in Asia and Oceania (SABRAO).

Pal, B. P. 1966. *Wheat.* ICAR Monograph. New Delhi, India: Indian Council for Agricultural Research (ICAR).

Rana, R. S. 1977. Wheat variability for tolerance to salt affected soils. In *Genetics and Wheat Improvement: Proceedings of the 1st National Seminar of Genetics and Wheat Improvement, 1977, Ludhiana, India.*

Rana, R. S. 1986. Genetic diversity for salt-stress resistance of wheat in India. *Rachis* 5(1): 32-37.

Rana, R. S., and Singh, K. N. 1977. Evaluation of plant materials and breeding for crop varieties suited to saline/sodic conditions. In *Annual Report, Central Soil Salinity Research Institute (CSSRI).* Karnal, Haryana, India: CSSRI.

Singh, K. N. 1983. Screening of wheat germplasm for salinity tolerance. In *Exploration, Conservation, Evaluation, Documentation of Indigenous Wheat Genetic Resources.* New Delhi, India: National Bureau of Plant Genetic Resources (NBPGR).

## Chapter 20

No references

## Chapter 21

Blum, A., Ebercon, A., Sinmena, B., Goldernberg, H., and Gerchter-Amitai, Z. K., and Grama, A. 1983. Drought resistance reactions of wild emmer, *Triticum dicoccoides*, and wild emmer x wheat derivatives. In Sakamoto, S. (ed.), *Proceedings of the 6th International Wheat Genetics Symposium*. Kyoto, Japan: Plant Germplasm Institute, University of Kyoto.

Bor, N. L. 1968. Gramineae. In Townsend, C. C., et al. (eds.), *Flora of Iraq*. (Vol. 9) Baghdad, Iraq: Ministry of Agriculture.

Frankel, O., and Soulé, M. 1981. *Conservation and Evaluation*. Cambridge, UK: Cambridge University Press.

Jain, S. K., Qualset, C. O., Bhatt, G. M., and Wu, K. K. 1975. Geographical patterns of phenotypic diversity in a world collection of durum wheat. *Crop Science* 15: 700-704.

Jaradat, A. A. 1989. Ecotypes and genetic divergence among sympatrically distributed populations of *Hordeum vulgare* and *Hordeum spontaneum* from the xeric regions of Jordan. Submitted to *Theoretical and Applied Genetics*.

Jaradat, A. A., and Jana, S. 1987. Collection of wild emmer wheat in Jordan. *Plant Genetic Resources Newsletter* 69:19-22.

Jaradat, A. A., Jaradat, T. T., and Jana, S. 1988. Genetic diversity of wild wheat, *Triticum dicoccoides*, in Jordan. In Miller, T. E., and Koebner, R. M. D. (eds.), *Proceedings of the 7th International Wheat Genetics Symposium*. Cambridge, UK: Institute of Plant Science Research (IPSR).

Kushner, U., and Halloran, G. M. 1982. Variation in vernalization and photoperiod response in tetraploid wheat, *Triticum dicoccoides*, ecotypes. *Journal of Applied Ecology* 19: 545-554.

Kushner, U., Du Cros, D. L., Lagudah, E., and Halloran, G. M. 1984. Origin and variation of gliadin proteins associated with pasta quality in durum wheat. *Euphytica* 33: 289-294

Moseman, J. G., Nevo, E., Murshidi, M. A., and Zohary, D. 1984. Resistance of *Triticum dicoccoides* to infection with *Erysiphe graminis tritici*. *Euphytica* 33: 273-280.

Mouterde, P. 1966. *Nouvelle Flore du Laban et de la Syrie*. (Vol. 1) Beirut, Lebanon: L'Imprimerie Catholique.

Nevo, E., Beiles, A., Gutterman, Y., Storch, N., and Kaplan, D. 1984. Genetic resources of wild cereals in Palestine and vicinity. I. Phenotypic variation within and between populations of wild wheat, *Triticum dicoccoides*. *Euphytica* 33: 717-735.

Plucknett, D. L., Smith, N., Williams, J. T., and Anishetty, N. 1983. Crop germplasm conservation and developing countries. *Science* 220: 163-169.

Poiarkova, H. 1988. Morphology, geography and infraspecific taxonomics of *Triticum dicoccoides* Koern: A retrospective of 80 years of research. *Euphytica* 38: 11-23.

Post, G. E., and Dinsmore, J. E. 1933. *Flora of Syria, Palestine and Sinai.* (Vol. 2) Beirut, Lebanon: American University of Beirut Press.

Srivastava, J. P., Damania, A. B., and Pecetti, L. 1988. Landraces, primitive forms and wild progenitors of macaroni wheat, *Triticum durum*: Their use in dryland agriculture. In Miller, T. E., and Koebner, R. M. D. (eds.), *Proceedings of the 7th International Wheat Genetics Symposium.* Cambridge, UK: Institute of Plant Science Research (IPSR).

Wilkinson, G. N., Eckert, S. R., Hancock, T. W., and Mayo, O. 1983. Nearest neighbour (NN) analysis of field experiments. *Journal of the Royal Statistics Society* 45: 151-211.

Witcombe, J. R., Bourgois, J. J., and Rifai, R. 1981. Germplasm collections from Syria and Jordan. *Plant Genetic Resources Newsletter* 50: 2-8.

## Chapter 22

Axford, D. W. E., McDermott, E. E., and Redman, D. G. 1979. Note on the sodium dodecyl sulphate test of bread-making quality: Comparison with Pelshenke and Zeleny tests. *Cereal Chemistry* 56: 582.

Boggini, G., and Pogna, N. E. 1989. The bread-making quality and storage protein composition of Italian durum wheat. *Journal of Cereal Science* 9: 131-138.

Boggini, G., Dal Belin Peruffo, A., Mellini, F., and Pogna, N. E. 1987. Storage protein composition morpho-physiological and quality characters of 24 old durum wheat varieties from Sicily. *Rachis* 1: 30-35.

Boggini, G., Dal Belin Peruffo, A., Tealdo, F., and Pogna, N. E. 1985. Agronomic characteristics and gliadin composition heterogeneity in old durum wheat varieties from Sicily. In *Proceedings of the Conference on Breeding Methodologies in Durum Wheat and Triticale, 1985, Viterbo, Italy.*

Bogyo, T. P., Porceddu, E., and Perrino, P. 1980. Analysis of sampling strategies for collecting genetic material. *Economic Botany* 34 (2): 160-174.

Ceccarelli, S., and Mekni, M. S. 1985. Barley breeding for areas receiving less than 250 mm annual rainfall. *Rachis* 4(2): 3-9.

D'Egidio, M. G., De Stefanis, E., Fortini, S., Galterio,G., Nardi, S., Sgruletta, D., and Bozzini, A. 1982. Standardization of cooking quality analysis in macaroni and pasta products. *Cereal Food World* 27: 367-368.

Dal Belin Peruffo, A., Pallavicini, C., and Cuniberti, T. 1980. Identificazione di varietà di frumento geneticamente affini mediante elettroforesi su gel di poliacrilamide. *Tecnica Molitoria* 31: 95-101.

De Cillis, V. 1942. *I Frumenti Siciliani.* Stazione Sperimentale di Granicoltura per la Sicilia. Publ. No. 9. Catania, Sicily, Italy: Stazione Sperimentale di Granicoltura per la Sicilia.

De Pace, C., Spagnoletti-Zeuli, P. L., Porceddu, E., and Scarascia Mugnozza, G. T. 1983. Variation in durum wheat population from different geographical origins. V. Linkage studies. In Sakamoto, S. (ed.), *Proceedings of the 6th International Wheat Genetics Symposium.* Kyoto, Japan: Plant Germplasm Institute, University of Kyoto.

Duwaryi, M., Tell, A. M., and Shgaidaf, F. 1987. Breeding for improving yield in moisture-limiting areas: The experience of Jordan. In *Proceedings of the Seminar on Improving Winter Cereals for Moisture Limiting Environment, 1987, Capri, Italy.*

Payne, P. I., Corfield, K.G., Holt, L. M., and Blackman, J. A. 1981. Correlation between the inheritance of certain high-molecular-weight subunits of glutenin and bread-making quality in progenies of six crosses of bread wheat. *Journal of Food Science and Agriculture* 32: 51-60.

Payne, P. I., Law, C. N., and Mudd, E. E. 1980. Control by homoeologous group 1 chromosomes of the high-molecular-weight subunits of glutenin, a major protein of wheat endosperm. *Theoretical and Applied Genetics* 58: 113-120.

Perini, D., and Verona, O. 1954. Suila diffusione in Italia delle vecchie e nuove razze di frumento. *Annali dell'Istituto Sperimentale per l'Agraria* 8(1) : 44-77.

Perrino, P., and Martignano, F. 1974. *The Presence and Distribution of Old Wheat Varieties in Sicily.* Bari, Italy: Annali Facoltà Agraria.

Pogna, N. E., Mellini, F., and Dal Belin Peruffo, A. 1985. Il ruolo della elettroforesi su gel di poliacrilamide nella identificazione varietale e nello sviluppo di nuove varietà di grano duro con buona qualità pastificatoria. *Genetica Agraria, Monografia* 7: 199-212.

Porceddu, E. 1979. Genetic variability in durum wheat germplasm of different origins. *Genetica Agraria, Monografia* 4: 39-70.

Porceddu, E., and Bennett, E. 1971. Primi risultati di una spedizione di esplorazione-raccolta di vecchie varietà di frumento in Sicilia. *Ecologia Agraria* 7 (4) : 3-18.

Spagnoletti-Zeuli, P. L., De Pace, C., and Porceddu, E. 1984. Variation in durum wheat populations from three geographical origins. I. Material and spike characteristics. *Euphytica* 33: 563-575.

Spagnoletti-Zeuli, P. L., De Pace, C., and Porceddu, E. 1985. Variation in durum wheat population from different geographical origins. III. Assessment of genetic diversity for breeding purposes. *Pflanzenzuchtung* 94 (3): 177-191.

Spagnoletti-Zeuli, P. L., De Pace, C., Porceddu, E., Scarascia Mugnozza, G. T., and Volpe, N. 1983. Variability in durum wheat population from different geographical origins. IV. Genetic analysis of correlated sequential characters. In Sakamoto, S. (ed.), *Proceedings of the 6th International Wheat Genetics Symposium.* Kyoto, Japan: Plant Germplasm Institute, University of Kyoto.

Spagnoletti-Zeuli, P. L., Porceddu, E., Volpe, N., and Perrino, P. 1988. Una nuova varietà di frumento duro: Il Norba. *L'Informatore Agrario* 44(4): 57-58.

## Chapter 23

No references

## Chapter 24

Aamodt, O. S., and Johnston, W. H. 1936. Studies on drought resistance in spring wheat. *Canadian Journal of Research* 14C:122-152.

Baker, R. J. 1987. Differential response to environmental stress. In Weir, B. S., et al. (eds.), *Proceedings of the 2nd International Conference on Quantative Genetics*. Sunderland, Massachusetts, USA: Sinauer Association.

Clarke, J. M., and McCaig, T. N. 1982. Excised-leaf water retention capability as an indicator of drought resistance of *Triticum* genotypes. *Canadian Journal of Plant Science* 62: 571-578.

Clarke, J. M., and Townley-Smith, T. F. 1986. Heritibility and relationship to yield of excised-leaf water retention in durum wheat. *Crop Science* 26: 289-292

Clarke, J. M., Romagosa, I., Jana, S., Srivastava, J. P., and McCaig, T. N. (in press) Relationship of excised-leaf water loss rate and yield of durum wheat in diverse environments. *Canadian Journal of Plant Science*.

Duwayri, M., Tell, A. M., and Shqaidef, F. 1987. Breeding for improved yield in moisture-limiting areas: The experience of Jordan. In Srivastava, J. P., et al. (eds.), *Improving Winter Cereals for Moisture Limiting Environments*. New York, USA: John Wiley.

Finlay, K. W., and Wilkinson, G. N. 1963. The analysis of adaptation in a plant-breeding programme. *Australian Journal of Agricultural Research* 14: 742-754.

Hurd E. A., Townley-Smith, T. F., Patterson, L. A., and Owen, C. H. 1972. Wascana, a new durum wheat. *Canadian Journal of Plant Science* 52: 687-688.

Hurd, E. A. 1964. Root study of three wheat varieties and their resistance to drought and damage by soil cracking. *Canadian Journal of Plant Science* 44: 240-248.

Hurd, E. A. 1969. A method of breeding for yield of wheat in semi-arid climates. *Euphytica* 18: 217-226.

Hurd, E. A., Townley-Smith, T. F., Mallough, D., and Patterson, L. A. 1973. Wakooma durum wheat. *Canadian Journal of Plant Science* 53: 261-262.

Joppa, L. R., and Williams, N. D. 1988. Genetics and breeding of durum wheat in the United States. In Fabriani, G., and Lintas, C. (eds.), *Durum Wheat: Chemistry and Technology*. St Paul, Minnesota, USA: American Association Cereal Chemistry.

Lin, C. S., Binns, M. R., and Lefkovitch, L. P. 1986. Stability analysis: Where do we stand? *Crop Science* 26: 894-900.

Porceddu, E., Ceoloni, C., Lafiandra, D., Tanzarella, O. A., and Scarascia Mugnozza, G.T. 1968. Genetic resources and plant breeding: Problems and prospects. In Miller, T. E., and Koebner, R. M. D. (eds.), *Proceedings of the 7th International Wheat Genetics Symposium*. Cambridge, UK: Institute of Plant Science Research (IPSR).

Prairie Pools Inc. 1988. *Prairie Grain Variety Survey*. Saskatchewan, Canada: Prairie Pools Inc.

Srivastava, J. P., Damania, A. B., and Pecetti, L. 1988. Landraces, primitive forms and wild progenitors of macaroni wheat, *Triticum durum*: Their use in dryland agriculture. In Miller, T.E., and Koebner, R.M.D. (eds.), *Proceedings of the 7th International Wheat Genetics Symposium*. Cambridge, UK: Institute of Plant Science Research (IPSR).

Townley-Smith, T. F., and Hurd, E. A. 1979. Testing and selecting for drought resistance in wheat. In Mussell, H., and Staples, R. (eds.), *Stress Physiology in Crop Plants*. New York, USA: John Wiley.

Townley-Smith, T. F., Hurd, E. A., and McBean, D. S. 1974. Techniques of selection for yield in wheat. In Sears, E. R., and Sears, L. M. S. (eds.), *Proceedings of the 4th International Wheat Genetics Symposium*. Columbia, Missouri, USA: Agricultural Experiment Station, University of Missouri.

## Chapter 25

Ceoloni, C. 1987. Current methods of chromosome engineering in wheat. In *Proceedings of the European Wheat Aneuploid Cooperative (EWAC) Conference, 1987, Martonvasar, Hungary.*

Ceoloni, C. 1988. Transfer of alien genes into cultivated wheat and triticale genotypes by the use of homoeologous pairing mutants. In *Proceedings of the Food and Agriculture Organization (FAO)/International Atomic Energy Agency (IAEA) Meeting, 1985, Rome, Italy*. Vienna, Austria: IAEA.

Ceoloni, C., and Feldman, M. 1987. Effect of *Ph2* mutants promoting homoeologous pairing on spindle sensitivity to colchicine in common wheat. *Genome* 29: 658-663.

Ceoloni, C., Del Signore, G., Pasquini, M., and Testa, A. 1988. Transfer of mildew resistance from *Triticum longissimus* into wheat by induced homoeologous recombination. In Miller, T. E., and Koebner, R. M. D. (eds.), *Proceedings of the 7th International Wheat Genetics Symposium*. Cambridge, UK: Institute of Plant Science Research (IPSR).

Ceoloni, C., Strauss, I., and Feldman, M. 1986. Effect of different doses of group-2 chromosomes on homoeologous pairing in intergeneric wheat hybrids. *Canadian Journal of Genetics and Cytology* 28: 240-246.

Chen, K. C., and Dvorak, J. 1984. The inheritance of genetic variation in *Triticum speltoides* affecting heterogenetic chromosome pairing in hybrids with *Triticum aestivum*. *Canadian Journal of Genetics and Cytology* 26: 279-287.

Dover, G., and Riley, R. 1972. Variation at two loci affecting homoeologous meiotic chromosome pairing in *Triticum aestivum* x *Aegilops mutica* hybrids. *Nature and New Biology* 235: 61-62.

Dvorak, J. 1987. Chromosomal distribution of genes in diploid *Elytrigia elongata* that promote or suppress pairing of wheat homoeologous chromosomes. *Genome* 29: 34-40.

Evans, G. M. 1988. Genetic control of chromosome pairing in polyploids. In *Proceedings of the 3rd Kew Chromosome Conference, 1988, London, UK.*

Feldman, M. 1979. Genetic resources of wild wheats and their use in breeding. *Genetica Agraria, Monografia* 4: 9-26.

Feldman, M. 1988. Cytogenetic and molecular approaches to alien gene transfer in wheat. In Miller, T. E., and Koebner, R. M. D. (eds.), *Proceedings of the 7th International Wheat Genetics Symposium*. Cambridge, UK: Institute of Plant Science Research (IPSR).

Feldman, M., and Mello-Sampayo, T. 1967. Suppression of homoeologous pairing in hybrids of polyploid wheats x *Triticum speltoides*. *Canadian Journal of Genetics and Cytology* 9: 307-13.

Giorgi, B. 1978. A homoeologous pairing mutant isolated in *Triticum durum* cv. Cappelli. *Mutation Breeding Newsletter* 11: 4-5.

Giorgi, B., and Ceoloni, C. 1985. A ph1 hexaploid triticale: Production, cytogenetic behaviour and use for intergeneric gene transfer. In *Proceedings of the Eucarpia Meeting on Genetics and Breeding of Triticale,1984, Clermont-Ferrand, France.*

Jones, G. H. 1984. The control of chiasma distribution. In Evans, C.W., and Dickinson, H.G. (eds.), *Controlling Events in Meiosis*. Cambridge, UK: Society for Experimental Biology.

Koebner, R. M. D., and Shepherd, K. W. 1985. Induction of recombination between rye chromosome IRL and wheat chromosomes. *Theoretical and Applied Genetics* 71: 208-15.

Mello-Sampayo, T., and Canas, A.P. 1973. Suppressors of meiotic chromosome pairing in common wheat. In Sears, E. R., and Sears, L. M. S. (eds.), *Proceedings of the 4th International Wheat Genetics Symposium*. Columbia, Missouri, USA: Agricultural Experiment Station, University of Missouri.

Miller, T. E., and Reader, S. M. 1985. The effect of increased dosage of wheat chromosomes on chromosome pairing and an analysis of the chiasma frequencies of individual wheat bivalents. *Canadian Journal of Genetics and Cytology* 27: 421-425.

Riley, R. and Chapman, V. 1958. Genetic control of the cytologically diploid behaviour of hexaploid wheat. *Nature* 182: 713-15.

Schlegel, R. 1979. The effect of heterochromatin on chromosome pairing in different amphidiploid wheat-rye hybrids. *Cereal Research Communications* 7(4): 319-327.

Sears E. R. 1977. An induced mutant with homoeologous pairing in common wheat. *Canadian Journal of Genetics and Cytology* 19: 585-593.

Sears, E. R. 1972. Chromosome engineering in wheat. In *Proceedings of the 4th Stadler Genetics Symposium*. St Louis, Missouri, USA: University of Missouri.

Sears, E. R. 1976. Genetic control of chromosome pairing in wheat. *Annual Review of Genetics* 10: 31-51.

Sears, E. R. 1982. A wheat mutation conditioning an intermediate level of homoeologous chromosome pairing. *Canadian Journal of Genetics and Cytology* 24: 715-19.

Sears, E. R. 1983. The transfer to wheat of interstitial segments of alien chromosomes. In Sakamoto, S. (ed.), *Proceedings of the 6th International Wheat Genetics Symposium*. Kyoto, Japan: Plant Germplasm Institute, University of Kyoto.

Sears, E. R. 1984. Mutations in wheat that raise the level of meiotic chromosome pairing. In Gustafson, J. P. (ed.), *Proceedings of the 16th Stadler Genetics Symposium*. New York, USA and London, UK: Plenum Press.

Sears, E. R., and Okamoto, M. 1958. Intergenomic chromosome relationships in hexaploid wheat. In *Proceedings of the 10th International Congress on Genetics, 1958, Montreal, Canada* (Vol 2).

## Chapter 26

Borlaug, N. E. 1988. Challenges for global food and fiber production. *Journal of the Royal Swedish Academy of Agriculture and Forestry (Supplement)* 21: 15-55.

Chang, T. T. 1989. The case for large collections. In Brown, A. H. D., et al. (eds.) *The Use of Plant Genetic Resources*. Cambridge, UK: Cambridge University Press.

Chapman, C. G. D. 1986. The role of genetic resources in wheat breeding. *Plant Genetic Resources Newsletter* 65: 2-5

Creech, J. L., and Reitz, L. P. 1971. Plant germplasm now and for tomorrow. *Advances in Agronomy* 23: 1-49.

Dalrymple, D. G. 1986. *Development and Spread of High-Yielding Wheat Varieties in Developing Countries*. Washington DC, USA: Bureau for Science and Technology, Agency for International Development.

Day, P. R., and Lupton, F. G. H. 1987. Future prospects. In Lupton, F. G. H. (ed.), *Wheat Breeding: Its Scientific Basis*. London, UK: Chapman and Hall.

Donald, C. M. 1981. Competitive plants, communal plants, and yield in wheat crops. In Ewans, L. T., and Peacock, W. J. (eds.), *Wheat Science — Today and Tomorrow*. Cambridge, UK: Cambridge University Press.

Ewans, L. T. 1981. Yield improvement in wheat: Empirical or analytical. In Ewans, L. T., and Peacock, W. J. (eds.), *Wheat Science — Today and Tomorrow*. Cambridge, UK: Cambridge University Press.

Frankel, O. H. 1977. Natural variation and its conservation. In Muhammed, A., and Von Botstel, R. C. (eds.), *Genetic Diversity of Plants*. New York, USA: Plenum Press.

Frankel, O. H. 1989. Perspectives on genetic resources. In *Centro Internacional de Mejoramiento de Maíz y Trigo (CIMMYT) 1988 Annual Report: Delivering Diversity*. Mexico, DF: CIMMYT.

Frankel, O. H., and Brown, A. H. D. 1984. Plant genetic resources today: A critical appraisal. In Holden, J.H.W., and Williams, J.T. (eds.), *Crop Genetic Resources: Conservation and Evaluation*. London, UK: George Allen and Unwin.

Gale, M. D., and Miller, T. E. 1987. The introduction of alien genetic variation in wheat. In Lupton, F. G. H. (ed.), *Wheat Breeding: Its Scientific Basis*. London, UK: Chapman and Hall.

Gale, M. D., and Youssefian, S. 1986. Dwarfing genes in wheat. In Russell, G. E. (ed.), *Progress in Plant Breeding*. London, UK: Butterworths.

Harlan, J. R. 1984. Evaluation of wild relatives of crop plants. In Holden, J. H. W., and Williams, J. T. (eds.), *Crop Genetic Resources: Conservation and Evaluation*. London, UK: George Allen and Unwin.

Kihara H. 1983. Origin and history of 'Daruma', a parental variety of Norin 10. In Sakamoto, S. (ed.), *Proceedings of the 6th International Wheat Genetics Symposium*. Kyoto, Japan: Plant Germplasm Institute, University of Kyoto.

Kimber, G., and Feldman, M. 1987. *Wild Wheat: An Introduction*. Special Report 353. Missouri, Columbia, USA: College of Agriculture, University of Missouri.

Kronstad, W. E. 1986. Germplasm: The key to past and future wheat improvement. In *Genetic Improvement in Yield of Wheat*. Crop Science Society of America (CSSA) Special Publication 13. Madison, Wisconsin, USA: CSSA.

Krull, C. F., and Borlaug, N. E. 1970. The utilization of collections in plant breeding and production. In Frankel, O.H., and Bennett, E. (eds.), *Genetic Resources in Plants: Their Exploration and Conservation*. Oxford, UK: Blackwell Scientific Publications.

Law, C. N., and Krattiger, A. F. 1987. Genetics of grain quality in wheat. In Morton, I. D. (ed.), *Cereals in a European Context: European Conference on Food Science and Technology*. Florida, USA: VCH Publishers.

Lucken, K. A. 1987. *Hybrid Wheat. Wheat and Wheat Improvement.* Agronomy Monograph 13 (2nd edn). Madison, Wisconsin, USA: Crop Science Society of America (CSSA).

Mackey, J., and Qualset, C. O. 1986. Conventional methods of wheat breeding. In *Genetic Improvement in Yield of Wheat*. Crop Science Society of America (CSSA) Special Publication 13. Madison, Wisconsin, USA: CSSA.

Marshall, D. R. 1989. Limitations to the use of germplasm collections. In Brown, A. H. D., et al. (eds.), *The Use of Plant Genetic Resources*. Cambridge, UK: Cambridge University Press.

McIntosh, R. A. A. 1983. A catalogue of gene symbols for wheat. In Sakamoto, S. (ed.), *Proceedings of the 6th International Wheat Genetics Symposium*. Kyoto, Japan: Plant Germplasm Institute, University of Kyoto.

Mujeeb-Kazi, A. 1988. Wide crosses. *Centro Internacional de Mejoramiento de Maíz y Trigo (CIMMYT) Report on Wheat Improvement 1985-86*. Mexico, DF: CIMMYT.

Parlevliet, J. E. 1988. Strategies for the utilization of partial resistance for control of cereal rusts. In Simmonds, N.W., and Rajaram, S. (eds.), *Breeding Strategies for Resistance to the Rusts of Wheat*. Mexico, DF: Centro Internacional de Mejoramiento de Maíz y Trigo (CIMMYT) .

Payne, P. I. 1987. Genetics of wheat storage proteins and the effect of allelic variation on bread-making quality. *Annual Review of Plant Physiology* 38: 141-153.

Peeters, J. P. and Williams, J. T. 1984. Towards better use of genebanks with special reference to information. *Plant Genetic Resources Newsletter* 60: 22-32.

Roelfs, A. P. 1988a. Resistance to leaf and stem rust in wheat. In Simmonds, N. W., and Rajaram, S. (eds.), *Breeding Strategies for Resistance to the Rusts of Wheat*. Mexico, DF: Centro Internacional de Mejoramiento de Maíz y Trigo (CIMMYT) .

Roelfs, A. P. 1988b. Advances and understanding of rust resistance. Lecture presented at Centro de Investigaciones Agricolas del Novoest (CIANO), Mexico.

Stakman, E. C., and Harrar, J. G. 1957. *Principles of Plant Pathology*. New York, USA: The Ronald Press Company.

Williams, P. H. 1989. Screening for resistance to diseases. In Brown, A. H. D., et al. (eds.), *The Use of Plant Genetic Resources*. Cambridge, UK: Cambridge University Press.

## Chapter 27

Acevedo, E., and Ceccarelli, S. (in press). Role of physiologist-breeder in a program for drought resistance conditions. In *Proceedings of the Symposium on Drought Resistance in Cereals: Theory and Practice, 1988, Cairo, Egypt.*

Ashraf, M., and McNeilly, T. 1988. Variability in salt tolerance of nine spring wheat cultivars. *Journal of Agronomy and Crop Science* 160: 14-21.

Boursier, P., Lynch, J., Läuchli, A., and Epstein, E. 1987. Chloride partitioning in leaves of salt-stressed sorghum, maize, wheat and barley. *Australian Journal of Plant Physiology* 14: 463-473.

Dewey, D. R. 1984. The genomic system of classification as a guide to intergeneric hybridization with the perennial Triticeae. In Gustafson, J. P. (ed.), *Gene Manipulation in Plant Improvement.* New York, USA: Plenum Press.

Dvorak, J., and Ross, K. 1986. Repression of tolerance of $Na^+$, $K^+$, $Mg^{2+}$, $Cl^-$ $SO_4^{2-}$ ions and sea water in the amphiploid of *Triticum aestivum* x *Elytrigia elongata. Crop Science* 26: 658-660.

Dvorak, J., Edge, M., and Ross, K. 1988. On the evolution of the adaptation of *Lophopyrum elongatum* to growth in saline environments. *Proceedings of the National Academy of Sciences* 85: 3805-3809.

Epstein, E. 1985. Salt tolerant crops: Origin, development, and prospects of the concept. *Plant and Soil* 89: 187-198.

Flowers, T. J., and Läuchli, A. 1983. Sodium versus potassium: Substitution and compartmentation. In Läuchli, A., and Bielski, R. L. (eds.), *Encyclopedia of Plant Physiology.* (Vol. 15B) Berlin, Germany FR: Springer Verlag.

Flowers, T. J., Hajibagheri, M. A., and Clipson, N. J. W. 1986. Halophytes. *Quarterly Review of Biology* 61: 313-336.

Forster, B. P., Gorham, J., and Miller, T. E. 1987. Salt tolerance of an amphiploid between *Triticum aestivum* and *Agropyron junceum. Plant Breeding* 98: 1-8.

Forster, B. P., Miller, T. E., and Law, C. N. 1988. Salt tolerance of two wheat-*Agropyron junceum* disomic addition lines. *Genome* 30: 559-564.

Gorham, J., and Wyn Jones, R. G. (in press) A physiologist's approach to improving the salt tolerance of wheat. Submitted to *Rachis.*

Gorham, J., Forster, B. P., Budrewicz, E., Wyn Jones, R. G., Miller. T. E., and Law, C. N. 1986. Salt tolerance in the Triticeae: Solute accumulation and distribution in an amphididiploid derived from *Triticum aestivum* cv. Chinese Spring and *Thinopyrum bessarabicum. Journal of Experimental Botany* 37: 1435-1449.

Gorham, J., Hardy, C., Wyn Jones, R. G., Joppa, L. R., and Law, C. N. 1987. Chromosomal location of a K/Na discrimination character in the D genome of wheat. *Theoretical and Applied Genetics* 74: 584-588.

Gorham, J., Hughes, Ll., and Wyn Jones, R. G. 1980. Chemical composition of salt-marsh plants from Ynys Môn (Anglesey): The concept of physiotypes. *Plant Cell and Environment* 3: 309-318.

Gorham, J., McDonnell, E., Budrewicz, E., and Wyn Jones, R. G., 1985. Salt tolerance in the Triticeae: Growth and solute accumulation in leaves of *Thinopyrum bessarabicum. Journal of Experimental Botany* 36: 1021-1031.

Greenway, H. 1962. Plant responses to saline substrates. I. Growth and ion uptake of several varieties of *Hordeum* during and after sodium chloride treatment. *Australian Journal of Biological Sciences*15: 16-38.

Greenway, H., and Munns, R. 1980. Mechanisms of salt tolerance in non-halophytes. *Annual Review of Plant Physiology* 31: 149-190.

Kingsbury, R. W., and Epstein, E. 1986. Salt sensitivity in wheat. A case for specific ion toxicity. *Plant Physiology* 80: 651-654.

Kinzel, H. 1982. *Pflanzenökologie und Mineralstoffwechsel.* Stuttgart, Germany FR: Verlag Eugen Ulmer.

Leigh, R. A., and Wyn Jones, R. G. 1986. Cellular compartmentation in plant nutrition: The selective cytoplasm and the promiscuous vacuole. In Tinker, P. B., and Läuchli, A. (eds.), *Advances in Plant Nutrition.* New York, USA: Praeger.

McGuire, P. E., and Dvorak, J. 1981. High salt tolerance potential in wheatgrasses. *Crop Science* 21: 702-705.

Miller, T. E. 1987. Systematics and evolution. In Lupton, F. G. H. (ed.), *Wheat Breeding: Its Scientific Basis.* London, UK: Chapman and Hall .

Munns, R., Greenway, H., and Kirst, G.O. 1983. Halotolerant eukaryotes. In Lange, O.L., et al. (eds.), *Encyclopedia of Plant Physiology.* (New series: Vol. 12C) Berlin, Germany: Springer Verlag.

Qureshi, R. H., Ahmad, R., Ilyas, M., and Aslam, Z. 1980. Screening of wheat, *Triticum aestivum* L., for salt tolerance. *Pakistan Journal of Agricultural Science* 17: 19-25.

Richards, R. A. 1983. Should selection for yield in saline regions be made on saline or non-saline soils? *Euphytica* 32: 431-438.

Richards, R. A., Bennett, C. W., Qualset, C. O., Epstein, E., Norlyn, J. D., and Winslow, M. D. 1987. Variation in yield of grain and biomass in wheat, barley and triticale in a salt-affected field. *Field Crops Research* 15: 277-288.

Shah, S. H., Gorham, J., Forster, B. P., and Wyn Jones, R. G. 1987. Salt tolerance in the Triticeae: The contribution of the D genome to cation selectivity in wheat. *Journal of Experimental Botany* 38: 254-269.

Storey, R., Graham R. D., and Shepherd, K. W. 1985. Modification of the salinity response of wheat by the genome of *Elytrigia elongatum. Plant and Soil* 83: 327-330.

Wyn Jones, R. G. 1981. Salt tolerance. In Johnson, C. B. (ed.), *Physiological Processes Limiting Plant Productivity.* London, UK: Butterworth.

Wyn Jones, R. G., and Gorham, J. (1989) Physiological effects of salinity. In Acevedo, E., Fereres, E. and Srivastava, J. P. (eds.) *Scope for Genetic Improvement: Proceedings of the International Symposium on Improving Winter Cereals under Temperature and Salinity Stresses, 1987.* Cordoba, Spain: Technical School for Agricultural Engineers.

Yeo, A. R., and Flowers, T. J. 1986. Salinity resistance in rice, *Oryza sativa* L., and a pyramiding approach to breeding varieties for saline soils. *Australian Journal of Plant Physiology* 13: 161-173.

Yeo, A. R., Yeo, M. E., and Flowers, T. J. 1988. Selection of lines with high and low sodium transport from within varieties of an inbreeding species: Rice, *Oryza sativa* L. *New Phytologist* 110: 13-19.

## Chapter 28


Blanco, A., Orecchia, C., and Simeone, R. 1983. Cytology, morphology and fertility of the amphiploid *Triticum durum* Desf. *Haynaldia villosa* L. Schur. In Sakamoto, S. (ed.), *Proceedings of the 6th International Wheat Genetics Symposium*. Kyoto, Japan: Plant Germplasm Institute, University of Kyoto.

Blanco, A., Simeone, R., and Resta, P. 1987. The addition of *Dasypyrum villosum* (L.) Candargy chromosomes to durum wheat, *Triticum durum* Desf. *Theoretical and Applied Genetics* 74: 328-33.

Cervato, A., Marudelli, M., and Piva, C. 1985. Growth analysis and yields in spring barley vs. pea in the central Po plain . (In Italian, with English summary). *Rivista di Agronomia* 19: 347-357.

De Pace, C. 1987. Genetic variability in natural populations of *Dasypyrum villosum* (L.) Candargy. PhD dissertation. University of California, Davis, California, USA.

De Pace, C., Montebove, L., Delre, V., Jan, C. C., Qualset, C. O., and Scarascia Mugnozza, G.T. 1988. Biochemical versatility of amphiploids derived from crossing *Dasypyrum villosum* (L.) Candargy and wheat: Genetic control and phenotypical aspects. *Theoretical and Applied Genetics* 76: 513-529.

Della Gatta, C., Tanzarella, O.A., Resta, P., and Blanco, A. 1984. Protein content in a population of *Haynaldia villosa* and electrophoretic pattern of the amphiploid *Triticum durum* x *H. villosa*. In Porceddu, E. (ed.), *Breeding Methodologies in Durum Wheat and Triticale*. Viterbo, Italy: University of Tuscia.

Halloran, C. M. 1966. Hybridization of *Haynaldia villosa* with *Triticum aestivum*. *Australian Journal of Botany* 14: 355-359.

Hyde, B. B., 1953. Addition of individual *Haynaldia villosa* chromosomes to hexaploid wheat. *American Journal of Botany* 40: 174-182.

Jan, C. C., De Pace, C., McGuire, P. E., and Qualset, C. O., 1986. Hybrids and amphiploids of *Triticum aestivum* L. and *Triticum turgidum* L. with *Dasypyrum villosum* (L.) Candargy. *Pflanzenzüchtung* 96: 97-106.

Kostoff, D. 1937. Chromosome behaviour in *Triticum* hybrids and allied genera. III. *Triticum* x *Haynaldia* hybrids. *Zeitschrift Für Züchtung (now Pflanzenzüchtung) Reihe A Band* 21: 380-382.

Massantini, F., Baonari, E., Masoni, A., and Ercoli, L. 1985. Comparative growth analysis of durum wheat and alfalfa. (In Italian, with English summary) *Rivista di Agronomia* 19: 358-365.

Nakajima, G. 1966. Karyogenetical studies on the intergeneric F1 hybrids raised between *Triticum* and *Haynaldia*. *La Kromosomo* 64: 2083-2100.

Qualset, C.O., De Pace, C., Jan, C.C., Scarascia Mugnozza, G.T., and Tanzarella, O.A. 1981. *Haynaldia villosa* : A species with potential use in wheat breeding. In *Agronomy Abstracts 1981, Meeting of the American Society of Agronomy (ASA)*. Madison, Wisconsin, USA: ASA.

Sando, W. J. 1935. Intergeneric hybrids of *Triticum* and *Secale* with *Haynaldia villosa*. *Journal of Agricultural Research* 51: 759-800.

Scott, P. R. 1981. Variation in host susceptibility. In Asher, M. J. C., and Shipton, P. J. (eds.), *Ecology and Control of Take-All*. London, UK: Academic Press
Sears, E. R. 1953. Addition of the genome of *Haynaldia villosa* to *Triticum aestivum*. *American Journal of Botany* 40: 168-174.
Strampelli, N. 1932. *Origins, developments, work and results.* (In Italian) Rome, Italy: Istituto Nazionale di Genetica per la Cerealicoltura.
Tschermak, E. 1930. Neue Beobachtungen am fertilen Artbastard *Triticum turgidovillosum.* In *Berichte der Deutschen Botanischen Gesellschaft, Band XLVIII* (9): 400-402.
Watson, D. J., Thorne, G. N., and French, S. A. W. 1963. Analysis of growth and yield of winter and spring wheats. *Annals of Botany* 27: 1-22.
Wittmer, G., Iannucci, A., De Santis, G., Baldelli, G., De Stefanis, E., Brando, A., Ciocca, L., and Rascio, A. 1982. Adaptability and physiological factors limiting the yielding ability of durum wheat. (In Italian, with English summary) *Rivista di Agronomia* 16: 61-70.

## Chapter 29

No references

## Chapter 30

Germplasm Resource Laboratory of Food Crop Institute, Jiangsu Academy of Agricultural Sciences (JAAS). 1983. Documentation and analysis of wheat germplasm resources. *Journal of Crop Germplasm Resources* 1: 7-11.
Jing Shanbao. 1983. *Chinese Wheat Cultivars and Their Genealogies*. Beijing, China: Agriculture Publishing House.
Zeng Xueqi, En Zaicheng, and Guan Lihua. 1987. Performance and evaluation of exotic wheat germplasm in Yunnan province. *Journal of Crop Germplasm Resources* 4: 23-24
Zheng Diansheng. 1988. Wheat introduction in China. *World Agriculture* 11: 25-28.

## Chapter 31

Allan, R. E., Heyne, E. G., Jones, E. T., and Johnston, C. O. 1959. Genetic analyses of ten sources of Hessian fly resistance, their interrelationships and association with leaf rust reaction in wheat. In *Kansas Agricultural Experiment Station and USA-Agricultural Research Service (ARS)Technical Bulletin 104*.
Anon. 1939. *La Cécidomyie Destructive, Mayetiola destructor (Say) Diptera Cecidomyilidae*. Memento No. 8, Defense des vegetaux. Rabat, Morocco.
Barnes, H. F. 1956. *Gall Midges of Economic Importance*. London, UK: Crosly Lockwood.

Cartwright, W. B., and La Hue, D. W. 1944. Testing wheat in the greenhouse for Hessian fly resistance. *Journal of Economic Entomology* 37: 385-387.

Dahms, R. G. 1967. Insects attacking wheat. *Agronomy* 13: 428-430.

El Bouhssini, M., Amri, A., and Hatchett, J. H. 1988. Wheat genes conditioning resistance to the Hessian fly (Diptera: Cecidomyiidae) in Morocco. *Journal of Economic Entomology* 81: 709-712.

Foster, J. E., Araya, J. E., Safranski, G. G., Cambron, S. E., and Taylor, P. L. 1988. Rearing Hessian fly for use in the development of resistant wheat. *Purdue University Agricultural Experiment Station Bulletin 536.*

Gallagher, L. W., Benyassine, A., Benlhabib, O., and Obanni, M. 1987. Sources of resistance to *Mayetiola destructor* in breadwheat and durum wheat in Morocco. *Euphytica* 36: 591-602.

Gallun, R. L. 1983. Genetics of host-parasite interaction in Hessian fly, *Mayetiola destructor* (Say) and wheat. In *Proceedings of the 15th International Congress on Genetics, 1983, New Delhi, India.*

Gallun, R. L., and Reitz, L. P. 1971. *Wheat Cultivars Resistant to Hessian fly.* United States Department of Agriculture (USDA)-ARS Research Report 134. Washington DC, USA: US Government Printers.

Gill, B. S., Hatchett, J. H., and Raupp, W. J. 1987. Chromosomal location of Hessian fly-resistance gene H13 in the D genome of wheat. *Journal of Heredity* 78: 97-100.

Hatchett, J. H., and Gill, B. S. 1983. Expression and genetics of resistance to Hessian fly in *Triticum tauschii* (Coss) Schmal. In Sakamoto, S. (ed.), *Proceedings of the 6th International Wheat Genetics Symposium.* Kyoto, Japan: Plant Germplasm Institute, University of Kyoto.

Lebsock, K. L., Quick, J.S., and Joppa, R. L. 1972. Registration of D6647, D6654, D6659, and D6660 durum wheat germplasm (Reg. no. GP 29 to GP 32). *Crop Science* 12: 721.

Maas, F. B. III., Patterson, F. L. , Foster, J. E, and Hatchett, J. H. 1987. Expression and inheritance of resistance of Marquillo wheat to Hessian fly biotype D. *Crop Science* 27:49-52.

Maas, F. B. III., Patterson, F. L., Foster, J. E. , and Ohm, H. W. (in press) Expression and inheritance of Hessian fly resistance of ELS 6404-160 durum wheat to Hessian fly. *Crop Science.*

Obanni, M., Ohm, H. W., Foster, J. E. , and Patterson, F. L. (in press) Genetics of resistance of PI 422297 durum wheat to Hessian fly. *Crop Science.*

Obanni, M., Patterson, F. L. , Foster, J. E. , and Ohm, H. W. 1988. Genetic analyses of resistance of durum wheat PI 428435 to Hessian fly. *Crop Science* 28: 223-226.

Oellermann, C. M., Patterson, F. L., and Gallun, R. L. 1983. Inheritance of resistance in Luso wheat to Hessian fly. *Crop Science* 23: 221-224.

Patterson, F. L., and Gallun, R. L. 1973. Inheritance of resistance of Seneca wheat to race E of Hessian fly. In Sears, E. R. , and Sears, L. M. S. (eds.), *Proceedings of the 4th International Wheat Genetics Symposium.* Missouri, Columbia, USA: University of Missouri.

Patterson, F. L., Foster, J. E., and Ohm, H. W. 1988. Gene *H16* in wheat for resistance to Hessian fly. *Crop Science* 28: 652-654.

Stebbins, N. B., Patterson, F. L., and Gallun, L. 1983. Inheritance of resistance of PI 94587 wheat to biotypes B and D of Hessian fly. *Crop Science* 23: 251-253.

Steel, R. G. D., and Torrie, J. H. 1980. *Principles and Procedures of Statistics: A Biometrical Approach.* (2nd edn) New York, USA: McGraw-Hill.

Tyler, J. M., and Hatchett, J. H. 1983. Temperature influence on expression of resistance to Hessian fly (Diptera: Cecidomyiidae) in wheat derived from *Triticum tauschii. Journal of Economic Entomology* 76: 323-326.

## Chapter 32

Kerber, E. R., and Dyck, P. L. 1973. Inheritance of stem rust resistance transferred from diploid wheat, *Triticum monococcum*, to tetraploid and hexaploid wheat and chromosome location of the gene involved. *Canadian Journal of Genetics and Cytology* 15: 397-409.

Rashid, G., Quick, J. S., and Statler, G. D. 1976. Inheritance of leaf rust resistance in three durum wheats. *Crop Science* 16: 294-296

Statler, G. D. 1973. Inheritance of leaf rust resistance in Leeds durum wheat. *Crop Science* 13: 116-117

Statler, G. D. 1982. Inheritance of virulence of *Puccinia recondita* f. sp. *tritici* on durum and spring wheat cultivars. *Phytopathology* 72: 210-213

## Chapter 33

Avivi, L. 1979. Utilization of *Triticum dicoccoides* for the improvement of grain protein quantity in cultivated wheats. *Genetica Agraria, Monografia* 4:27-38.

Avivi, L., Levy, A. A., and Feldman M. 1983. Studies on high protein durum wheat derived from crosses with the wild tetraploid wheat, *Triticum turgidum* var. *dicoccoides*. In Sakamoto, S. (ed.), *Proceedings of the 6th International Wheat Genetics Symposium.* Kyoto, Japan: Plant Germplasm Institute, University of Kyoto.

Gerechter-Amitai, Z. K., and Stubbs, R. W. 1970. A valuable source of yellow rust resistance in Palestinian populations of wild emmer, *Triticum dicoccoides. Euphytica* 18:12-21.

Gill, K. S., Gill, B. S., and Synder, E. B. 1988. *Triticum araraticum* chromosome substitutions in common wheat, *Triticum aestivum* cv. Wachita. In Miller, T. E., and Koebner, R. M. D. (eds.), *Proceedings of the 7th International Wheat Genetics Symposium.* Cambridge, UK: Institute of Plant Science Research (IPSR).

Islam, A. K. M. R., and Shepherd, K. W. (1988) Incorporation of barley chromosomes into wheat. In Bajaj, Y. P. S (ed.), *Biotechnology in Agriculture and Forestry.* Berlin, Germany FR: Springer Verlag.

Levy A. A., and Feldman, M. 1987. Increase of grain protein percentage in high yielding common wheat breeding lines by genes from wild tetraploid wheat. *Euphytica* 26: 253-259.

Nagle, K. 1980. Utilization of a tetraploid high protein mutant in cross breeding for protein improvement of hexaploid wheats. In *Proceedings of Food and Agriculture Organization (FAO)/ Swedish International Development Agency (SIDA)/ Swedish Agency for Research Cooperation with Developing Countries (SAREC) Seminar on Nutritional Quality of Barley and Spring Wheat, 1980, Ankara, Turkey.*

Riley, R., and Kimber, G. 1966. The transfer of alien genetic variation to wheat. *Annual Report of Plant Breeding Institute (PBI), Cambridge*: 6-36.

Sharma, H. C., and Gill, B. S. 1983. Current status of wide hybridization in wheat. *Euphytica* 32: 17-31.

Tahir, M. (1986) Evaluation and utilization of *Triticum turgidum* L. var. *dicoccoides* for the improvement of durum wheat. In *Proceedings of the 5th International Wheat Conference, 1986, Rabat, Morocco.*

Tahir, M. 1983. Genetic variability in protein content of *Triticum aestivum*, *T. durum* and *T. dicoccoides*. *Rachis* 2:14-15.

Tahir, M. 1985. Evaluation and utilization of *Triticum dicoccoides*. *Annual Report, International Center for Agricultural Research in the Dry Areas (ICARDA)*. Aleppo, Syria: ICARDA.

## Chapter 34

Approval Methods of the Association (AMA). 1983. *Method No. 46-12*. St Paul, Minnesota, USA: American Association of General Chemists.

Avivi, L. 1978. High grain protein content in wild tetraploid wheat, *Triticum dicoccoides* Korn. In *Proceedings of the 5th International Wheat Genetics Symposium*. New Delhi, India: Indian Society of Genetics and Plant Breeding.

Damania, A. B., Tahir, M., and Somaroo, B. H. 1988. Improvement of durum wheat proteins utilizing wild gene resources of *Triticum dicoccoides* Koern. at the International Center for Agricultural Research in the Dry Areas (ICARDA). In Miller, T. E., and Koebner, R. M. D. (eds.), *Proceedings of the 7th International Wheat Genetics Symposium*. Cambridge, UK: Institute of Plant Science Research (IPSR).

El-Haramein, F., Williams, P. C., Sayegh, A., Nachit, M., and Srivastava, J. P. 1986. Selecting durum wheats on the basis of flat bread quality. *Cereal Food World* 31: 589 (Abstract No. 83).

International Center for Agricultural Research in the Dry Areas (ICARDA), Aleppo, Syria. 1986. *Technical Manual* 14: 77-85.

Joppa, L. R., Khan, K., and Williams, N. D. 1983. Chromosomal location of genes for gliadin polypeptides in durum wheat *Triticum turgidum* L. *Theoretical and Applied Genetics* 64: 289-293.

Kushnir, U., Du Cros, D. I., Lagudah, E., and Halloran, G. M. 1984. Origin and variation of gliadin proteins associated with pasta quality in durum wheat. *Euphytica* 33: 289-294

Lafiandra, D., Colaprico, G., Kasarda, D. D., and Porceddu, E. 1987. Null alleles for gliadin blocks in bread and durum wheat cultivars. *Theoretical and Applied Genetics* 74: 610-616.

McDermott, E. E., and Redman, D. G. 1977. Small-scale tests for bread-making quality. *Flour Milling and Baking Research Association Bulletin* 6: 200-213

Tkachuk, R., and Mellish, V. J. 1980. Wheat cultivar identification by high voltage gel electrophoresis. *Annales de Technologie Agricole* 209: 207-212.

Williams, P. C., El-Haramein, F., Nelson, W., and Srivastava, J. P. 1988. Evaluation of wheat quality by baking Syrian-type two-layered flat breads. *Journal of Cereal Science* 7: 195-207.

# *Index*

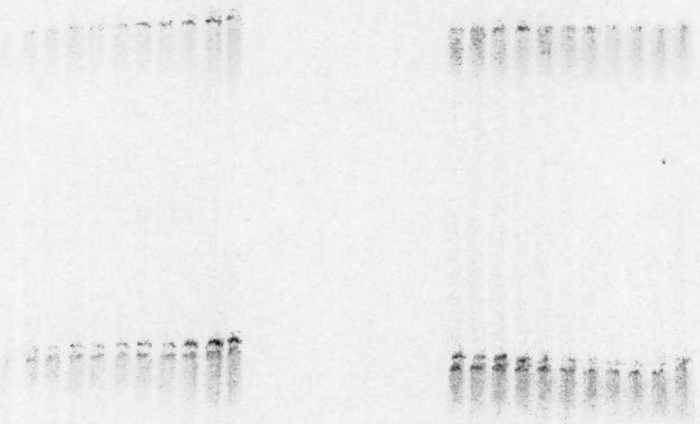